Thomas M. Klapötke
Chemistry of High-Energy Materials

Also of interest

Energetic Materials Encyclopedia.
Klapötke, 2018
ISBN 978-3-11-044139-0, e-ISBN 978-3-11-044292-2

Energetic Compounds.
Methods for Prediction of Their Performance
Keshavarz, Klapötke, 2017
ISBN 978-3-11-052184-9, e-ISBN 978-3-11-052186-3

The Properties of Energetic Materials.
Determination and Prediction of Combustion Properties
Keshavarz, Klapötke, 2018
ISBN 978-3-11-057220-9, e-ISBN 978-3-11-057222-3

Chemical Kinetics.
For Engineers
Csavdari, 2021
ISBN 978-3-11-047515-9, e-ISBN 978-3-11-047517-3

Industrial Inorganic Chemistry.
Benvenuto, 2017
ISBN 978-3-11-049446-4, e-ISBN 978-3-11-049447-1

High Temperature Materials and Processes.
Hiroyuki Fukuyama (Editor-in-Chief)
ISSN 0334-6455, e-ISSN 2191-0324

Thomas M. Klapötke

Chemistry of High-Energy Materials

5th Edition

DE GRUYTER

Author
Prof. Dr. Thomas M. Klapötke
Ludwig-Maximilians University Munich
Department of Chemistry
Butenandtstr. 5–13 (Building D)
81377 Munich, Germany
tmk@cup.uni-muenchen.de

ISBN 978-3-11-062438-0
e-ISBN (PDF) 978-3-11-062457-1
e-ISBN (EPUB) 978-3-11-062469-4

Library of Congress Control Number: 2019946369

Bibliographic information published by the Deutsche Nationalbibliothek
The Deutsche Nationalbibliothek lists this publication in the Deutsche Nationalbibliografie; detailed bibliographic data are available on the Internet at http://dnb.dnb.de.

© 2019 Walter de Gruyter GmbH, Berlin/Boston
Typesetting: Meta Systems Publishing & Printservices GmbH, Wustermark
Printing and binding: CPI books GmbH, Leck
Cover image: antorti / E+ / Getty Images

www.degruyter.com

"We will not waver; we will not tire; we will not falter;
And we will not fail. Peace and freedom will prevail."

G. W. Bush,
Presidential Address to the Nation,
October 7th 2001

Preface to this 5th English edition

Everything which has been said in the preface to the first German, as well as in the first to fourth English editions still holds true, and essentially does not need any addition or correction. In this revised fifth edition in English, the manuscript has been updated and various recent aspects of energetic materials have been added:

(i) Some errors which were unfortunately present in the fourth edition have been corrected and the references have also been updated where appropriate.//
(ii) Chapters on the critical diameter, colored smokes, delay compositions, sensitivities, gun erosion, TKX-50, liners, thermobaric weapons and nitrocellulose have been updated.//
(iii) Four new chapters on Aging (6.7), Air Bags (13.6), Spent Acid (10.4) and Toxicity Measurements (13.7) have been added.

In addition to the people thanked in the German and first four English editions, the author would like to thank Dr. Manfred Bohn (ICT), Dr. Johann Glück (ZF), Dr. Paul Wanninger (PWX), Mr. Moritz Mühlemann (BIAZZI), Mr. Marcel Holler (Bayern Chemie), Ms. Cornelia Unger (LMU) and Ms. Teresa Küblböck (LMU) for many inspired discussions.

Munich, July 2019   Thomas M. Klapötke

Preface to the 4th English edition

Everything which has been said in the preface to the first German, as well as in the first to third English editions, still holds and essentially does not need any addition or correction. In this revised fourth edition in English, the manuscript has been updated and various recent aspects of energetic materials have been added:
(i) Some errors, which were unfortunately present in the third edition, have been corrected, and the references have also been updated where appropriate.
(ii) The chapters on shaped charges, visible pyrotechnics, smokes (addition of P_3N_5), primary explosives (MTX-1), composite and double-base propellants, and Gurney energy have been updated.
(iii) Four new chapters on Electric Detonators (5.4), Detonation velocity from Laser Induced Air Shock (9.7), Thermally Stable Explosives (9.8), and Explosive Welding (13.5) have been added.

In addition to the people thanked in the German and first three English editions, the author would like to thank Professor Dr. Mohammad H. Keshavarz, Dr. Tomasz Witkowski, Col. Dr. Ahmed Elbeih, Cpt. Mohamed Abd-Elghany, M.Sc. and Lt. Andreea Voicu, M.Sc. for many inspired discussions.

Munich, July 2017   Thomas M. Klapötke

Preface to the 3rd English edition

Everything which has been said in the preface to the first German and first and second English editions still holds and essentially does not need any addition or correction. In this revised third edition in English the manuscript has been updated and various recent aspects of energetic materials have been added:
(i) some errors which unfortunately occurred in the first and second editions have been corrected and the references have also been updated where appropriate.
(ii) The chapters on critical diameters, delay compositions, visible light (blue) pyrotechnics, polymer-bonded explosives (PBX), HNS, thermodynamic calculations, DNAN, smoke (yellow) formulations and high-nitrogen compounds have been updated.
(iii) Five new short chapters on Ignition and Initiation (chapters 5.2 and 5.3), the Plate Dent Test (chapter 7.4), Underwater Explosions (chapter 7.5) and the Trauzl Test (chapter 6.6) have been added.

In addition to the people thanked in the German and first and second English editions, the author would like to thank Dr. Vladimir Golubev and Tomasz Witkowski (both LMU) for many inspired discussions concerning hydrocode calculations. The author is also indebted to and thanks Dr. Manuel Joas (DynITEC, Troisdorf, Germany) for his help with the preparation of chapters 5.2 and 5.3.

Munich, October 2015   Thomas M. Klapötke

Preface to the 2nd English edition

Everything said in the preface to the first German and first English editions still holds and essentially does not need any addition or correction. In this revised second edition in English we have up-dated the manuscript and added some recent aspects of energetic materials:

(i) We have tried to correct some mistakes which can not be avoided in a first edition and also updated the references where appropriate.

(ii) The chapters on Ionic Liquids, Primary Explosives, NIR formulations, Smoke Compositions and High-Nitrogen Compounds were updated.

(iii) Two new short chapters on Co-Crystallization (9.5) and Future Energetic Materials (9.6) have been added.

In addition to the people thanked in the German and first English edition, the author would like to thank Dr. Jesse Sabatini and Dr. Karl Oyler (ARDEC, Picatinny Arsenal, NJ) for many inspired discussions concerning pyrotechnics.

Munich, May 2012
Thomas M. Klapötke

Preface to the first English edition

Everything said in the preface to the first German edition remains valid and essentially does not need any addition or correction. There are several reasons for translating this book into English:
- The corresponding lecture series at LMU is now given in English in the postgraduate M.Sc. classes, to account for the growing number of foreign students and also to familiarize German students with the English technical terms.
- To make the book available to a larger readership world-wide.
- To provide a basis for the author's lecture series at the University of Maryland, College Park.

We have tried to correct some omissions and errors which can not be avoided in a first edition and have also updated the references where appropriate. In addition, five new chapters on Combustion (Ch. 1.4), NIR formulations (Ch. 2.5.5), the Gurney Model (Ch. 7.3), dinitroguanidine chemistry (Ch. 9.4) and nanothermites (Ch. 13.3) have been included in the English edition. The chapter on calculated combustion parameters (Ch. 4.2.3) has been extended.

In addition to the people thanked in the German edition, the author would like to thank Dr. Ernst-Christian Koch (NATO, MSIAC, Brussels) for pointing out various mistakes and inconsistencies in the first German edition. For inspired discussions concerning the Gurney model special thanks goes to Joe Backofen (BRIGS Co., Oak Hill). Dr. Anthony Bellamy, Dr. Michael Cartwright (Cranfield University), Neha Mehta, Dr. Reddy Damavarapu and Gary Chen (ARDEC) and Dr. Jörg Stierstorfer (LMU) are thanked for ongoing discussions concerning secondary and primary explosives.

The author also thanks Mr. Davin Piercey, B.Sc. for corrections and for writing the new chapter on nanothermites, Dr. Christiane Rotter for her help preparing the English figures and Dr. Xaver Steemann for his help with the chapter on detonation theory and the new combustion chapter. The author thanks the staff of de Gruyter for the good collaboration preparing the final manuscript.

Munich, January 2011   Thomas M. Klapötke

Preface to the first German edition

This book is based on a lecture course which has been given by the author for more than 10 years at the Ludwig-Maximilian University Munich (LMU) in the postgraduate Master lecture series, to introduce the reader to the chemistry of highly energetic materials. This book also reflects the research interests of the author. It was decided to entitle the book "Chemistry of High-Energy Materials" and not simply "Chemistry of Explosives" because we also wanted to include pyrotechnics, propellant charges and rocket propellants into the discussion. On purpose we do not give a comprehensive historical overview and we also refrained from extensive mathematical deductions. Instead we want to focus on the basics of chemical explosives and we want to provide an overview of recent developments in the research of energetic materials.

This book is concerned with both the civil applications of high-energy materials (e.g. propellants for carrier or satellite launch rockets and satellite propulsion systems) as well as the many military aspects. In the latter area there have been many challenges for energetic materials scientists in recent days some of which are listed below:

- In contrast to classical targets, in the on-going global war on terror (GWT), new targets such as tunnels, caves and remote desert or mountain areas have become important.
- The efficient and immediate response to time critical targets (targets that move) has become increasingly important for an effective defense strategy.
- Particularly important is the increased precision ("we want to hit and not to miss the target", Adam Cumming, DSTL, Sevenoaks, U.K.), in order to avoid collateral damage as much as possible. In this context, an effective coupling with the target is essential. This is particularly important since some evil regimes often purposely co-localize military targets with civilian centers (e.g. military bases near hospitals or settlements).
- The interest in insensitive munitions (IM) is still one of the biggest and most important challenges in the research of new highly energetic materials.
- The large area of increasing the survivability (for example by introducing smokeless propellants and propellant charges, reduced signatures of rocket motors and last but not least, by increasing the energy density) is another vast area of huge challenge for modern synthetic chemistry.
- Last but not least, ecological aspects have become more and more important. For example, on-going research is trying to find suitable lead-free primary explosives in order to replace lead azide and lead styphnate in primary compositions. Moreover, RDX shows significant eco- and human-toxicity and research is underway to find suitable alternatives for this widely used high explosive. Finally, in the area of rocket propulsion and pyrotechnical compositions, replacements for toxic ammonium perchlorate (replaces iodide in the thyroid

gland) which is currently used as an oxidizer are urgently needed. Despite all this, the performance and sensitivity of a high-energy material are almost always the key-factors that determine the application of such materials – and exactly this makes research in this area a great challenge for synthetically oriented chemists.

The most important aspect of this book and the corresponding lecture series at LMU Munich, is to prevent and stop the already on-going loss of experience, knowledge and know-how in the area of the synthesis and safe handling of highly energetic compounds. There is an on-going demand in society for safe and reliable propellants, propellant charges, pyrotechnics and explosives in both the military and civilian sector. And there is no one better suited to provide this expertise than well trained and educated preparative chemists.

Last but not least, the author wants to thank those who have helped to make this book project a success. For many inspired discussions and suggestions the authors wants to thank the following colleagues and friends: Dr. Betsy M. Rice, Dr. Brad Forch and Dr. Ed Byrd (US Army Research Laboratory, Aberdeen, MD), Prof. Dr. Manfred Held (EADS, TDW, Schrobenhausen), Dr. Ernst-Christian Koch (NATO MSIAC, Brussels), Dr. Miloslav Krupka (OZM, Czech Republic), Dr. Muhamed Sucesca (Brodarski Institute, Zagreb, Croatia), Prof. Dr. Konstantin Karaghiosoff (LMU Munich), Prof. Dr. Jürgen Evers (LMU Munich), as well as many of the past and present co-workers of the authors research group in Munich without their help this project could not have been completed.

The author is also indebted to and thanks Dipl.-Chem. Norbert Mayr (LMU Munich) for his support with many hard- and soft-ware problems, Ms. Carmen Nowak and Ms. Irene S. Scheckenbach (LMU Munich) for generating many figures and for reading a difficult manuscript. The author particularly wants to thank Dr. Stephanie Dawson (de Gruyter) for the excellent and efficient collaboration.

Munich, July 2009   Thomas M. Klapötke

Contents

Preface to this 5th English edition —— vii

Preface to the 4th English edition —— viii

Preface to the 3rd English edition —— ix

Preface to the 2nd English edition —— x

Preface to the first English edition —— xi

Preface to the first German edition —— xii

1	**Introduction —— 1**	
1.1	Historical Overview —— 1	
1.2	New Developments —— 11	
1.2.1	Polymer-Bonded Explosives —— 11	
1.2.2	New High (Secondary) Explosives —— 15	
1.2.3	New Primary Explosives —— 25	
1.2.4	New Oxidizers for Solid Rocket Motors —— 35	
1.2.5	Initial Characterization of New Energetic Materials —— 41	
1.3	Definitions —— 42	
1.4	Combustion, Deflagration, Detonation – A Short Introduction —— 49	
1.4.1	Fire and Combustion —— 49	
1.4.2	Deflagration and Detonation —— 51	
2	**Classification of Energetic Materials —— 55**	
2.1	Primary Explosives —— 55	
2.2	High (Secondary) Explosives —— 60	
2.3	Propellant Charges —— 69	
2.4	Rocket Propellants —— 72	
2.4.1	Chemical Thermal Propulsion (CTP) —— 89	
2.5	Pyrotechnics —— 91	
2.5.1	Detonators, Initiators, Delay Compositions and Heat-Generating Pyrotechnics —— 91	
2.5.2	Light-Generating Pyrotechnics —— 95	
2.5.3	Decoy Flares —— 104	
2.5.4	Smoke Munitions —— 111	
2.5.5	Near-Infrared (NIR) Compositions —— 122	

3	Detonation, Detonation Velocity and Detonation Pressure —— 125	
4	Thermodynamics —— 137	
4.1	Theoretical Basis —— 137	
4.2	Computational Methods —— 143	
4.2.1	Thermodynamics —— 143	
4.2.2	Detonation Parameters —— 147	
4.2.3	Combustion Parameters —— 151	
4.2.4	Example: Theoretical Evaluation of New Solid Rocket Propellants —— 160	
4.2.5	Example: EXPLO5 Calculation of the Gun Propellant Properties of Single, Double and Triple Base Propellants —— 167	
4.2.6	Semiempirical Calculations (EMDB) —— 169	
5	Initiation —— 173	
5.1	Introduction —— 173	
5.2	Ignition and Initiation of Energetic Materials —— 175	
5.3	Laser Ignition and Initiation —— 177	
5.4	Electric Detonators —— 183	
6	Experimental Characterization of Explosives —— 187	
6.1	Sensitivities —— 187	
6.2	Long-Term Stabilities —— 214	
6.3	Insensitive Munitions —— 216	
6.4	Gap Test —— 217	
6.5	Classification —— 218	
6.6	Trauzl Test —— 220	
6.7	Aging of Energetic Materials —— 223	
7	Special Aspects of Explosives —— 235	
7.1	Shaped Charges —— 235	
7.2	Detonation Velocities —— 244	
7.3	Gurney Model —— 250	
7.3.1	Example: Calculation of the Gurney Velocity for a General Purpose Bomb —— 255	
7.4	Plate Dent Tests vs. Fragment Velocities —— 256	
7.5	Underwater explosions —— 262	
8	Correlation between the Electrostatic Potential and the Impact Sensitivity —— 269	
8.1	Electrostatic Potentials —— 269	
8.2	Volume-Based Sensitivities —— 272	

9	**Design of Novel Energetic Materials — 275**
9.1	Classification — 275
9.2	Polynitrogen Compounds — 277
9.3	High-Nitrogen Compounds — 283
9.3.1	Tetrazole and Dinitramide Chemistry — 284
9.3.2	Tetrazole, Tetrazine and Trinitroethyl Chemistry — 291
9.3.3	Ionic Liquids — 296
9.4	Dinitroguanidine Derivatives — 300
9.5	Co-Crystallization — 302
9.6	Future Energetics — 303
9.7	Detonation Velocities from Laser-induced Air Shock — 317
9.8	Thermally Stable Explosives — 323
10	**Synthesis of Energetic Materials — 331**
10.1	Molecular Building Blocks — 331
10.2	Nitration Reactions — 332
10.3	Processing — 337
10.4	Safe Handling of Spent Acid — 338
11	**Safe Handling of Energetic Materials in the Laboratory — 343**
11.1	General — 343
11.2	Protective Equipment — 344
11.3	Laboratory Equipment — 347
12	**Energetic Materials of the Future — 349**
13	**Related Topics — 357**
13.1	Thermobaric Weapons — 357
13.2	Agent Defeat Weapons — 360
13.3	Nanothermites — 362
13.3.1	Example: Iron Oxide / Aluminum Thermite — 369
13.3.2	Example: Copper Oxide / Aluminum Thermite — 370
13.3.3	Example: Molybdenum Trioxide / Aluminum Thermite — 371
13.4	Homemade Explosives — 371
13.5	Explosive Welding — 372
13.6	Gas Generators (Air Bags) — 373
13.7	Toxicity Measurements — 376
14	**Study Questions — 381**
15	**Literature — 385**

16	**Appendix** —— 397	
	Important reaction types in organic nitrogen chemistry —— 397	
	Curing of a diol (HTPB) with isocyanate binder —— 401	
	Important reaction types in inorganic nitrogen chemistry —— 401	
	Collection of formulas —— 411	

Author —— 417

Index —— 419

1 Introduction

1.1 Historical Overview

In this chapter we do not want to be exhaustive in scope, but rather to focus on some of the most important milestones in the chemistry of explosives (Tab. 1.1). The development of energetic materials began with the accidental discovery of **blackpowder** in China (~ 220 BC). In Europe this important discovery remained dormant until the 13th and 14th centuries, when the English monk Roger Bacon (1249) and the German monk Berthold Schwarz (1320) started to research the properties of blackpowder. At the end of the 13th century, blackpowder was finally introduced into the military world. However, it was not until 1425 that Corning greatly improved the production methods and blackpowder (or gunpowder) was then introduced as a propellant charge for smaller and later also for large calibre guns. Even today, ca. 100,000 pounds of blackpowder are still used in the US military per year.

One application in which BP is still in use is in "time blasting fuses". They were invented as early as 1831, and contain approx. 4.7 g/m BP in the core (see Fig. 1.1a). The burn rate is about 135 m/s, and since they are RF safe (radio frequency), they are very popular with the military.

Fig. 1.1a: Time Blasting Fuse.

Tab. 1.1: Historical overview of some important secondary explosives.

substance	acronym	development	application	density/g cm^{-3}	explosive power[b]
blackpowder	BP	1250–1320	1425–1900	ca. 1.0	
nitroglycerine	NG	1863	in propellant charges	1.60	170
dynamite	Dy	1867	civil/commercial only	varies	varies
picric acid	PA	1885–1888	WW I	1.77	100
nitroguanidine	NQ	1877	most in TLPs	1.71	99
trinitrotoluene	TNT	1880	WW I	1.64	116
nitropenta	PETN	1894	WW II	1.77	167
hexogen	RDX	1920–1940	WW II	1.81	169
octogen	HMX	1943	WW II	1.91 (β polymorph)	169
hexanitrostilbene	HNS	1913	1966	1.74	
triaminotrinitro-benzene	TATB	1888	1978	1.93	
HNIW	CL-20	1987	under evaluation	2.1 (ε polymorph)	

[a] rel. to PA

The next milestone was the first small-scale synthesis of **nitroglycerine (NG)** by the Italian chemist Ascanio Sobrero (1846). Later, in 1863 Imanuel Nobel and his son Alfred commercialized NG production in a small factory near Stockholm (Tab. 1.1). NG is produced by running highly concentrated, almost anhydrous, and nearly chemically pure glycerine into a highly concentrated mixture of nitric and sulfuric acids (HNO_3 / H_2SO_4), while cooling and stirring the mixture efficiently. At the end of the reaction, the nitroglycerine and acid mixture is transferred into a separator, where the NG is separated by gravity. Afterwards, washing processes using water and alkaline soda solution remove any residual acid.

Nitroglycerin (NG) showed itself to be extremely dangerous for blasting. Alfred Nobel scrambled to find a solution, and in doing so, invented dynamite. His solution was to solidify the NG by adding diatomaceous earth (mineral powder). It was not understood why this worked. It is now known that liquid HEs are sensitized by *cavitation bubble collapse*. Liquid HEs are dangerous, because they are often insensitive in pure form, but highly sensitized by aeration.

Initially NG was very difficult to handle because of its high impact sensitivity and unreliable initation by blackpowder. Among many other accidents, one explosion in 1864 destroyed the Nobel factory completely, killing Alfred's brother Emil. In the same year, Alfred Nobel invented the metal blasting cap detonator, and replaced blackpowder with **mercury fulminate (MF)**, $Hg(CNO)_2$. Although the Swedish-German Scientist Johann Kunkel von Löwenstern had described $Hg(CNO)_2$

Fig. 1.1b: Molecular structure of mercury fulminate, $Hg(CNO)_2$.

NG

NC

Fig. 1.2: Molecular structures of nitroglycerine (NG) and nitrocellulose (NC).

as far back as in the 17th century, it did not have any practical application prior to Alfred Nobel's blasting caps. It is interesting to mention that it was not until the year 2007 that the molecular structure of $Hg(CNO)_2$ was elucidated by the LMU research team (Fig. 1.1b) [1, 2]. Literature also reports the thermal transformation of MF, which, according to the below equation, forms a new mercury containing explosive product which is reported to be stable up to 120 °C.

$$3\,Hg(CNO)_2 \rightarrow Hg_3(C_2N_2O_2)_3$$

After another devastating explosion in 1866 which completely destroyed the NG factory, Alfred Nobel focused on the safe handling of NG explosives. In order to reduce the sensitivity, Nobel mixed NG (75%) with an absorbent clay called "Kieselguhr" (25%). This mixture called "Guhr Dynamite" was patented in 1867. Despite the great success of dynamite in the civil sector, this formulation has never found significant application or use in the military sector.

One of the great advantages of NG (Fig. 1.2) in comparison to blackpowder (75% KNO_3, 10% S_8, 15% charcoal) is that it contains both the fuel and oxidizer in the same molecule which guarantees optimal contact between both components, whereas in blackpowder, the oxidizer (KNO_3) and the fuel (S_8, charcoal) have to be physically mixed.

At the same time as NG was being researched and formulated several other research groups (Schönbein, Basel and Böttger, Frankfurt-am-Main) worked on the nitration of cellulose to produce **nitrocellulose (NC)**. In 1875 Alfred Nobel discovered that when NC is formulated with NG, they form a gel. This gel was further refined to produce blasting gelatine, gelatine dynamite and later in 1888 ballistite (49% NC, 49% NG, 2% benzene and camphor), which was the first smokeless powder. (Cordite which was developed in 1889 in Britain, had a very similar composition.) In 1867 it was proven that mixtures of NG or dynamite and ammonium nitrate (AN) showed enhanced performance. Such mixtures were used in the civil sector. In 1950 manufacturers started to develop explosives which were waterproof and solely contained the less hazardous AN. The most prominent formulation was ANFO (Ammonium Nitrate Fuel Oil) which found extensive use in commercial areas (mining, quarries etc.). Since the 1970s aluminium and monomethylamine were added to such formulations to produce gelled explosives which could detonate more easily. More recent developments include production of emulsion explosives which contain suspended droplets of a solution of AN in oil. Such emulsions are water proof, yet readily detonate because the AN and oil are in direct contact. Generally, emulsion explosives are safer than dynamite and are simple and cheap to produce.

Picric acid (PA) was first reported in 1742 by Glauber, however it was not used as an explosive until the late 19th century (1885–1888), when it replaced blackpowder in nearly all military operations world-wide (Fig. 1.3). PA is prepared best by dissolving phenol in sulfuric acid and the subsequent nitration of the resulting of phenol-2,4-disulfonic acid with nitric acid. The direct nitration of phenol with nitric acid is not possible because the oxidizing HNO_3 decomposes the phenol molecule. Since the sulfonation is reversible, the $-SO_3H$ groups can then be replaced with $-NO_2$ groups by refluxing the disulfonic acid in concentrated nitric acid. In this step the third nitro group is introduced as well. Although pure PA can be handled safely, a disadvantage of PA is its tendency to form impact sensitive metal salts (picrates, primary explosives) when in direct contact with shell walls. PA was used as a grenade and as mine filling.

Tetryl was developed at the end of the 19th century (Fig. 1.3) and represents the first explosive of the nitroamino (short: nitramino) type. Tetryl is best obtained by dissolving monomethylaniline in sulfuric acid and then pouring the solution into nitric acid, while cooling the process.

The above mentioned disadvantages of PA are overcome by the introduction of **trinitrotoluene (TNT)**. Pure 2,4,6-TNT was first prepared by Hepp (Fig. 1.3) and its structure was determined by Claus and Becker in 1883. In the early 20th century TNT almost completely replaced PA and became the standard explosive during WW I. TNT is produced by the nitration of toluene with mixed nitric and sulfuric acid. For military purposes TNT must be free of any isomer other than the 2,4,6-isomer. This is achieved by recrystallization from organic solvents or from 62% nitric acid. TNT is still one of the most important explosives for blasting charges today. Charges are produced through casting and pressing. However, cast charges of TNT often show sensitivity issues and do not comply with the modern insensitive munition requirements (IM). For this reason alternatives to TNT have been suggested. One of these replacements for TNT is NTO (filler) combined with 2,4-dinitroanisole (DNAN, binder).

Nitroguanidine (NQ) was first prepared by Jousselin in 1887 (Fig. 1.3). However, during WW I and WW II it only found limited use, for example in formulations

Fig. 1.3: Molecular structures of picric acid (PA), tetryl, trinitrotoluene (TNT), Nitroguanidine (NQ), pentaerythritol tetranitrate (PETN), hexogen (RDX), octogen (HMX), hexanitrostilbene (HNS) and triaminotrinitrobenzene (TATB).

with AN in grenades for mortars. In more recent days NQ has been used as a component in triple-base propellants together with NC and NG. One advantage of the triple-base propellants is that unlike double-base propellants the muzzle flash is reduced. The introduction of about 50% of NQ to a propellant composition also results in a reduction of the combustion temperature and consequently reduced erosion and increased lifetime of the gun. NQ can be prepared from dicyandiamide and ammonium nitrate via guanidinium nitrate which is dehydrated with sulfuric acid under the formation of NQ:

$$\underset{H_2N}{\overset{H_2N}{>}}C=\underset{N}{\overset{N}{>}}C\equiv N \xrightarrow[-H_2N-CN]{NH_4NO_3} C(NH_2)_3{}^{\oplus}NO_3{}^{\ominus} \xrightarrow[-H_2O]{H_2SO_4} H_2N-C\underset{N-NO_2}{\overset{NH_2}{<}}$$

The most widely used explosives in WW II other than TNT were hexogen (RDX) and pentaerythritol tetranitrate (nitropenta, PETN) (Fig. 1.3). Since PETN is more sensitive and chemically less stable than RDX, RDX was (and is) the most commonly used high explosive. PETN is a powerful high explosive and has a great shattering effect (brisance). It is used in grenades, blasting caps, detonation cords and boosters. PETN is not used in its pure form because it is too sensitive. A formulation of 50 % TNT and 50 % PETN is known as "pentolite". In combination with plasticized nitrocellulose PETN is used to form polymer bonded explosives (PBX). The military application of PETN has largely been replaced by RDX. PETN is prepared by introducing pentaerythritol into concentrated nitric acid while cooling and stirring the mixture efficiently. The then formed bulk of PETN crystallizes out of the acid solution. The solution is then diluted to about 70 % HNO_3 in order to precipitate the remaining product. The washed crude product is purified by recrystallization from acetone.

Hexogen (RDX) was first prepared in 1899 by Henning for medicinal use. (*N.B.* NG and PETN are also used in medicine to treat angina pectoris. The principal action of these nitrate esters is vasodilation (i.e. widening of the blood vessels). This effect arises because in the body the nitrate esters are converted to nitric oxide (NO) by mitochondrial aldehyde dehydrogenase, and nitric oxide is a natural vasodilator.) In 1920 Herz prepared RDX for the first time by the direct nitration of hexamethylene tetramine. Shortly afterwards Hale (Picatinny Arsenal, NJ) developed a process that formed RDX in 68 % yield. The two processes most widely used in WW II were
1. the Bachmann process (KA process) in which hexamethylene tetramine dinitrate reacts with AN and a small amount of nitric acid in an acetic anhydride medium to form RDX (type B RDX). The yields are high, however, 8–12 % of HMX form as a side product.
2. the Brockman process (type A RDX) essentially produces pure RDX.

After WW II **octogen (HMX)** started to become available. Until today, most high explosive compositions for military use are based on TNT, RDX and HMX (Tab. 1.2).

Since 1966 **hexanitrostilbene (HNS)** and since 1978 **triaminotrinitrobenzene (TATB)** are produced commercially (Fig. 1.3). Both secondary explosives show excellent thermal stabilities and are therefore of great interest for the NAVY (fuel fires) and for hot deep oil drilling applications (Fig. 1.3). Especially HNS is known as a heat- and radiation-resistant explosive which is used in heat-resistant explosives in the oil industry. The brisance of HNS is lower than that of RDX, but the melting point of approx. 320 °C is much higher. HNS can directly be prepared from

Tab. 1.2: Composition of some high explosive formulations.

name	composition
Composition A	88.3 % RDX, 11.7 % non-energetic plasticizers
Composition B	60 % RDX, 39 % TNT, 1 % binder (wax)
Composition C4	90 % RDX, 10 % polyisobutylene
octol	75 % HMX, 25 % RDX
torpex[a]	42 % RDX, 40 % TNT, 18 % aluminum
PBXN-109	64 % RDX, 20 % aluminum, 16 % binder and plasticizer
OKFOL	96.5 % HMX, 3.5 % wax

[a] An Australian improved development of torpex is known under the name H6 and also contains hexogen (RDX), trinitrotoluene (TNT) and aluminum. H6 was used as a high explosive formulation in the MOAB bomb (Massive Ordnance Air Blast bomb). MOAB (also known as GBU-43/B) is with a load of approx. 9500 kg high explosive formulation (30 % TNT, 45 % RDX, 20 % AL, 5 % wax) one of the largest conventional bombs ever used.

Tab. 1.2a: Desired properties of potential HNS replacements

Thermal Stability	No changes after 100 h at 260 °C
Detonation Velocity	> 7500 m/s
Specific Energy *	> 975 kJ/kg
Impact Sensitivity	> 7.4 J
Friction Sensitivity	> 235 N
Total Costs	< 500 Euro/kg
Critical diameter	≥ HNS

* specific energy. $F = p_e \cdot V = n \cdot R \cdot T$

trinitrotoluene through oxidation with sodium hypochlorite in a methanol/THF solution:

$$2\ C_6H_2(NO_2)_3CH_3 + 2\ NaOCl \rightarrow C_6H_2(NO_2)_3-CH=CH-C_6H_2(NO_2)_3 + 2\ H_2O + 2\ NaCl$$

Since oil deposits which are located closer to the surface are becoming rare, deeper oil reserves now have to be explored where (unfortunately) higher temperatures are involved. Therefore, there is an ongoing search for explosives which are even more thermally stable (decomposition temperatures > 320 °C) than HNS, but at the same time show better performance (Tab. 1.2a). Higher thermal stabilities usually result in compounds with lower sensitivities which are therefore safer to handle.

According to J. P. Agrawal, new energetic materials with high thermal stabilities can be achieved by incorporating the following points in the compounds:

- Salt formation
- Introduction of amino groups
- Introduction of conjugation
- Condensation with a triazole ring.

Fig. 1.3a: Molecular structures of PATO and PYX.

Fig. 1.3b: Synthetic route for PYX.

Two possible replacements for HNS which are presently under investigation are PYX and PATO.

Various picryl and picrylamino substituted 1,2,4–triazoles which were formed by condensing 1,2,4-triazole or amino-1,2,4-triazole with picryl chloride (1-chloro-2,4,6-trinitrobenzene) were studied in detail by Coburn & Jackson. One of these molecules is PATO (3-picrylamino-1,2,4-triazole), a well known, thermally stable explosive, which is obtained by the condensation of picryl chloride with 3-amino-1,2,4-triazole (Fig. 1.3a). Another promising candidate for a high-temperature explosive is PYX (Fig. 1.3a). The synthesis for PYX is shown in Fig. 1.3b.

Fig. 1.3c: Molecular structure of BTDAONAB.

Tab. 1.2b: Comparative data of the thermal and explosive properties of TATB, HNS, BTDAONAB and BeTDAONAB.

Property	TATB	HNS	BTDAONAB	BeTDAONAB
density / g/cc	1.94	1.74	1.97	1.98
Sensitivity to temperature / °C	360	318	350	260
DTA (exo) / °C	360	353	550	275
DSC (exo) / °C	371	350		268
Ω_{CO} / %	−18.6	−17.8	−6.8	−5.9
IS / J	50	5	21	21
FS / N	> 353	240	353	362
VoD / m s^{-1}	7900	7600	8600	8700
p_{C-J} / kbar	273	244	341	354

Agrawal et al. reported the synthesis of BTDAONAB (Fig. 1.3c) which does not melt below 550 °C and is considered to be a better and thermally more stable explosive than TATB. According to the authors, this material has a very low impact (21 J), no friction sensitivity (> 360 N) and is thermally stable up to 550 °C. These reported properties makes BTDAONAB superior to all of the nitro-aromatic compounds which have been discussed. BTDAONAB has a VoD of 8300 m/s while TATB is about 8000 m/s [Agrawal et al., *Ind. J. Eng. & Mater Sci.*, **2004**, *11*, 516–520; Agrawal et al., Central *Europ. J. Energ. Mat.* **2012**, *9(3)*, 273–290.]

Moreover, recently another nitro-aromatic compound (BeTDAONAB), similar to Agrawal's BTDAONAB has been published by Keshavaraz et al., which is also very insensitive (Fig 1.3d). In this compound, the terminal triazole moieties have been replaced by two more energetic (more endothermic) tetrazole units [Keshavaraz et al., Central *Europ. J. Energ. Mat.* **2013**, *10(4)*, 455; Keshavaraz et al., *Propellants, Explos. Pyrotech.*, DOI: 10.1002/prep.201500017]. Table 1.2b shows a comparison of the thermal and explosive properties of TATB, HNS, BTDAONAB and BeTDAONAB.

TATB is obtained from trichloro benzene by nitration followed by a reaction of the formed trichlorotrinitro benzene with ammonia gas in benzene or xylene solution.

As shown above, the number of chemical compounds which have been used for high explosive formulations until after WW II is relatively small (Tab. 1.1 and 1.2). As

Fig. 1.3d: Synthetic route for the synthesis of BeTDAONAB.

we can also see from Table 1.1 and 1.2 the best performing high explosives (RDX and HMX; TNT is only used because of its melt-cast applications) possess relatively high densities and contain oxidizer (nitro and nitrato groups) and fuel (C—H back bone) combined in one and the same molecule. One of the most powerful new high explosive is **CL-20** which was first synthesized in 1987 by the Naval Air Warfare Center (NAWF) China Lake (Fig. 1.7, Tab. 1.1). CL-20 is a cage compound with significant cage strain which also contains nitramine groups as oxidizers and possesses a density of about 2 g cm^{-3}. This already explains the better performance in comparison with RDX and HMX. However, due to the relatively high sensitivity of the (desirable) ε polymorph as well as possible phase transition problems and high production costs (ca. $ 1,000/Lb; cf. RDX ca. $ 20.—/kg; GAP ca. $ 500/kg) so far CL-20's wide and general application has not been established.

1.2 New Developments

1.2.1 Polymer-Bonded Explosives

Since about 1950 polymer-bonded (or plastic-bonded) explosives (PBX) have been developed in order to reduce sensivity and to facilitate safe and easy handling. PBX also show improved processibility and mechanical properties. In such materials the crystalline explosive is embedded in a rubber-like polymeric matrix. One of the most prominent examples of a PBX is **Semtex**. Semtex was invented in 1966 by Stanislav Brebera, a chemist who worked for VCHZ Synthesia in Semtin (hence the name Semtex), a suburb of Pardubice in the Czech Republic. Semtex consists of varying ratios of PETN and RDX. Usually polyisobutylene is used for the polymeric matrix, and phthalic acid n-octylester is the plasticizer. Other polymer matrices which have been introduced are polyurethane, polyvinyl alcohol, PTFE (teflon), Viton, Kel-F, PBAN, HTPB and various polyesters. PBAN (polybutadiene acrylonitrile) was used in the space shuttle composite propellant.

Often, however, problems can arise when combining the polar explosive (RDX) with the non-polar polymeric binder (*e.g.* polybutadiene or polypropylene). In order to overcome such problems, additives are used to facilitate mixing and intermolecular interactions. One of such polar additives is dantacol (DHE) (Fig. 1.4).

Fig. 1.4: Structure of Dantacol (DHE).

One disadvantage of the polymer-bonded explosives of the first generation, is that the non-energetic binder (polymer) and plasticizer lessened the performance. To overcome this problem energetic binders and plasticizers have been developed. The most prominent examples for **energetic binders** are (Fig. 1.5, a):
- poly-GLYN, poly(glycidyl)nitrate
- poly-NIMMO, poly(3-nitratomethyl-3-methyl-oxetane)
- GAP, glycidylazide polymer
- poly-AMMO, poly(3-azidomethyl-3-methyl-oxetane),
- poly-BAMO, poly(3,3-bis-azidomethyl-oxetane).

Examples for **energetic plasticizers** are (Fig. 1.5, b):
- NENA derivatives, alkylnitratoethylnitramine,
- EGDN, ethyleneglycoldinitrate,

Fig. 1.5: Energetic binders (a) and energetic plasticizers (b). Synthesis of the NENA compound, ANTTO (c) (ANTTO = azido nitrato trinito triaza octane).

Fig. 1.5a: Ring-opening reactions (ROP) for the synthesis of energetic polymers.

- MTN, metrioltrinitrate,
- BTTN, butane-1,2,4-trioltrinitrate.

Fig. 1.5a shows a general synthesis route for various energetic polymers using the ring-opening polymerization (ROP) strategy).

For binders in particular – but also for plasticizers – it is important to know the glass transition temperature. The value of the glass transition temperature should be as low as possible but at least −50 °C. If the temperature of a polymer drops below T_g, it behaves in an increasingly brittle manner. As the temperature rises above T_g, the polymer becomes more rubber-like. Therefore, knowledge of T_g is essential in the selection of materials for various applications. In general, values of T_g well below room temperature correspond to elastomers and values above room temperature to rigid, structural polymers.

In a more quantitative approach for the characterization of the liquid-glass transition phenomenon and T_g, it should be noted that in cooling an amorphous material from the liquid state, there is no abrupt change in volume such as that which occurs on cooling a crystalline material below its freezing point, T_f. Instead, at the glass transition temperature, T_g, there is a change in the slope of the curve of specific volume vs. temperature, moving from a low value in the glassy state to

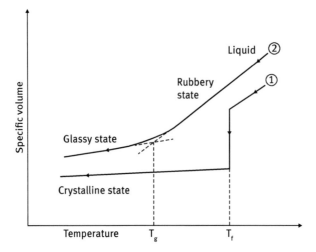

Fig. 1.5b: Specific volume vs. temperature plot for a crystalline solid and a glassy material with a glass transition temperature (T_g).

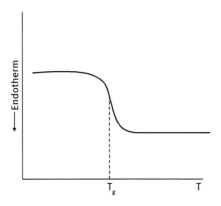

Fig. 1.5c: DSC plot illustrating the glass transition process for a glassy polymer which does not crystallize and is being slowly heated from below T_g.

a higher value in the rubbery state over a range of temperatures. This comparison between a crystalline material (1) and an amorphous material (2) is illustrated in the figure below. Note that the intersection of the two straight line segments of curve (2) defines the quantity T_g (Fig. 1.5b).

Differential scanning calorimetry (DSC) can be used to determine experimentally the glass transition temperature. The glass transition process is illustrated in Fig. 1.5c for a glassy polymer which does not crystallize and is being slowly heated from a temperature below T_g. Here, the drop which is marked T_g at its midpoint, represents the increase in energy which is supplied to the sample to maintain it at the same temperature as the reference material. This is necessary due to the relatively rapid increase in the heat capacity of the sample as its temperature is increases pass T_g. The addition of heat energy corresponds to the endothermal direction.

The typical formulation of a cast polymer-bonded explosive is shown below (reproduced with kind permission of Dr. Paul Wanninger):

percentage	ingredient	comment
85–90 %	AP, RDX, HMX, Al	
10–15 %	HTPB, plasticizer	HTPB is the binder
0.2 %	Lecithin	Lecithin is used to reduce the rheological yield point
0.2 %	Antioxidants	An antioxidant is added for long-life stability
10^{-3}–10^{-4} %	Catalyst	The amount of catalyst determines the curing time
0.05–0.1 %	bonding agent	Different bonding agents for RDX, HMX and AP or Al are recommended to improve the mechanical properties

Typical properties of melt cast, cast, and pressed PBX formulations are shown below (reproduced with kind permission of Dr. Paul Wanninger):

	melt cast PBX	cast PBX	pressed PBX
Ingredients	TNT, RDX, HMX, NTO	RDX, HMX + polymer	RDX, HMX + thermoplast
Technical process	melting and casting	mixing under vacuum and casting under vacuum, polymerization	HE + Binder + solvent, pressing at up to 1500 bar
Density (in % of TMD)	98	99.5	97–98
Density gradient	high	very low	varies
Vulnerability	high	very low	medium
Homogenicity	low	high	medium

1.2.2 New High (Secondary) Explosives

New secondary explosives which are currently under research, development or testing include 5-nitro-1,2,4-triazol-3-one (NTO), 1,3,3-trinitroazetidine (TNAZ), hexanitrohexaazaisowurtzitane (HNIW, CL-20) and octanitrocubane (ONC) (Fig. 1.7). **NTO** has already found application as a very insensitive compound in gas generators for automobile inflatable air bags and in some polymer-bonded explosive formulations. (*N.B.* Initially NaN_3 was used in air bag systems, however, nowadays guanidinium nitrate is often used in combination with oxidizers such as AN in some non-azide automotive inflators.) It is used to enhance burning at low flame temperatures. Low flame temperatures are desired in order to reduce the formation of NO_x

Fig. 1.6: Synthesis of 1,3,3-trinitroazetidine (TNAZ).

gasses in inflators. NTO is usually produced in a two-step process from semicarbazide hydrochloride with formic acid via the intermediate formation of 1,2,4-triazol-5-one (TO) and subsequent nitration with 70 % nitric acid:

Another interesting new and neutral high explosive is BiNTO, which can be synthesized as shown in the below equation from commercially available NTO in a one-step reaction.

TNAZ was first synthesized in 1983 and has a strained four-membered ring backbone with both C-nitro and nitramine (N—NO$_2$) functionalities. There are various routes that yield TNAZ all of which consist of several reaction steps. One possible synthesis of TNAZ is shown in Figure 1.6. It starts from epichlorohydrine and ^{t}Bu-amine. As far as the author of this book is aware, there has been no wide-spread use for TNAZ so far.

CL-20 (1987, A. Nielsen) and ONC (1997, Eaton) are without doubt the most prominent recent explosives based on molecules with considerable cage-strain. While CL-20 is now already produced in 100 kg quantities (e.g. by SNPE, France or Thiokol, USA, ca. $ 1000.—/Lb) on industrial pilot scale plants, ONC is only available on a mg to g scale because of its very difficult synthesis. Despite the great

Fig. 1.7: Molecular structures of 5-nitro-1,2,4-triazol-3-one (NTO), 1,3,3-trinitroazetidine (TNAZ), hexanitrohexaazaisowurtzitane (CL-20), octanitrocubane (ONC) and 4,10-dinitro-2,6,8,12-tetraoxa-4,10-diazaisowurtzitane (TEX).

Fig. 1.8: Synthesis of hexanitrohexaazaisowurtzitane (CL-20).

enthusiasm for CL-20 since its discovery over 20 years ago it has to be mentioned that even today most of the high explosive formulations are based on RDX (see Tab. 1.2). There are several reasons why CL-20 despite its great performance has not yet been introduced successfully:

- CL-20 is much more expensive than the relatively cheap RDX.
- CL-20 has some sensitivity issues (see insentitive munitions).
- CL-20 exists in several polymorphic forms and the desired ε polymorph (because of its high density and detonation velocity) is thermodynamically not the most stable one.

Interconversion of the ε form into a more stable but perhaps also more sensitive other polymorph would result in a loss of performance and an increase in sentitivity.

CL-20 is obtained by the condensation of glyoxal with benzylamine in an acid catalyzed reaction to yield hexabenzylhexaazaisowurtzitane (Fig. 1.8). Afterwards

Fig. 1.9: Molecular structures of FOX-7 and FOX-12 (a). Two alternative synthetic routes for the synthesis of FOX-7 (b).

the benzyl groups are replaced under reducing conditions (Pd-C catalyst) by easily removable acetyl substituents. Nitration to form CL-20 takes place in the final reaction step.

Another very insensitive high explosive which is structurally closely related to CL-20 is 4,10-dinitro-2,6,8,12-tetraoxa-4,10-diazaisowurtzitane (**TEX**, see Fig. 1.7), which was first described by Ramakrishnan and his co-workers in 1990. It displays one of the highest densities of all nitramines (2.008 g cm^{-3}) [1c].

Tab. 1.3: Characteristic performance and sensitivity data of FOX-7 and FOX-12 in comparison with RDX.

	FOX-7	FOX-12	RDX
detonation pressure, p_{C-J} / kbar	340	260	347
detonation velocity, D / m s^{-1}	8870	7900	8750
impact sensitivity / J	25	> 90	7.5
friction sensitivity / N	> 350	> 352	120
ESD / J	ca. 4.5	> 3	0.2

Fig. 1.10: DSC-Plot of FOX-7.

The chemist N. Latypov of the Swedish defense agency FOI developed and synthesized two other new energetic materials. These two compounds have become known as **FOX-7** and **FOX-12** (Fig. 1.9, a). FOX-7 or DADNE (diamino dinitro ethene) is the covalent molecule 1,1-diamino-2,2-dinitro ethene: $(O_2N)_2C=C(NH_2)_2$. The synthesis of FOX-7 always includes several reaction steps. Two alternative ways to prepare FOX-7 are shown in Figure 1.9 (b). FOX-12 or GUDN (guanylurea dinitramide) is the dinitramide of guanylurea: $[H_2N-C(=NH_2)-NH-C(O)-NH_2]^+[N(NO_2)_2]^-$.

It is interesting that FOX-7 has the same C/H/N/O ratio as RDX or HMX. Although neither FOX-7 nor (and in particular not) FOX-12 meet RDX in terms of performance (detonation velocity and detonation pressure). Both compounds are much less sensitive than RDX and might be of interest due to their insensitive munition (IM) properties. Table 1.3 shows the most characteristic performance and sensitivity data of FOX-7 and FOX-12 in comparison with RDX.

FOX-7 exists in at least three different polymorphic forms (α, β and γ). The α modification converts reversibly into the β form at 389 K (Fig. 1.10) [2]. At 435 K the β polymorph converts into the γ phase and this interconversion is not reversible. The γ form can be quenched at 200 K. When heated the γ form decomposes at 504 K. Structurally, the three polymorphs are closely related and quite similar, with the planarity of the individual FOX-7 layers increasing from α via β to γ (i.e. γ possesses the most planar layers) (Fig. 1.11).

Another member of the family of nitramine explosives is the compound dinitroglycoluril (DINGU) which was first reported as early as 1888. The reaction between glyoxal (O=CH—CH=O) and urea yields glycoluril which can be nitrated with 100 % nitric acid to produce DINGU. Further nitration with a mixture of HNO_3/N_2O_5 yields the corresponding tetramine SORGUYL. The latter compound is of interest because of its high density (2.01 g cm^{-3}) and its high detonation velocity (9150 m s^{-1}) (Fig. 1.12). SORGUYL belongs to the class of cyclic dinitroureas. These compounds generally show a higher hydrolytic activity and may therefore be of interest as "self-remediating" energetic materials.

A new neutral nitrimino-functionalized high explosive which was first mentioned in 1951 (*J. Am. Chem. Soc.* **1951**, *73*, 4443, see also Damavarapu et al., OPR & D, **2016**, 683–686) and which was recently suggested as a RDX replacement in C4 and Comp.B by Damavarapu (ARDEC) is bisnitraminotriazinone (DNAM). This compound has a melting point of 228 °C and a remarkably high density of 1.998 g/cc. Due to the high density and the not too negative enthalpy of formation (ΔH_f° = –111 kJ mol^{-1}) DNAM has a detonation velocity of 9200 m s^{-1} but still desirably low sensitivities (IS = 82.5 cm, FS = 216 N, ESD = 0.25 J). The synthesis of DNAM can be achieved in 50–60 % yield by nitration of melamine using in-situ generated $AcONO_2$ as the effective nitrating agent (Fig. 1.13a) or by direct nitration of melamine. One possible concern about DNAM is that the compound hydrolyzes rapidly at 80° with liberation of nitrous oxide. At room temperature, the hydrolysis requires one to two days and is acid catalyzed.

The reaction of DNAM with $NaHCO_3$, CsOH and $Sr(OH)_2 \cdot 8 H_2O$ yields the corresponding mono-deprotonated salts NaDNAM, CsDNAM and Sr(DNAM)$_2$, respectively.

Pyrazine derivatives are six-membered heterocyclic compounds containing two nitrogen atoms in the ring system. As high-nitrogen heterocylic compounds, they have an ideal structure for energetic materials (EMs). Some of them have a high formation enthalpy, fine thermal stability and good safety characteristics. The basic structure of energetic pyrazine compounds is that of 3,5-dinitro-2,6-diaminopyrazine (**I** in Fig. 1.13). One of the most prominent members in this family is the 1-oxide 3,5-dinitro-2,6-pyrazinediamine-1-oxide (also known as **LLM-105**, Fig. 1.13). LLM-105 has a high density of 1.92 g cm^{-3} and it shows a detonation velocity of 8730 m s^{-1} and a detonation pressure of 359 kbar which are comparable to those of RDX (density = 1.80 g cm^{-3}, exptl. values: VoD = 8750, P_{C-J} = 347 kbar). LLM-105 is a lot less impact sensitive than RDX and is not sensitive towards electrostatics and friction [1d].

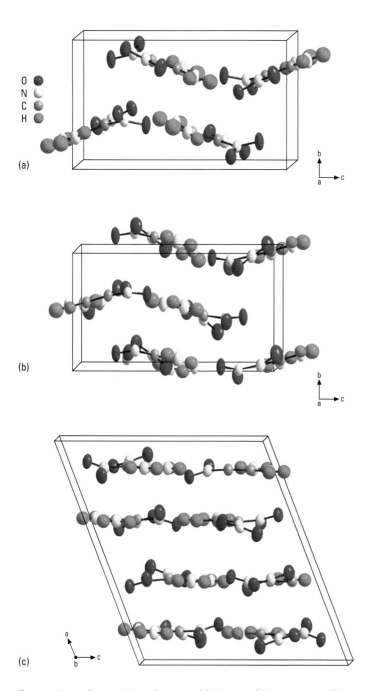

Fig. 1.11: Crystalline packing of α-FOX-7 (a), β-FOX-7 (b) and γ-FOX-7 (c).

Fig. 1.12: Molecular structures of dinitroglycoluril (DINGU) and the corresponding tetramine (SORGUYL).

Fig. 1.13: Synthesis of LLM-105 starting from 2,6-dichloropyrazine. 3,5-dinitro-2,6-diaminopyrazine is oxidized to 3,5-dinitro-2,6-pyrazinediamine-1-oxide (LLM-105) in the final step.

Fig. 1.13a: Synthesis of DNAM.

The synthesis of LLM-116 is shown in Fig. 1.13c.

Another N-oxide which has recently been suggested by Chavez et al. (LANL) as an insensitive high explosive is 3,3′-diaminoazoxy furazan (DAAF). Though the detonation velocity and detonation pressure of DAAF are rather low (7930 m s^{-1}, 306 kbar @ 1.685 g/cc), the low sensitivity (IS > 320 cm, FS > 360 N) and a critical diameter of < 3 mm make this compound promising. The synthesis of DAAF is shown in Fig 1.13b [see also E.-C. Koch, Propellants, Explosives, Pyrotech, **2016**, *41*, 529–538].

Fig. 1.13b: Synthesis of DAAF.

Fig. 1.13c: Synthesis of LLM-116.

There are various methods to prepare LLM-105. Most methods start from commercially available 2,6-dichloropyrazine (Fig. 1.13) and oxidize dinitropyrazinediamine in the final step to the 1-oxide (LLM-105).

Organic peroxides are another class of explosives which has been researched recently. This class of explosives (organic, covalent peroxides) includes the following compounds:
- H_2O_2
- peracids, R–C(O)–OOH
- peresters, R–C(O)–OO–R′
- perethers, R–O–O–R′
- peracetals, $R_2'C-(OOR)_2$.
- peroxyanhydride, R–C(O)–O–O–C(O)–R′

Triacetone triperoxide (TATP, Fig. 1.14) is formed from acetone in sulfuric acid solution when acted upon by 45 % (or lower concentration) hydrogen peroxide (the acid acts as a catalyst). Like most other organic peroxides TATP has a very high impact (0.3 J), friction (0.1 N) and thermal sensitivity. TATP has the characteristics of a primary explosive. For this reason and because of its tendency to sublime (high volatility) it is not used in practice (apart from terrorist and suicide bomber activities).

Fig. 1.14: Molecular structures of triacetone triperoxide (TATP), hexamethylene triperoxide diamine (HMTD), methyl ethyl ketone peroxide (MEKP) and diacetone diperoxide (DADP).

Because of the use of TATP by terrorists, a reliable and fast detection of this material is desirable. In addition to conventional analytical methods such as mass spectrometry and UV (ultra violet) spectroscopy specially trained explosive detection dogs (EDD) play an important role in the detection of organic peroxides. However, fully trained EDDs are expensive (up to $ 60 k) and can only work for 4 h per day. Although the high vapor pressure helps the dogs to detect the material, it is also a disadvantage because of the limited time-span in which the dog is able to find it (traces may sublime and disappear forever). Matrices in which the compounds can be imbedded are sought after for safe training of explosive detection dogs. These matrices should not have any volatility or any characteristic smell for the dogs. In this respect zeolites may be of interest [1e, f]. The ongoing problem with zeolites is that they need to be loaded with solutions and the solvents (e.g. acetone) may not completely vaporize before the peroxide.

Typical organic peroxides, which have been or may be used by terrorists are so-called homemade explosives (HMEs): triacetone triperoxide (TATP), hexamethylene triperoxide diamine (HMTD), methyl ethyl ketone peroxide (MEKP) and diacetone diperoxide (DADP) (Fig. 1.14).

The following class of N-oxide compounds is considerably more stable than the above mentioned peroxides. For example, the oxidation of 3,3'-azobis(6-amino-1,2,4,5-tetrazine) in H_2O_2 / CH_2Cl_2 in the presence of trifluoroacetic acid anhydride yields the corresponding N-oxide (Fig. 1.15). This compound has a desirable high density and only modest impact and frictiteon sensitivity.

Another oxidizing reagent that has proven useful at introducing N-oxides is commercially available Oxone® (2 $KHSO_5 \cdot KHSO_4 \cdot K_2SO_4$). The active ingredient in this oxidizing agent is potassium peroxomonosulfate, $KHSO_5$, which is a salt of Caro's acid, H_2SO_5. Examples of oxidation reactions involving Oxone® are shown in Figure 1.16, including the interconversion of an amine (R_3N) into an N-oxide. (*N.B.* Sometimes, mCPBA [meta-chloro perbenzoic acid] or CF_3COOH are also used as an oxidizing agent for form N-oxides.)

Another tetrazine derivative, 3,6-bis(1 H-1,2,3,4-tetrazole-5-ylamino)-s-tetrazine, has recently been prepared from (*bis*(pyrazolyl)tetrazine (Fig. 1.15). It is interesting to note that the tetrazine derivatives potentially form strong intermolecular interac-

Fig. 1.15: Synthesis of an N-oxide (a) and preparation of 3,6-bis(1H-1,2,3,4-tetrazole-5-ylamino)-s-tetrazine (b).

Fig. 1.16: Oxidation reactions with Oxone® as the oxidant (see for example: B. R. Travis, M. Sivakumar, G. O. Hollist, B. Borhan, Org. Lett. **2003**, 5, 1031–1034).

tions via π-stacking. This can influence many of the physical properties in a positive way, for example by reducing the electrostatic sensitivity.

1.2.3 New Primary Explosives

In early days Alfred Nobel already replaced mercury fulminate (MF, see above), which he had introduced into blasting caps, with the safer to handle primary explosives lead azide (LA) and lead styphnate (LS) (Fig. 1.17). However, the long-term use of LA and LS has caused considerable lead contamination in military training grounds which has stimulated world-wide activities in the search for replacements that are heavy-metal free. In 2016 the US military still used ca. 2000–3000 kg LA per year. In 2006 Huynh und Hiskey published a paper proposing iron and copper complexes of the type $[cat]_2^+[M^{II}(NT)_4(H_2O)_2]$ ($[cat]^+ = NH_4^+$, Na^+; M = Fe, Cu;

Fig. 1.17: Molecular structures of lead styphnate (LS), lead azide (LA), an iron and copper nitrotetrazolate complexes as well as copper(I) 5-nitrotetrazolate (DBX-1) and potassium-7-hydroxy-6-dinitrobenzofuroxane (KDNP).

NT = 5-nitrotetrazolate) as environmentally friendly, "green" primary explosives (Fig. 1.17) [3].

In 2007 the LMU Munich research group reported on the compound copper bis(1-methyl-5-nitriminotetrazolate) with similarly promising properties (Fig. 1.17) [4]. Because they have only been discovered recently, none of the above mentioned complexes has found application yet, but they appear to have substantial potential as lead-free primary explosives.

Another environmentally compatible primary explosive is copper(I) 5-nitrotetrazolate (Fig. 1.17). This compound has been developed under the name of **DBX-1** by Pacific Scientific EMC and is a suitable replacement for lead azide. DBX-1 is thermally stable up to 325 °C (DSC). The impact sensitivity of DBX-1 is 0.04 J (ball-drop instrument) compared with 0.05 J for LA. The compound is stable at 180 °C

Fig. 1.18: Typical design of a stab detonator; 1: initiating charge, stab mix, e.g. NOL –130 (LA, LS, tetrazene, Sb$_2$S$_3$, Ba(NO$_3$)$_2$); 2: transfer charge (LA); 3: output charge (RDX).

for 24 hrs in air and for 2 months at 70 °C. DBX-1 can be obtained from NaNT and Cu(I)Cl in HCl/H$_2$O solution at a higher temperature. However, the best preparation for DBX-1 in a yield of 80–90 % is shown in the following equation where sodium ascorbate, NaC$_6$H$_7$O$_6$, is used as the reducing agent:

$$\text{CuCl}_2 + \text{NaNT} \xrightarrow{\text{reducing agent, H}_2\text{O, 15 min, }\Delta T} \text{DBX-1}$$

The mercury salt of NT, Hg(NT)$_2$, is also known under the name DXN-1 or DXW-1.

A possible replacement for lead styphnate is potassium-7-hydroxy-6-dinitrobenzofuroxane (**KDNP**) (Fig. 1.17). KDNP is a furoxane ring containing explosive and can best be prepared from commercially available bromo anisol according to the following equation. The KN$_3$ substitutes the Br atom in the final reaction step and also removes the methyl group:

A typical stab detonator (Fig. 1.18) consists of three main components:
1. initiating mixture or initiating charge (initiated by a bridgewire),
2. transfer charge: primary explosive (usually LA),
3. output charge: secondary explosive (usually RDX).

A typical composition for the initiating charge is:
20 % LA,
40 % LS (basic),
 5 % tetrazene,
20 % barium nitrate,
15 % antimony sulfide, Sb_2S_3.

It is therefore desirable, to find suitable heavy metal-free replacements for both lead azide and lead styphnate. Current research is addressing this problem. The following replacements in stab detonators are presently being researched:
1. initiating charge   LA   →   DBX-1
    LS   →   KDNP
2. transfer charge: LA   →   triazine triazide (TTA) or APX
3. output charge: RDX →   PETN or BTAT

Primary explosives are substances which show a very rapid transition from deflagration to detonation and generate a shock-wave which makes transfer of the detonation to a (less sensitive) secondary explosive possible. Lead azide and lead styphnate are the most commonly used primary explosives today. However, the long-term use of these compounds (which contain the toxic heavy metal lead) has caused considerable lead contamination in military training grounds. Costly clean-up operations require a lot of money that could better be spent improving the defense capability of the country's forces. A recent article published on December 4th 2012 in the Washington Post (http://www.washingtonpost.com/blogs/federal-eye/wp/2012/12/03/new-report-warns-of-high-lead-risk-for-military-firing-range-workers/) entitled "Defense Dept. Standards On Lead Exposure Faulted" stated: "… it has found overwhelming evidence that 30-year-old federal standards governing lead exposure at Department of Defense firing ranges and other sites are inadequate to protect workers from ailments associated with high blood lead levels, including problems with the nervous system, kidney, heart and reproductive system."

Devices using lead primary explosives – from primers for bullets to detonators for mining – are manufactured in the tens of millions every year in the United States. In the US alone, over 750 lbs. of lead azide are consumed every year for military use.

Researchers from LMU Munich have now synthesized in collaboration with ARDEC at Picatinny Arsenal, N.J. a compound named K_2DNABT (Fig. 1.18a), a new heavy metal-free primary explosive which has essentially the same sensitivity (impact, friction and electrostatic sensitivity) as that of lead azide, but does not contain toxic lead. Instead of lead, it contains the ecologically and toxicologically benign element potassium instead. Preliminary experimental detonation tests (dentplate tests) and high-level computations have shown that the performance of K_2DNABT even exceeds that of lead azide. Therefore, there is great hope that toxic

Fig. 1.18a: Chemical structures of K$_2$DNABT and DBX-1.

lead azide and / or lead styphnate can be replaced in munitions and detonators with this physiologically and ecologically benign compound.

In theory, unprotected 1,1′-diamino-5,5′-bistetrazole can be nitrated. However, the amination of 5,5′-bistetrazole is a procedure which results in only low yields and also requires considerable effort, therefore an alternative route was developed. The bisnitrilimine would appear to be a suitable precursor, however, unfortunately unprotected bisnitrilimine is not known and only the corresponding diphenyl derivative is known. Therefore another derivative was prepared which contains a more easily removable protecting group than the phenyl group. The synthetic sprocess for the synthesis of K$_2$DNABT starts from the easily preparable dimethyl carbonate. This is reacted with hydrazine hydrate to form the carbazate **1**. The subsequent condensation reaction with half an equivalent of glyoxal forms compound **2**, which is subsequently oxidized with NCS (N-chlorosuccinimide) to the corresponding chloride. Substitution with sodium azide offers the diazide (in only 38 % yields) which then is cyclized with hydrochloric acid in ethereal suspension. The carboxymethyl protected 1,1′-diamino-5,5′-bistetrazole is then gently nitrated with N$_2$O$_5$ (Fig. 1.18b).

An alkaline aquatic work-up with KOH precipitates dipotassium 1,1′-dinitramino-5,5′-bistetrazolate. The products of the individual stages can be purified by recrystallization, or used as obtained. No column chromatography must be used. Fortunately K$_2$DNABT shows low water solubility, which (i) facilitates its isolation and purification and (ii) avoids future toxicity problems due to potential ground water pollution.

A primary explosive is an explosive that is extremely sensitive to stimuli such as impact, friction, heat or electrostatic discharge. Only very small amounts of energy are required to initiate such a material. Generally, primary explosives are considered to be materials which are more sensitive than PETN. Primary explosives are described as being the initiating materials that initiate less sensitive energetic materials such as secondary explosives (e.g. RDX/HMX) or propellants. A small quantity – usually only milligrams – is required to initiate a larger charge of explo-

Fig. 1.18b: Synthetic pathway for the formation of K$_2$DNABT.

sive which is safer to handle. Primary explosives are widely used in primers, detonators and blasting caps. The most commonly used primary explosives are lead azide and lead styphnate. Lead azide is the more powerful of the two and is used as a pure substance typically in detonators as a transfer charge, or in formulations for initiation mixes (e.g. NOL-130). Lead styphnate is mostly used in formulations for initiation and primer mixes and is rarely found as a neat material in applications.

Although lead azide has been used extensively for decades, it is a very poisonous material that reacts with copper, zinc or alloys containing these metals, forming other azides that can be highly sensitive and dangerous to handle. Furthermore, lead-based materials have been clearly found to cause environmental and health related problems. Lead-based materials are included on the EPA Toxic Chemical List (EPA List of 17 Toxic Chemicals); they are additionally regulated under the Clean Air Act as Title II Hazardous Air Pollutants, as well as classified as toxic pollutants under the Clean Water Act, and are on the Superfund list of hazardous substances. Under the Clean Air Act, the USEPA (US Environmental Protection Agency) revised the National Ambient Air Quality Standard (NAAQS) to 0.15 µg/m³, which is ten times more stringent than the previous standard. Lead is both an acute and chronic toxin, and the human body has difficulty in removing it once it has been absorbed and dissolved in the blood. Consequently, a chief concern is the absorption of lead by humans who are exposed to the lead-containing constituents of the initiating mix, as well as the combustion by-products of lead-based compositions. The health effects of lead are well documented however, recent studies have shown that there are no safe exposure levels for lead, particularly for children. There is a direct correlation between lead exposure and develop-

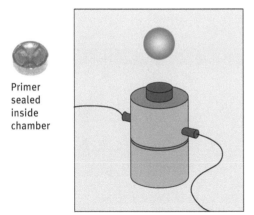

Fig. 1.18c: Primer performance test setup.

ment, including IQ loss (even at the revised lead NAAQS, exposure levels are consistent with an IQ loss of over 2 points), behavioral issues and even hearing loss. Their use during training and testing deposits heavy metals on ranges and can impact sustainable use of these ranges.

These initiator and transfer charge compositions also require expensive handling procedures during production and disposal. The manufacturing of any lead-based primary explosives, such as lead azide or lead styphnate, results in the production of significant quantities of highly toxic, hazardous waste. In addition, the handling and storage of these compounds is also a concern.

Due to its environmental and health impact, there is a need to develop green primary explosives, to replace lead-based compounds. Current research seeks to replace lead azide and lead-based formulations in commonly used detonators, and in the M55 stab detonators and percussion primers, such as M115, M39, M42, etc. in particular. One important strategy for developing lead-free alternatives focuses on the use of high-nitrogen compounds. High-nitrogen compounds are widely considered to be viable, environmental-friendly energetic, since the predominant detonation product is non-toxic nitrogen (N_2) gas. At the same time, high-nitrogen compounds possess high enthalpies of formation (ΔH_f), which lead to high energy outputs. In developing/testing viable lead replacements, the following criteria must be considered for new compounds. The material:
- must be safe to handle and possess a rapid deflagration to detonation transition
- be thermally stable to temperatures above 150 °C and must have a melting point greater than 90 °C
- should possess high detonation performance and sensitivity
- should have long term chemical stability
- should not contain toxic heavy metals or other known toxins
- be easy to synthesize and affordable.

Fig. 1.18d: Performance of K$_2$DNABT/ RDX after initiation by an electrical igniter.

Tab. 1.3a: Comparison of the sensitivities of K$_2$DNABT with other primary explosives.

sample	Impact / in	Friction / N	ESD / mJ	Density / g cm^{-3}	VoD / m s^{-1}
LA	7–11	0.1	4.7	4.8	5300
LS	5	0.1	0.2	3.00	4900
DBX-1	4	0.1	3.1	2.58	ca. 7000
K$_2$DNABT	2	0.1	0.1	2.2	ca. 8330

For primer performance testing, primers are sealed in an air-tight test apparatus and initiated by dropping an 8 or 16 oz steel ball onto the primary explosive. Pressure transducers measure the output (Fig. 1.18c).

The sensitivity of K$_2$DNABT was tested and was found to be very sensitive to impact, friction and ESD, as all primary explosives are. K$_2$DNABT is more sensitive compared to lead azide, lead styphnate and DBX-1 (as shown in Tab. 1.3a).

To compare the performance of K$_2$DNABT with that of lead azide, 40 mg of K$_2$DNBT was loaded in a small aluminum holder and 1 g of RDX was pressed in a standard detonator copper shell. This was initiated by an electrical igniter. The performance of K$_2$DNABT/RDX was shown to pass the requirements expected of a primary explosive mixture as can be seen in Fig. 1.18d.

A modified Small Scale Shock Reactivity Test (SSRT) was performed with K$_2$DNABT as well as with lead azide for comparison. In these tests, 500 mg of each compound was ignited by an igniter and it was found that K$_2$DNABT showed more indentation than lead azide on the witness plate (Fig 1.18e).

When performed electrically, K$_2$DNABT is comparable to lead styphnate in performance. Primer mixes have been also been formulated and require testing.

Future work requires optimization of the system in order for K$_2$DNABT to be considered as a replacement for lead azide in detonators. From the initial results, K$_2$DNABT appears to be a good possible replacement for lead styphnate when initi-

Fig. 1.18e: Witness plate of SSRT of lead azide (left) and K$_2$DNABT (right).

Fig. 1.18f: Fischer test of K$_2$DNABT.

ated via a hot-wire or electric bridgewire. Formulations will also need to be optimized in the future, with respect to finding the correct particle size.

An image of the Fischer test is shown in Fig. 1.18f.
Another promising and thermally stable (Tab. 1.3b) lead-free primary explosive is copper(II) 5-chlorotetrazolate (Cu ClT, PSI & LMU). The synthesis is achieved in a one-step reaction (Fig. 1.18a) starting from commercially available aminotetrazole. Cu ClT can then be further converted into the synthetically useful compounds sodium chlorotatrazolate and chlorotetrazole (Fig. 1.18h).

In the area of metal-free primary explosives, covalently bound azides are often advantageous. Although these compound soften do not show the high thermal stability of metal complexes, some may have an application as LA replacements in

Tab. 1.3b: Energetic properties of copper(II) chlorotetrazolate (CuClT).

	CuClT
formula	$C_2Cl_2CuN_8$
molecular mass [g mol^{-1}]	270.53
impact sensitivity [J]	1
friction sensitivity [N]	< 5
ESD-test [J]	0.025
N [%]	41.42
$T_{dec.}$ [°C]	289

Fig. 1.18g: Synthesis of diazidoglyoxime (DAGL).

Fig. 1.18h: Synthesis of copper(II) chlorotetrazolate (Cu ClT).

Tab. 1.3c: Energetic properties of TTA and DAGL in comparison with LA.

	TTA	DAGL	LA
formula	C_3N_{12}	$C_2H_2N_8O_2$	N_6Pb
molecular mass [g mol^{-1}]	204.1	170.1	291.3
impact sensitivity [J]	1.3	1.5	2.5–4
friction sensitivity [N]	< 0.5	< 5	0.1–1
ESD [J]	< 0.36 (?)	0.007	0.005
$T_{dec.}$ [°C]	187 (m.p. 94)	170	315

transfer charges (see Fig. 1.18). Two of the most promising candidates are triazido triazine (triazine triazide, TTA, ARDEC, see above) and diazidoglyoxime (DAGL, LMU, Tab. 1.3c). The latter one can be prepared according to Fig. 1.18h.

1.2.4 New Oxidizers for Solid Rocket Motors

Solid propellants of essentially all solid rocket boosters are based on a mixture of aluminum (Al, fuel) and ammonium perchlorate (AP, oxidizer).

Ammonium perchlorate (AP) has applications in munitions, primarily as an oxidizer for solid rocket and missile propellants. It is also used as an air bag inflator in the automotive industry, in fireworks, and appears as a contaminant of agricultural fertilizers. As a result of these uses and ammonium perchlorate's high solubility, chemical stability, and persistence, it has become widely distributed in surface and ground water systems. There is little information about the effects of perchlorate on the aquatic life, but it is known that perchlorate is an endocrine disrupting chemical that interferes with normal thyroid function which impacts both growth and development in vertebrates. Because perchlorate competes for iodine binding sites in thyroids, adding iodine to culture water has been examined in order to determine if perchlorate effects can be mitigated. Finally, perchlorate is known to affect normal pigmentation of amphibian embryos. In the US alone the cost for remediation is estimated to be several billion dollars. That money that is urgently needed in other defense areas [5–7].

The currently most promising chlorine-free oxidizers which are being researched at present are ammonium dinitramide (ADN), which was first synthesized in 1971 in Russia (Oleg Lukyanov, Zelinsky Institute of Organic Chemistry) and is being commercialized today by EURENCO, as well as the nitroformate salts hydrazinium nitroformate (HNF, APP, Netherlands) and triaminoguanidinium nitroformate (Germany) (Fig. 1.19) [8]. The salt hydroxyl ammonium nitrate, $HO-NH_3^+\ NO_3^-$ (HAN) is also of interest. However, all four compounds possess relatively low decomposition temperatures and TAGNF only has a positive oxygen balance with respect to CO (not to CO_2).

$NH_4^\oplus\ \ ^\ominus N(NO_2)_2$ ADN

$N_2H_5^\oplus\ \ C(NO_2)_3^\ominus$ HNF

$C(NH-NH_2)_3^\oplus\ \ C(NO_2)_3^\ominus$ TAGNF

Fig. 1.19: Molecular structures of ammonium dinitramide (ADN), hydrazinium nitroformate (HNF) and triaminoguanidinium nitroformate (TAGNF).

While ADN has the best oxygen balance (Ω_{CO_2} = 25.8%, cf. AP 34.0%) of all presently discussed AP replacements it still has some stability issues with respect to binder compatibility and thermal stability ($T_{dec.}$ = 127 °C). Thermal decomposition of ADN is observed at 127 °C after complete melting at 91.5 °C. The main decomposition pathway is based on the formation of NH_4NO_3 and N_2O followed by the thermal decomposition of NH_4NO_3 to N_2O and H_2O at higher temperatures. Side reactions forming NO_2, NO, NH_3, N_2 and O_2 are described and a mechanism for the acid-catalyzed decomposition of hydrogen dinitramide, dissociation product of ADN, is proposed.

Alternatively, ammonium nitrate (AN, Ω_{CO_2} = 20.0%, begins decomposition at m.p. = 169.9 °C, complete decomposition at 210 °C) has been discussed, however this compound has severe burn rate issues. Furthermore, AN is hygroscopic and shows phase transitions from one polymorph to another at 125.2 °C, 84.2 °C, 32.3 °C and –16.9 °C. Phase stabilized ammonium nitrate (PSAN) and spray crystallized AN (SCAN) are special qualities provided by ICT.

Another recently suggested organic oxidizer is TNC-NO_2 which has an oxygen balance of Ω_{CO_2} = 14.9% and a thermal stability of up to 153 °C. TNC-NO_2 can be synthesized by direct nitration of TNC (2,2,2-trinitroethyl carbamate) using mixed acid:

The further new nitroethyl compounds based on boron esters are tris-(2-nitroethyl) borate and tris-(2,2,2-trinitroethyl) borate. Especially the trinitroethyl derivative is a suitable candidate for high energy density oxidizers and for smoke-free, green coloring agents in pyrotechnic compositions. Tris-(2-nitroethyl) borate and tris-(2,2,2-trinitroethyl) borate can be obtained from boron oxide with 2-nitroethanol and 2,2,2-trinitroethanol, respectively:

The oxygen balance of **2** is –59.70% and of **3** is +13.07%. The density for **3** was determined to 1.982 g cm^{-3}, which is a quite high value. DSC measurements revealed an exothermic decomposition at 216 °C for **2** and 161 °C for **3**.

The starting material 2,2,2-trinitroethanol (**1**) was prepared from the reaction of trinitromethane with formaldehyde (see Fig. 9.18).

Sensitivity data for **3**.

grain size	< 100 μm
impact sensitivity	15 J
friction sensitivity	144 N
electrostatic sensitivity	0.5 J

In general, elemental boron and boron compounds have been the subject of many studies involving propulsion systems due to the very high heats of combustion. However, due to some combustion problems, the theoretical benefits of boron and its compounds have rarely been delivered. The main limitations for boron combustion are (i) ignition delay due to the presence of an oxide layer on the surface of the boron particle (only for elemental boron) and (ii) the energy release during the combustion process in hydrogen containing gases is significantly lowered due to the formation of HBO_2 or "HOBO" as commonly referred to. The slow oxidation of "HOBO" to the preferred product of combustion B_2O_3 often results in the particles leaving the propulsion-system prior to releasing all of their energy thus resulting in less than expected performance.

Nitroform (trinitromethane, NF) is a valuable starting material for the production of propellant and explosive components due to its high oxygen content. It was first obtained as the ammonium salt by Shiskov in 1857. The acidic hydrogen atom facilitates the formation of derivatives, since deprotonation can readily occur with bases to form, for example, hydrazinium nitroformate (HNF) [69]. HNF possesses many useful properties, which makes it useful for application: it can be used as an oxidizer in rocket propellants, and it is also chlorine-free, which is a desirable property in modern propellant compositions [70, 71].

NF is also a highly useful starting material in the synthesis of many other chlorine-free oxidizers, such as $TNC-NO_2$ and tris(trinitroethyl) borate. Whereas HNF is directly synthesized from NF, the oxidizers $TNC-NO_2$ and tris(trinitroethyl) borate are prepared from trinitroethanol (TNE). TNE can easily be synthesized in two steps from an aqueous solution of nitroform with formaldehyde:

$$H-C(NO_2)_3 + CH_2O \rightarrow (NO_2)_3C-CH_2-OH + 8.56 \text{ kcal mol}^{-1} \qquad (1)$$

NF participates in many reactions such as addition reactions [72], condensation reactions [73], and substitution reactions [74]. NF reacts with unsaturated compounds to form multi-nitroalkane derivatives, and therefore, it can be used for the synthesis of an explosive composition containing the nitroform group such as 1-(2,2,2-trinitroethylamino)-2-nitroguanidines [75] or 1,1,1,3-tetranitro-3-azabutane [76].

Since it has proven itself to be a crucially important raw material in the synthesis of many energetic compounds, synthetic routes for the formation of NF have attracted considerable attention and have been explored, for example, by the nitra-

Fig. 1.20: Synthetic routes for the preparation of NF.

tion of various substrates including acetylene, acetic anhydride, pyrimidine-4,6-diol, and isopropanol (Fig. 1.20).

The most investigated preparative method of nitroform in the literature is the nitration of acetylene with nitric acid. However, this method requires the use of expensive and toxic mercury nitrate as a catalyst, which is clearly detrimental from the economic and environmental point of view. Other preparative methods for nitroform, which have been reported in the literature, include the nitration of other suitable compounds, such as acetone or isopropanol, the hydrolysis of tetranitromethane, or the nitration-hydrolysis of pyrimidine-4,6-dione [77, 78]. Due to the volatility and flammability of acetone, methods involving its use are dangerous, whereas the method involving tetranitromethane requires a water vapor distillation process which makes it difficult to prepare the raw materials.

For the large-scale production of NF, the use of isopropanol has certain advantages, e.g. safety, cost, and sustainability (environmental aspects). The synthesis of NF starting from isopropanol was first reported by Frankel [79] in 1978, and is

Fig. 1.21: Synthesis of NF from isopropanol in fuming nitric acid.

Fig. 1.22: Synthesis of TNE via potassium nitroformate.

based on the oxidation, nitration, and hydrolysis of isopropanol with fuming nitric acid (Fig. 1.21).

Alternatively, on a laboratory scale, NF can be obtained from tetranitromethane (TNM) which decomposes under alkaline conditions to yield HNO3 and nitroformate salts. TNM itself can be prepared in the laboratory by the nitration of acetic anhydride with fuming nitric acid (Fig. 1.20).

$$C(NO_2)_4 + KOH \rightarrow HNO_3 + K^+C(NO_2)_3^- \tag{2}$$

Another convenient synthesis which can be used to obtain potassium nitroformate (which can be subsequently converted into TNE) on a laboratory scale is the nitration of barbituric acid as shown in Fig. 1.22.

All of the problems mentioned above with the known synthetic routes for the preparation of NF demanded new, safer and more efficient methods to be developed for the synthesis of NF. In the last few years, several new methods have been reported. In 2003, A. Langlet et al published a patent [80] concerned with the preparation of nitroform (NF) by nitrating a starting material which is dissolved or suspended in sulfuric acid. A nitrating agent consisting of nitric acid, nitrate salts or nitrogen pentoxide is added to the sulfuric acid solution/suspension at temperatures in the range of –10 °C to +80 °C. The starting material can be selected from one of the following classes of compounds:

a) a gem-dinitroacetyl urea; X, in which X is H or a group, wherein Y is an alkoxy or amino group, and salts thereof;
b) gem-dinitroacetyl guanidine
c) 4,6-hydroxypyrimidine

The sulfuric acid can have a concentration of 70–100 %, preferably 95 %, and the nitric acid is added as a concentrated acid (85–100 %). The molar ratio between the nitrating acid and the substrate may be in the range 2.0–6.0 : 1. Good yields of NF have been obtained with a molar ratio as low as 3 : 1, which (in combination with the reuse of sulfuric acid) makes the method economically attractive. After completion of the nitration, the product (the gem-trinitro compound) is hydrolyzed and nitroform is spit off by mixing the reaction with an aqueous medium (e.g. crushed ice).

The nitroform which is prepared can be extracted from the reaction mixture using a suitable polar extracting agent, such as methylene chloride or diethyl ether. A neutralizing agent can then be added to precipitate the corresponding nitroform salt. Extraction of the gem-trinitro compound directly from the nitration mixture is also possible, and nitroform can then be subsequently obtained using a hydrolysis procedure, for instance, by adding a base. The sulfuric acid may then be used several times.

In 2015, Hong-Yan Lu et al. published an article which presented a new method to synthesize nitroform by the nitrolysis of cucurbituril. Cucurbituril (CB[6]) is a cyclic oligomer of 6 units of glycoluril linked by 12 methylene bridges [81]. In this study, CB[n] (n = 5–8) was nitrolyzed with fuming nitric acid in acetic anhydride to produce nitroform. This process requires only mild reaction conditions, reduces risk, is inexpensive, and therefore must be regarded as a new and useful preparative route for the formation of nitroform.

Fig. 1.23: Nitrolysis of CB.

In short, there is a reason why AP has been used as the most important oxidizer for composite propellants for decades. It has the advantage of being completely convertible to gaseous reaction products and has an oxygen balance of $\Omega_{CO_2} = 34\%$. It can easily be prepared by neutralizing ammonia with perchloric acid and is purified by crystallization. AP is stable at room temperature but decomposes at measurable rates at temperatures greater than about 150 °C. At decomposition temperatures below approximately 300 °C, AP undergoes an autocatalytic reaction which ceases after about 30 % decomposition. This is usually called the low-temperature reaction. The residue is quite stable at these temperatures unless rejuvenated by sublimation, recrystallization, or mechanical disturbance. At temperatures above 350 °C, the high-temperature decomposition occurs; this reaction is not autocatalytic and decomposition is complete. Concurrently with these decomposition reactions, AP also undergoes dissociative sublimation.

1.2.5 Initial Characterization of New Energetic Materials

Once a new energetic material (high explosive, primary explosive, oxidizer) has been synthesized in the lab, various characterization and evaluation methods need to get applied before one can consider up-scaling of the material and the preparation of formulations. The most important characterization and evaluation methods are summarized in Tab. 1.3d.

Tab. 1.3d: Synthesis, Characterization & Evaluation of New EMs (**before** formulation).

	Synthesis & Characterization:	
	– IS, FS, ESD (grain size) – CHN analysis – X-ray, density or pycnometry – ^{1}H, ^{13}C{1H}, ^{14}N/^{15}N NMR; IR/Raman – DSC (5 K/min), TGA, m.p., b.p., T_{dec} – calcd: VoD, p_{C-J}, I_{sp}, Q_{ex}, ... – Phase transitions / polymorphs (powder X-ray, DSC) – aquatic toxicity – better, cheaper ($/kg), higher yield, safer prep? – hydrolysis, degradation, detonation/combustion products – shape and habit of crystals (electron microscopy) – vapor pressure measurement	
Secondary Explosives:	**Primary Explosives:**	**Oxidizers:**
1. VST 2. compatibilities (DSC & VST) # 3. VoD (exptl.) 4. steel shell (Koenen) 5. Trauzl 6. under-water ex. 7. SSRT 8. FCO 9. SCO 10. GAP 11. isothermal long term 12. bomb calorimetry 13. ANSYS: fragment velocity, E_{kin}.	1. laser initiation 2. initiation cap. (RDX. PETN, HNS) 3. electromag. ball-drop primer test 4. M55 stab 5. compatibilities *	1. VST 2. compatibilities (DSC & VST) + 3. bomb calorimetry 4. FCO 5. SCO 6. isothermal long term

CuO, SS, Al, binder & plasticizer, DNAN, TNT, NTO, RDX, HMX
* RDX, PETN, Al, SS, Cu, CuO, LA, LS, Ba(NO$_3$)$_2$, Tetracene, Sb$_2$S$_3$
+ HTPB, binder, plasticizer, Al, SS, NC, NQ, GAP

1.3 Definitions

In accordance to the ASTM international definition (American Society for Testing and Materials: http://www.astm.org/), an **energetic material** is defined as a compound or mixture of substances which contains both the fuel and the oxidizer and reacts readily with the release of energy and gas. Examples of energetic materials are
– primary explosives
– secondary explosives
– gun propellant charges

- rocket fuels
- pyrotechnics, e.g. signal torches und illuminants, smoke and fog generators, decoys, incendiary devices, gas generators (air bags) and delay compositions.

Energetic materials can be initiated using thermal, mechanical or electrostatic ignition sources and do not need atmospheric oxygen to maintain the exothermic reaction.

A material which could be considered to be a potential explosive, is a chemical compound or a mixture of substances in a metastable state, which is capable of a quick chemical reaction, without requiring additional reaction partners (e.g. atmospheric oxygen). In order to estimate how explosive a substance is, the Berthelot-Rot value B_R can be used (equation 1). Here ρ_0 (in kg m^{-3}) is the density of the potential explosive, V_0 (in m^3 kg^{-1}) is the volume of smoke and Q_v (in kJ kg^{-1}) is the heat of explosion.

$$B_R [\text{kJ m}^{-3}] = \rho_0^2 \, V_0 \, Q_v \tag{1}$$

Table 1.4 shows the Berthelot-Rot values for some well-known explosives. In general, compounds with a B_R value higher or equal to that of the "Oppauer salt" (55 % NH$_4$NO$_3$, 45 % (NH$_4$)$_2$SO$_4$) – the detonation of which in 1921 caused a massive and catastrophic explosion with over 1000 fatalities – are considered potentially explosive.

In contrast to the potentially explosive substances discussed above, explosion hazards are defined as follows. Hazardous explosive substances are compounds or mixtures of substances which give a positive result to at least one of the following tests:
- steel-sleeve test (Koenen test) with a nozzle plate with a 2 mm diameter hole
- impact sensitivity (using the BAM drop-hammer); more sensitive than 39 J,
- friction sensitivity; more sensitive than 353 N.

Energetic materials which contain the fuel as well as the oxidizer ignite on raising the temperature past the ignition temperature (due to the supply of external heat), forming a flame which generates heat as a result of the exothermic chemical

Tab. 1.4: Berthelot-Rot values for potentially explosive substances.

potentially explosive substance	$\rho_0 / 10^3$ kg m^{-3}	$V_0 / $ m^3 kg^{-1}	$Q_v / $ kJ kg^{-1}	$B_R / $ kJ m^{-3}
HMX	1.96	0.927	5253	18707
NG	1.60	0.782	6218	12448
TNT	1.65	0.975	3612	9588
gun powder	1.87	0.274	3040	2913
hydrazine	1.00	1.993	1785	3558
Oppauer salt	1.10	0.920	1072	1193

Fig. 1.24: Schematic representation of the pressure-over-time diagram of a combustion, a deflagration and a detonation.
(This diagram is reproduced with slight modification from the original of Prof. Dr. Manfred Held, who is herewith thanked for his permission to reproduce this.)

reaction which is greater than the heat loss of the surroundings. In contrast to simple combustion, (e.g. a candle = fuel) which occurs without an increase in pressure and uses oxygen from the atmosphere, in a deflagration (oxidizer and fuel combined in the energetic material), pressure is formed (Fig. 1.24). The term **Deflagration** means that when using a mixture of fuel and oxidizer, a flame is propagated which is less than the speed of sound, but which still can be heard. The (linear) burn rate (r in m s^{-1}) is therefore the speed with which the flame (zone of reaction is on the surface) moves through the unreacted material. Since the temperature increases with increasing pressure (p), the burn rate is also pressure dependent (equation 2). It is strongly influenced by the composition of the energetic material, as well as the confinement. In this context, β is a coefficient ($\beta = f(T)$) and α is the index of the burn rate which describes the pressure dependency. The index α is < 1 for energetic materials which deflagrate (usually 0.2 ... 0.6 for propellants), and > 1 for detonating explosives.

$$r = \beta p^\alpha \qquad (2)$$

A flame in the gas above an energetic solid conducts and radiates heat to the solid surface, which evaporates fuel. The fuel in turn feeds the flame – a cooperative system. There is an equilibrium standoff distance, which represents a balance between these two effects. It is a function of pressure.

A key observation is that a flame will not enter a crack (δ) unless the latter is large enough to support the necessary physical reaction structure (Fig. 1.25a and 1.25b). The standoff distance decreases as P increases, allowing the flame to enter smaller spaces as P increases. This behavior is given by Balyaev's equation (empirical, but with basis in theory):

$$\text{If } \dot{m}_a = \alpha P^\beta, \text{ then } \delta^2 P_{cr}^{1+2\beta} = k$$

Under certain conditions (e.g. strong confinement), a deflagration can change to a **Detonation** (but not vice versa!). This occurs when the reaction front reaches the

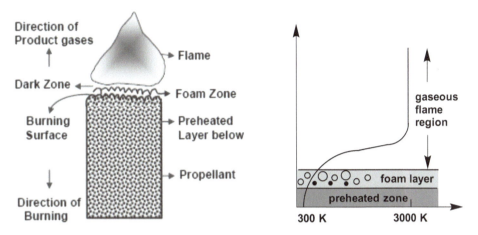

Fig. 1.25a: Self-sustained combustion in a stagnant environment (not to the scale). x-axis: T (K).

Fig. 1.25b: Flame intrusion into cracks.

speed of sound in the material (typical values for the speed of sound in different materials are: air 340 m s^{-1}, water 1484 m s^{-1}, glass 5300 m s^{-1}, iron 5170 m s^{-1}) and then propagates supersonically from the reacted into the unreacted material. The transition from a deflagration to a detonation is known as a **Deflagration-to-Detonation Transition** (DDT). The term detonation is used to describe the propagation of a chemical reaction zone through an energetic material accompanied by

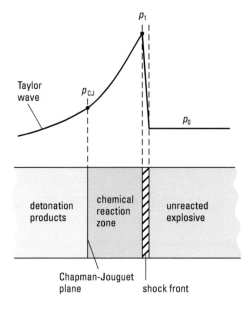

Fig. 1.26: Schematic representation of the detonation process and the structure of the detonation wave.

the influence of a shock-wave at a speed faster than the speed of sound. The detonation zone moves through the explosive at the speed of detonation D, perpendicular to the reaction surface and with a constant velocity. All of the properties of the system are uniform within the detonation zone. When these chemical reactions occur with the release of heat at a constant pressure and temperature, the propagation of the shock-wave becomes a self-sustaining process. Chemical substances which are able to undergo a DDT, are described as explosives and the corresponding self-sustaining process is called a detonation (Figs. 1.24 and 1.26).

Under the influence of the dynamic properties of the shock-wave, a thin layer of still unreacted explosive from the original specific volume V_0 ($V_0 = 1/\rho_0$) along with the shock adiabat of the corresponding explosive (or Hugoniot adiabat) is compressed to volume V_1 (Fig. 1.27). The pressure increases from p_0 to p_1 as a consequence of the dynamic compression, which subsequently results in an increase of the temperature in the thin compressed layer of the explosive (Figs. 1.26 and 1.27), which causes initiation of the chemical reaction. At the end of the chemical reaction, the specific volume and pressure have the values V_2 and p_2. At the end of the reaction zone the reaction products are in chemical equilibrium ($\Delta G = 0$) and the gaseous reaction products move at the local speed of sound, c. Mass, impulse and energy conservation apply. The requirement of a chemical equilibrium between the reaction products is known as the Chapman-Jouguet condition. This state corresponds to the point of the detonation products on the shock adiabat (Fig. 1.27 and 1.27a). At this point it is important to emphasize again, that in a deflagration the

Fig. 1.27: Shock adiabat for an explosive and the detonation products (detonation in stationary state).

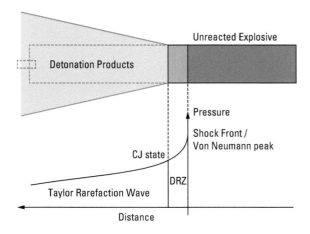

Fig. 1.27a: Structure of a stationary detonation wave (ideal detonation) in the pressure-distance plane according to the Zeldovich-Neumann-Doring (ZND) detonation model (DRZ = detonation reaction zone).

propagation of the reaction occurs as a result of thermal processes, whereas in the considerably faster detonation, a shock-wave mechanism is found.

Following the detonation model for the stationary state, the points (V_0, p_0), (V_1, p_1) and (V_2, p_2) lie on one line (Fig. 1.27), which is known as the **Rayleigh line**. The slope of the Rayleigh line is determined by the detonation velocity of the explosive. In accordance with Chapman and Jouguet's theory, the Rayleigh line is tangent to

Tab. 1.5: Reaction types of an energetic material with Q_{ex} = 1000 kcal kg^{-1}.

reaction type	reaction speed / m s^{-1}	mass flow / m^3 s^{-1}	gaseous products / m^3 s^{-1}	reaction time / s m^{-3}
Combustion	10^{-3}–10^{-2}	10^{-3}–10^{-2}	1–10	10^2–10^3
Deflagration	10^2	10^2	10^5	10^{-2}
Detonation	10^4	10^4	10^7	10^{-4}

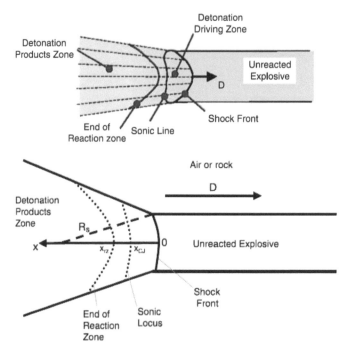

Fig. 1.28: Representation of a non-ideal detonation (see for example: S. Esen, Rock Mech.-Rock Engng **2008**, *41*, 46–497).

the shock adiabat of the detonation products and at this point corresponds to the end of the chemical reaction (V_2, p_2). This point is therefore also known as the **Chapman-Jouguet Point** (C-J point). At the C-J point an equilibrium is reached and the speed of the reaction products corresponds to that of the detonation velocity D. Now the gaseous products can expand and a dilution or Taylor wave occurs (Fig. 1.26). Concrete values for typical speeds of reactions and mass flow rates for combustion, deflagration and detonation are summarized in Table 1.5.

It is important to stress that the above description is a simplified one, and is only valid for ideal explosives. In an ideal explosive, there is a plane shock front (Fig. 1.26), whereas in a non-ideal detonation, the shock front is somewhat curved (Fig. 1.28), and the flow of the reaction products significantly diverge. A consequence of this is that reactions in a non-ideal detonation are never complete in the detonation driving zone (DZZ) – i.e. between the shock front and the sonic line. The detonation (a detonation is a reactive shock – shock followed by chemical reactions) driving reaction terminates at the "sonic line" and contributes to supporting the detonation process.

For an ideal explosive at the C-J point (at the end of the chemical reaction zone), conversion of the explosive into the products corresponding to the thermochemical equilibrium is complete (1.0), whereas for a non-ideal explosive the conversion is < 1. Thus, for a non-ideal explosive, the C-J point is also called the "sonic point" (end of DDZ) (Fig. 1.28), and where the Rayleigh line has the same slope as the Hugoniot of the detonation products formed at the sonic point.

For both ideal and non-ideal detonations, at the C-J point the "flow" (detonation velocity, D – particle velocity, Up) corresponds to sound velocity, Co:

$$D - U_p = C_o$$

1.4 Combustion, Deflagration, Detonation – A Short Introduction

Fire, combustion, deflagration, and detonation are all terms used to describe exothermic oxidation reactions. They are distinguishable from each other by their reaction velocities.

1.4.1 Fire and Combustion

Generally speaking, combustion is a heat-releasing chemical oxidation reaction, which is often accompanied by the occurrence of a flame. Flame temperatures differ widely from approx. 2000 K for an open flame to approx. 3000 K for the flame of an acetylene cutting torch. Remarkably, the heats of combustion of explosives is generally lower than those of common fuels (pine wood with 12.9 % water: 4.422 kcal/g, acetylene: 11.923 kcal/g, dynamite 75 %: 1.290 kcal/g.

Fire is a simple example of a combustion reaction, despite it being less predictable and slower than pure combustion reactions. There are four factors which decide how big a fire can get: fuel, oxygen, heat, and uninhibited chemical chain reactions. If one of these factors is restricted or even absent, the fire will become extinct. For example, lean fuel-air mixtures cannot be ignited because the ratio of fuel is too low and rich mixtures lack the sufficient oxygen ratio to be ignitable. Water as a fire extinguisher has a cooling effect and it therefore stops burning through heat restriction and heat transfer. During burning, a reducing agent (fuel, e.g. gasoline) reacts with oxygen. When heating an ignitable liquid, prior to reaching the actual boiling point of the liquid, a considerable amount of fumes can be observed on the surface of the liquid. As these gases mix with air, ignitable mixtures can develop. The temperature at which an ignitable fume layer is found directly on the liquid's surface is called the flash point of the substance. Liquids at temperatures exceeding the flash point are easily set on fire by a flame or spark. The flash point is usually found at temperatures lower than the boiling point. As the boiling liquid is heated further, the temperature of the heated fume-air mixture may become high enough to evoke a spontaneous oxidation. At this temperature, the fume-air mix will ignite itself spontaneously. The corresponding temperature is called the autoignition temperature (usually above 300 °C, for some compounds, e.g. diethyl ether, significantly lower). A comparable effect can be observed when water is added to overheated oil. As the oil is spilled in form of a finely dispersed cloud, an oil-air mixture is obtained at a temperature enabling rapid oxidation. The resulting hot oil cloud will ignite spontaneously. Thus when heating an ignitable liquid, the temperatures first reach the flash point, then the boiling point, and finally the autoignition point. In order to burn, liquid fuels must be evaporated to maximize contact with oxygen (air) and solid fuels must be sublimed first (e.g. camphor) or pyrolyzed into ignitable gases or fumes.

After a sufficient heat source starts the fire, it propagates through the fuel by heat carrying the uninhibited chemical chain reaction. Heat is transferred by convection, conduction and radiation. Convection is based on the lower densities of heated gases resulting in a gas flow. This is the base for the so-called Zone Model to predict heat flow and fire growth in rooms and spaces. In the Zone Model (e.g. CFAST program; Fire Growth and Smoke Transport Modeling with CFAST (version 6.1.1), National Institute of Standards and Technology (NIST), http://cfast.nist.gov), a room is divided into a cold gas layer (zone) at the floor and a hot zone at the ceiling. Heat transfer by conductivity is understood either as thin, when heat transfer within the object is faster than the rate at which the surface of the object is changed by heat transfer, or as thick, when there is a temperature gradient within the object and the unexposed surface has no significant influence on the heat transfer into the exposed surface. In thin materials (e.g. sheet metal), the temperature can be described by the one-dimensional heat transfer equation:

$$(\delta \rho C_p) \frac{\partial T}{\partial x} = \dot{q}''_{rad} + \dot{q}''_{con} + \dot{q}''_e$$

with the heat transfer rates $\dot{q}''_i$ (i = rad for radiation, con for convection and e for background loss) and the measurable material thickness δ, density ρ and heat capacity C_p. For thick materials, heat transfer within the material can be described one-dimensionally by

$$\rho C_p \frac{\partial T}{\partial t} = k \frac{\partial^2 T}{\partial x^2}$$

with the thermal conductivity k of the material. Heat transfer by radiation can be described by the model of black body radiation and Boltzmann's equation for the thermal energy emission of an ideal radiator

$$E_b = \sigma T^4$$

with the Boltzmann constant of σ. Unfortunately, in reality, most burning objects do not behave like ideal black body radiators and do not completely absorb thermal energy.

When a fire is started in the open the burning is fuel-limited. Since oxygen is abundant in the air, the rate at which fuel is supplied to the fire limits its growth. Even for fires in confined areas (rooms), the early stage of fire is fuel limited. The heat release of a fire is dependent on time. In the early stage of burning, the heat release will grow with the square of time until reaching a steady state. When the fire is extinguished, e.g. when all fuel or oxygen available is consumed, the heat release will decrease to zero. The heat release of a fire can be estimated from the height of its flames. In confined spaces, the initial stage of fuel limited burn will grow into a fully developed fire, wherein all burnable fuels in the room are burning. The fully developed fire will terminally be oxygen-limited as the amount of oxygen available to the chemical chain reactions will constrain the fire. The transition to a fuel-rich oxygen-limited fire will be indicated by a flashover, the transition to uniform burning of all burnable substances in the room, preceded by rollover, flaming-out and off-gassing. In the final oxygen-limited stage, more and more under-oxidized combustion products like CO, HCN and other potentially hazardous compounds are detected in the combustion product mixture.

Combustion reactions are excessively used in propellant systems. The impulse of the gaseous combustion products is used to propel a payload. For obvious technical reasons, the burning temperature of a rocket engine is of interest. The adiabatic flame temperature of combustion (T_{ad}) is the temperature at which reactants and products do not differ in enthalpy. The enthalpies of the components of the system have to be calculated from their standard enthalpy by adding of the enthalpy caused by heating to T_{ad}. This is where the specific heat capacity C_v comes into play. Unfortunately, the component ratios of the system are functions of the temperature, necessitating the use of iterative calculations of T_{ad}.

1.4.2 Deflagration and Detonation

Combustion reactions can be self-accelerating. The acceleration of chemical reactions is achieved by two routes: large increases in active particles and branching of exothermic chain reactions. Active particles, like radicals, are formed by the dissociation of molecules, a process expected to increasingly take place at elevated temperatures. The more active particles are generated, the more collisions between active particles can be observed which causes an accerleration of the overall reaction rate. Branching of chain reactions may occur whenever a single reaction produces more active particles than there were previously in this reaction. However, active particles are continuously de-activated by inevitable collisions. Chain branching then is in concurrence with chain termination. If the rate of chain branching is greater than the rate of chain termination, the reaction will be accelerated and explosion conditions can be obtained. If the energy release of the reaction exceeds the thermal heat loss, thermal explosion can be achieved. Since the loss of active particles does not require activation energy, the temperature's influence on the rate of chain branching is much bigger than it would be on chain termination. Taking the rates of active center creation, chain branching and chain termination into account, the critical temperature, below which explosion is impossible, is to be found.

When starting the reaction, there is an induction period, in which active particles are built up in the system. After the induction time, the number of active particles continues to rise without limit. From this time on, deflagration is observed which may develop into detonation.

Deflagration is the exothermic chemical reaction proceeding at subsonic speed, a detonation proceeds in a supersonic detonation wave. While deflagration can be initiated by mild energy release, e.g. a spark, detonation is triggered by localized explosions caused by the impact of a shock wave. Rapid release of a large amount of energy is typically needed to initiate detonations. However, there often is a transition from deflagration to detonation (a transition from detonation to deflagration is not apparent). This deflagration-to-detonation transition (DDT) is relevant because of its destructive potential. Experimental observations indicate that DDT rarely takes place in unconfined propagating flames. The presence of walls or other spatial confinements of the propagating flame greatly enhance the probability of DDT. In an unconfined burning, the pressure of the gaseous reaction products, which might form small shock waves, are able to expand freely without pressurizing the unburned composition. As the burning system becomes more and more confined, e.g. in a tube, the shock waves caused by the expansion of product gases can reflect from the walls, then form turbulences and fold to convulsions. These reflections, turbulences and convulsions may increase the flame velocity. The flame, which is thus accelerated by itself, can act with the flow of the product gases as piston in the refined system generating pressure and shock waves in the unreacted material. This may again contribute to an even faster and exponential self-acceleration of the flame until a miniature explosion prompts the onset of detonation. The shock wave must be strong enough to raise the temperature in the system up to roughly 1500 K in order to initiate the reaction. For Chapman-Jouguet type detonations, Mach numbers of $\sim 6/7$ are found for the leading shock wave; DDT takes place when the velocity of the reaction wave approaches the local speed of sound. Experimentally, a tulip shaped or spiked flame front appears shortly before the transition to detonation is achieved. As soon as the DDT has taken place, the reaction proceeds through the unreacted material in a self-sustained detonation.

Check your knowledge: See Chapter 14 for Study Questions.

2 Classification of Energetic Materials

Energetic materials which derive their energy from a chemical reaction (in contrast to a nuclear reaction) can be classified according to their use as shown in Figure 2.1.

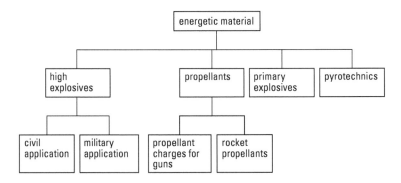

Fig. 2.1: Classification of energetic materials.

2.1 Primary Explosives

Primary explosives are substances which unlike secondary explosives show a very rapid transition from combustion (or deflagration) to detonation and are considerably more sensitive towards heat, impact or friction than secondary explosives. Primary explosives generate either a large amount of heat or a shockwave which makes the transfer of the detonation to a less sensitive secondary explosive possible. They are therefore used as initiators for secondary booster charges (e.g. in detonators), main charges or propellants. Although primary explosives (e.g. $Pb(N_3)_2$) are considerably more sensitive than secondary explosives (e.g. RDX), their detonation velocities, detonation pressures and heat of explosions are as a rule, generally lower than those of secondary explosives (Tab. 2.1).

Typical primary explosives are lead azide and lead styphnate (see Fig. 1.17). The latter one is less powerful than LA but easier to initiate. Tetrazene (Fig. 2.2) is often added to the latter in order to enhance the response (sensitizer). (N.B. mercury fulminate used to be used as a sensitizer). Tetrazene is an effective primer which decomposes without leaving any residue behind. It has been introduced as an additive to erosion-free primers based on lead trinitroresorcinate. Unfortunately, tetrazene is hydrolytically not entirely stable and in long term studies decomposes at temperatures above 90 °C. Diazodinitrophenol (Fig. 2.2) is also a primary explosive and is primarily used in the USA. However, the compound quickly darkens in sun-

Tab. 2.1: Typical sensitivity and performance data of primary and secondary explosives.

	typical primary explosives	Pb(N$_3$)$_2$	typical secondary explosives	RDX
sensitivity data				
impact sensitivity / J	≤ 4	2.5–4	≥ 4	7.4
friction sensitivity / N	≤ 10	< 1	≥ 50	120
ESD / J	0.002–0.020	0.005	≥ 0.1	0.2
performance data				
detonation velocity / m s^{-1}	3500–5500	4600–5100	6500–9000	8750
detonation pressure / kbar		343	210–390	347
heat of explosion / kJ kg^{-1}	1000–2000	1639	5000–6000	5277 (H$_2$O(g))

Fig. 2.2: Molecular structure of tetrazene and diazodinitrophenol (DDNP).

light. SAcN (silver acetylide nitrate, Ag$_2$C$_2$ · 6 AgNO$_3$), mainly in combination with PETN, has also shown to be useful for the use in detonators [9].

Tetrazene (see above) is widely used in ordnance systems as a sensitizer of primer mixes and is used in both percussion and stab applications. It has low thermal stability (decomposition at 90 °C within six days, forming two equivalents of aminotetrazole, 5-AT) and hydrolytic stability compared with other components of primer mixes. Therefore, there currently exists the need to find a replacement for tetrazene, which shows enhanced stability characteristics. MTX-1 (1-[(2 E)-3-(1 H-tetrazol-5-yl)triaz-2-en-1-ylidene]methanediamine) meets these criteria and shows great promise as a tetrazene replacement [J. W. Fronabarger, M. D. Williams, A. G. Stern, D. A. Parrish, *Central European Journal of Energetic Materials*, **2016**, *13*(1), 33–52]. Preliminary testing of this material has confirmed that MTX-1 has safety and performance properties which are similar to those of tetrazene, but its chem-

Fig. 2.2a: Synthesis of MTX-1.

Fig. 2.3: Structures of silver nitraminotetrazolate and disilver(aminotetrazole) perchlorate.

ical properties (including thermal and hydrolytic stability) are better than those of tetrazene. MTX-1 has been successfully evaluated against tetrazene in a variety of chemical/output tests, including comparative testing in the PVU-12 primer.

HNS is a temperature stable secondary explosive, which is particularly useful for blasting in very hot oil deposits, because it is stable to approx. 320 °C. Problems in this area however relate to the initiator, since HNS is relatively difficult to initiate. The most useful initiator is cadmium azide, $Cd(N_3)_2$, ($T_{dec.}$ ca. 295 °C). However, since cadmium is toxic, alternatives are currently being sought after. The two most promising compounds to date to replace $Cd(N_3)_2$ are silver nitriminotetrazolate ($T_{dec.}$ = 366 °C) and di(silveraminotetrazole) perchlorate ($T_{dec.}$ = 319 °C) (Fig. 2.3) [10].

Calcium nitriminotetrazolate has shown huge potential due to its large thermal stability (T_{dec} = 360 °C) and relatively low sensitivity (impact sensitivity: 50 J, friction sensitivity 112 N, ESD = 0.15 J), but at the same time good initiation behavior with commercial pyroelectric initiators (Fig. 2.4). The corresponding cadmium salt is also a good primary explosive.

While the environmental impact of cadmium azide in deep oil deposits is relatively low, the long-term use of $Pb(N_3)_2$ and lead styphnate in military training grounds has resulted in considerable lead contamination (see Ch. 1.2.3, see Fig. 1.17). "On demand lead azide" (ODLA) is available from the reaction of lead acetate and sodium azide (Fig. 2.3a, Tab. 2.1a). The recently introduced iron and copper complexes of the type $[Cat]_2 [M^{II}(NT)_4(H_2O)_2]$ ($[Cat]^+$ = NH_4^+, Na^+; M = Fe, Cu; NT = 5-nitrotetrazolate) as "green" primary explosives [3] are relatively easily obtained and show similar initiator properties as those of lead azide (Tab. 2.2).

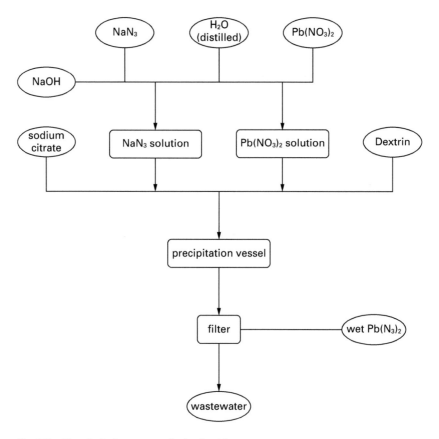

Fig. 2.3a: Manufacturing process for lead azide.

Fig. 2.4: Synthesis (a) and structure (b) of calcium (nitriminotetrazolate).

Tab. 2.1a: Key types of lead(II) azide suitable for industrial applications.

Type of lead azide	Product characteristics
Service (SLA)	Service lead azide (SLA) was developed by the British and does not involve a coating agent; instead, it makes use of the addition of acetic acid and sodium carbonate, which provides a nucleus to precipitate lead azide in a less-hazardous (compared to needle-shaped crystals) spherical morphology. SLA has a higher explosive performance than dextrinated lead azide (DLA), or RD1333/special purpose lead azide (SPLA), but is somewhat more sensitive.
Colloidal	Colloidal lead azide is the purest form of lead azide used in the United States, and is made by mixing dilute solutions of lead and azide salts. It has a very fine particle size of ~ 5 µm, making it ideal for its main application, which is as a coating on electric bridgewires for commercial electric detonators. It is known to be extremely sensitive to electrostatic discharge (ESD).
Dextrinated (DLA)	Dextrinated lead azide (DLA) is considered to be the safest-to-handle form and the most common type now used commercially. It was developed in the United States in 1931 as a solution to the numerous accidental explosions associated with attempts to manufacture pure lead azide. The key feature is the incorporation of dextrin (a short-chained, starch-based polysaccharide), which helps to desensitize the explosive by preventing the formation of large fragile crystals, albeit at the cost of performance and increased hygroscopicity.
RD1333 (SPLA)	RD1333 and special purpose lead azide (SPLA). Although involving somewhat different processing pathways, both (British developed) RD1333 and (the later U. S. developed) SPLA are similar in that they have nearly identical performance requirements/specifications and make use of the sodium salt of carboxymethylcellulose as the desensitizing agent. As such, in the United States, the two types are often used interchangeably. The materials were developed to meet the need for lead azide, which performed better than DLA (especially in smaller detonators), but which maintained some of its safe handling characteristics. SPLA makes up a large portion of the current U. S. military stockpile of lead azide, which has existed since the 1960s.
On-demand (ODLA)	On-demand lead azide (ODLA) is a much more recent development compared to the others described in this table. It is prepared by a military-qualified process, developed by the U. S. Army Armament Research, Development and Engineering Center, Picatinny Arsenal, NJ, which produces lead azide that meets RD1333 specifications and which is considered to be equivalent to RD1333 and SPLA. The main advantage is that it is produced in an on-demand, continuous fashion, therefore avoiding the hazards associated with handling large scale batches of the material. The small footprint and low cost of the processing equipment also means that it can be placed close to item production lines, which further reduces the need for the expensive and dangerous transport of lead azide on public roadways. ODLA was qualified by the U. S. Army in 2012 and is currently being evaluated in larger-scale loading operations.

Tab. 2.2: Properties of lead azide (LA) and lead styphnate (LS) in comparison to new "green" primary explosives.

primary explosive	T_{dec} / °C	impact sensitivity / J	ESD / J	density / g cm^{-3}	detonation velocity / m s^{-1}
NH$_4$FeNT	255	3	> 0.36	2.2	7700
NaFeNT	250	3	> 0.36	2.2	not determined
NH$_4$CuNT	265	3	> 0.36	2.0	7400
NaCuNT	259	3	> 0.36	2.1	not determined
LA	315	2.4	0.005	4.8	5500
LS	282	3.4	0.0002	3.0	5200

2.2 High (Secondary) Explosives

Secondary explosives (also known as "high explosives" = HEs) unlike primary explosives can not be initiated simply through heat or shock. In order to initiate primary explosives have to be used, whereby the shockwave of the primary explosive initiates the secondary. However, the performance of the secondary explosive is usually higher than that of the primary (see Tab. 2.1). Typical currently used secondary explosives are TNT, RDX, HMX, NQ and TATB (see also Tab. 1.2) and, for civil applications, HNS and NG e.g. in the form of dynamite for commercial use.

The current trends in the research of secondary explosives can be arranged into three branches (Fig. 2.5):
- Higher performance,
- Lower sensitivity (insensitive munition, IM),
- Lower toxicity of the explosives, their biological degradation products and the detonation products.

A higher performance for secondary explosives is of fundamental importance and always desired. The main performance criteria are:
1. heat of explosion Q (in kJ kg^{-1}),
2. detonation velocity D (m s^{-1}),
3. detonation pressure P (in kbar),

and less importantly,
4. explosion temperature T (K) and
5. volume of gas released V per kg explosive (in l kg^{-1}).

Depending on the location of the mission, one parameter can be more important than the other. For example, the performance (or brisance) of an explosive can not be described by one parameter alone. The term **brisance** describes the destructive

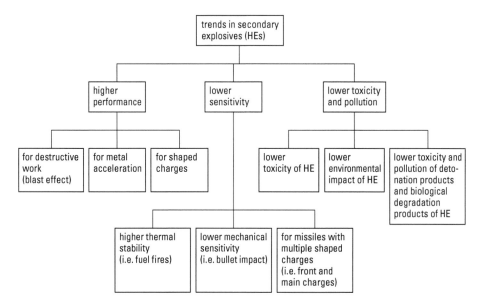

Fig. 2.5: Trends for new secondary explosives.

fragmentation effect of a charge on its immediate vicinity. The higher the charge density (corresponding to the energy per volume) and the detonation velocity (corresponding to the reaction rate), the more brisant the explosive is. Furthermore, the detonation velocity as well as the detonation pressure increase with increasing density. According to Kast, the brisance value (B) is defined as the product of the loading density (ρ), the specific energy (F for "force of an explosive") and the detonation velocity D:

$$\text{Brisance: } B = \rho \times F \times D$$

The specific energy ("force") of an explosive (F) on the other hand can be calculated according to the general equation of state for gases:

$$\text{Specific energy: } F = p_e V = nRT$$

Here p_e is the maximum pressure through the explosion (not to be confused with the much higher values for the detonation pressure p_{C-J} at the C-J point), V is the volume of detonation gases (in l kg^{-1} or m^3 kg^{-1}), n is the number of moles of gas formed by the explosion per kilogram of explosive, R is the gas constant and T is the temperature of the explosion. The specific energy also has the units J kg^{-1}. Consequently, the brisance can be given using the units kg s^{-3}.

Generally, the specific energy of secondary explosives is higher than that for propellant charges. This is due to the fact that in propellant charges the tempera-

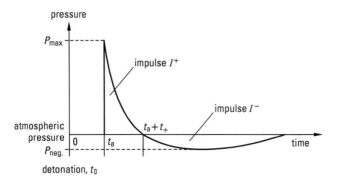

Fig. 2.6: Typical time dependent progression of a shock wave in the vicinity of a detonation. (This diagram is reproduced with slight modification from the original of Prof. Dr. Manfred Held, who is herewith thanked for his permission to reproduce this.)

ture of combustion is kept as low as possible in order to protect the barrel by preventing the formation of iron carbide (from the CO in the combustion). For shaped charges a high brisance and therefore a high specific energy and also high loading densities are particularly important. For typical general purpose bombs, a maximum heat of explosion (Q) and gaseous products (V) should be achieved.

As we have already seen above (see Fig. 1.24), the pressure vs. time diagram shows that the explosion pressure (not to be confused with the by magnitudes higher detonation pressure p_{C-J}) generated on detonation increases instantly and decreases exponentially with time. Figure 2.6 shows the typical time dependence of a blast wave. When the detonation occurs at time $t = 0$, the shock front reaches the "observer" i.e. the object, which is exposed to the shock front at time t_a. Afterwards, the pressure falls exponentially and for a certain period of time it falls beneath the pressure of its surroundings (atmospheric pressure).

As we can see in Figure 2.7, after a detonation, the shock wave naturally requires a certain amount of time t ($t_1 < t_2 < t_3 < t_4$), to reach a certain point. However, the maximum pressure p of the shockwave also decreases as the distance from the center of the detonation increases.

The shock front is formed by the fusion of the incident and reflected shock fronts from an explosion. The term is generally used with reference to a blast wave which is propagated in the air and reflected at the surface of the Earth. In an ideal case, the mach stem is perpendicular to the reflecting surface and slightly convex (forward). The mach stem is also sometimes referred to as the mach front.

If the explosion occurs above the ground (e.g. MOAB), when the expanding blast wave strikes the surface of the earth, it is reflected off the ground to form a second shock wave which travels behind the first (Fig. 2.7a). This reflected wave travels faster than the first (or incident) shock wave, since it is traveling through air already, moving at high speed due to the passage of the incident wave. The reflected blast wave merges with the incident shock wave to form a single wave,

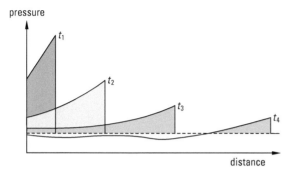

Fig. 2.7: Pressure-distance-dependence of a progressing shock wave at different points in time t ($t_1 < t_2 < t_3 < t_4$) after a detonation.
(This diagram is reproduced with slight modification from the original of Prof. Dr. Manfred Held, who is herewith thanked for his permission to reproduce this.)

Fig. 2.7a: Formation of a Mach Stem.

known as the Mach Stem (Fig. 2.7a). The overpressure at the front of the Mach wave is generally about twice as large as that at the direct blast wave front.

Generally, it can be said that the damaging effect of a shockwave produced by a detonation is proportional to its impulse (impulse = mass × velocity of the gaseous explosion products) and its maximum pressure, with the impulse being the most influential factor at smaller distances and the pressure being most important at larger distances. As a "rule of thumb", the distance D, which offers a chance of survival, is proportional to the cube route of the mass w of an explosive, while for typical secondary explosives at larger distances, the proportionality constant is approximately 2:

$$D = c\, w^{0.33} \approx 2\, w^{0.33}$$

It is important to note, that this approximation is only based on the pressure. The impulse of the shockwave and the fragment impact (e.g. from confined charges or fume-cupboard shields in lab) are *not* taken into consideration.

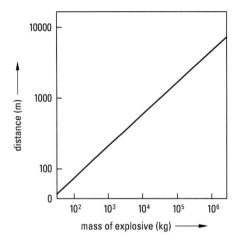

Fig. 2.8: "Safe" distance for occupied buildings from the detonation of a secondary explosive.

In order to work safely with highly energetic materials in the chemical laboratory the following rules must be obeyed:
- amounts used are to be kept as small as possible,
- the distance from the experiment is to be kept as large as possible (filled vessels are not to be transported by hand, but with well fitted tongs or clamps instead),
- mechanical manipulators are to be used when possible, particularly for larger quantities,
- vessels are never to be enclosed by hand (confinement),
- protective clothing (gloves, leather or Kevlar vest, ear protectors, face shield, antielectrostatic shoes) is to be worn at all times.

Figure 2.8 shows a correlation between the mass of the secondary explosive used and the "safe" distance for occupied buildings. Of course this is only a very rough guide and it is dependent on the building and the nature of the explosive used.

In addition to formulations of secondary explosives (see Tab. 1.2), metallized mixtures are sometimes used as well. Metals such as beryllium, magnesium or aluminum which are air resistant, but at the same time easily oxidized in very exothermic reactions are suitable. In practice, aluminum is used almost exclusively. Since most formulations possess a negative oxygen balance, the aluminum does not contribute to raising the heat of detonation in atmospheric explosions a lot, but it combusts afterwards (post-detonation combustion, often a large fireball) using the surrounding air which increases the visibility (psychological effect) considerably. This is different for underwater explosions, e.g. torpedoes (see Tab. 1.2, TORPEX), because here the aluminum obtains the oxygen from the (liquid) water, which contains oxygen in a high density. Metallized formulations are also suitable for special

Fig. 2.9: Safety (insensitivity) and performance.

missions (e.g. caves, tunnels e.g. in GWT), since in enclosed spaces the oxygen in the air needed to breathe for survival vanishes (see Ch. 13).

In attempts to strive for better performance, safety aspects (lower sensitivities) can not be ignored. For example, by using the formulations Composition B and octol (see Tab. 1.2) instead of TNT, the performance can be increased significantly (Fig. 2.1). However, the sensitivity increases as well which causes safety to decrease accordingly. The goal of current research is to develop considerably less sensitive secondary explosives, which offer maximum performance and a high safety standard (Fig. 2.9).

Particularly important in the area of explosives for torpedoes, missiles, war heads and bombs, which are transported by planes or submarine is, that if a fuel fire should occur, the charge is not thermally initiated. Insensitive, but high performing explosives also play a large role for shaped charges with several charges (front- and main charges) e.g. for operations against tanks, which are protected by Explosive Reactive Armour (ERA). Last but not least, insensitive also means that no accident can occur as a result of friendly fire or through enemy attack.

There still is a need for intensive research (Fig. 2.1) in the area of the latter aspect of new secondary explosives (and not only for those!) in terms of low toxicity and environmental impact. Conventional energetic materials (TNT, RDX), which are currently used by NATO armies for training purposes on their training grounds, have an extremely negative ecological impact, if they end up unwanted and uncontrolled in the environment as unexploded munition (UXOs = unexploded ordnance), or through low-order as well as high-order detonations. This necessitates expensive and very time-consuming remediation and detoxification operations. The development of alternative energetic materials, which are environmentally compatible, but still fulfil the performance and insensitivity requirements of the NATO armies (see above), i.e. safe handling, is an important step in the direction of developing ecologically compatible, non-toxic and sustainable highly energetic materials. For example, RDX is toxic when ingested or inhaled in large quantities, causing seizures, nausea and vomiting. For example, the EPA (Environmental Protection Agency) has suggested a limit of 2 µg l^{-1} for RDX in drinking water. However, this value is marginally overstepped in some areas in the vicinity of troop training grounds. Limits for RDX have also been determined for the workplace by NIOSH (National Institute for Occupational Safety and Health). These are in general 3 µg m^{-3} air and 1.5 µg m^{-3} air for a 40 hour working week.

Table 2.3 shows a summary of examples of some ecological and toxicologic problem areas of highly energetic materials.

Particularly in the USA, but also in Germany, great efforts are being made by different institutions to reduce the environmental impact caused by the use of highly energetic materials. Leaders in this area are SERDP (Strategic Environmental Research and Development Program: http://www.serdp.org/) as well as ESTCP (Environmental Security Technology Certification Program: http://www.estcp.org/). Some possible replacements for existing formulations based on DNAN are summarized in Tab. 2.3a.

In this context it should also be mentioned that 3,4-dinitropyrazole (DNP) is an interesting melt-cast explosive (m.p. 87 °C, $T_{dec.}$ = 276 °C) which can easily be prepared from nitropyrazole (Fig. 2.9a). Due to the relatively high density of 1.79 g cm^{-3}, DNP has good detonation parameters (VoD = 8115 m s^{-1}, p_{C-J} = 294 kbar) and may therefore be a potential replacement for Comp. B (RDX/TNT, VoD = 7969 m s^{-1}, p_{C-J} = 292 kbar). Therefore, DNP has great potential as a new melt-pour base. DNP (3,4-dinitropyrazole) is now technically mature, and has been consistently synthesized at BAE SYSTEMS on the 100-g scale with 65% crude yield and 99.9% HPLC purity but with ~5% remaining acids.

Another very promising melt-pour explosive is propyl nitroguanidine (PrNQ). The compound melts at 99 °C and has TNT like performance. The synthesis is easy and straight forward from commercially available NQ (Fig. 2.9b).

While traditionally general purpose bombs (GP bombs) are filled with sensitive tritonal (80% TNT, 20% Al powder) or H6 (44% RDX&NC, 29.5% TNT, 21% Al powder, 5% paraffin wax, 0.5% CaCl$_2$) and do not pass all IM requirements, PrNQ may emerge as a novel ingredient (melt-cast) for IM formulations for use in GP bombs.

Two of the most promising replacements for TNT and Comp B are IMX-101 (TNT replacement) and IMX-104 (Comp B replacement) (Tab. 2.3b). IMX stands for "Insensitive Munitions eXplosives". **IMX-101** is a high-performance insensitive high explosive composite mixture developed by BAE systems (Holston) and the US Army to replace TNT in artillery shells. **IMX-104** is an insensitive melt-pour explosive to replace Composition B for mortar applications and was also developed by the Holston Army Ammunition Plant (BAE). The main ingredients are:

IMX-101:   43.5% 2,4-dinitroanisole (DNAN), 19.7% 3-nitro-1,2,4-triazol-5-one (NTO) and 36.8% nitro-guanidine (NQ)
IMX-104:   32% DNAN, 15% RDX, 53% NTO.

Tab. 2.3: Ecological and toxicological problem areas of highly energetic materials.

class of energetic material	example	problem environment	problem human	possible solution
primary explosive	Pb(N$_3$)$_2$	Pb contamination of training grounds	heavy metal toxicity of Pb	Pb-free primary explosives, e.g. Fe
secondary explosive	RDX	RDX and degradation products are toxic for plants, microorganisms and microbes (earthworms)	kidney toxin	new high-N, high-O compounds
	TNT	TNT and degradation products are ecologically toxic		ew melt-castable high explosives, e.g. IMX-101: DNAN = dinitroanisole (binder) + NTO (filler) + NQ
	Comp.-B	contains RDX and TNT		IMX-104: DNAN (binder) + NTO and RDX (filler)
pyrotechnics	Ba (green)	heavy metal	heavy metal[b]	insoluble Ba compounds; Ba-free colorants, e.g. Cu
	ClO$_4^-$ as Chlorine source		decrease of thyroxin synthesis due to inhibition of iodine storage[b]	
pyrotechnics	Pb resorcynate burn rate catalyst in min smoke missiles	heavy metal	heavy metal	Cu, Bi based burn rate modifiers; reactive metal nano-powders as burn rate catalysts/modifiers
solid rocket propellant	AP	HCl in atmosphere, ozone layer	decrease of thyroxin synthesis due to inhibition of iodine storage	CaSi$_2$ additives as HCl scrubbers; new HEDOs, e.g. ADN
monopropellant	N$_2$H$_4$		carcinogenic	new propellants, e.g. DMAZ
bipropellant	MMH		carcinogenic	new bipropellants, e.g. DMAZ
	HNO$_3$	NO$_x$	toxic	e.g. 100% H$_2$O$_2$

[a] causes hypertension, muscular weakness, cardiotoxic effects, gastrointestinal effects
[b] EPA has recommended a max. level of 24.5 ppb ClO$_4^-$ in drinking water

Tab. 2.3a: Possible replacements for existing formulations based on DNAN.

Formulation	Ingredients	Replaces	Applications
IMX-101	DNAN 43 % NTO 20 % NQ 37 %	TNT	Artillery and other large caliber munitions
IMX-104	DNAN 32 % NTO 53 % RDX 15 %	Comp B	Mortar applications
PAX-21	RDX 36 % DNAN 34 % AP 30 %	Comp B	Mortar applications
PAX-25	RDX 20 % DNAN 60 % AP 20 %	Comp B	Mortar applications
PAX-26	DNAN, Al, AP, MNA	Tritonal	General purpose bomb
PAX-28	RDX 20 % DNAN 40 % Al 20 % AP 20 %	Comp B	warheads

Tab. 2.3b

Formulation	VoD / km s^{-1}	Critical diameter / inches
Composition B	7.98	0.169
HMX-104	7.4	0.875
TNT	cast 5.9 pressed 6.7	cast 0.5 pressed 1.6
IMX-101	6.9	2.6
MCX-6100	7.2	0.78

Fig. 2.9a: Synthesis of 3,4-Dinitropyrazole (DNP).

Fig. 2.9b: Synthesis of propyl nitroguanidine (PrNQ).

Other interesting new formulations include MCX-6100 (NTO 51%, DNAN 32%, RDX 17%; melt-cast for IM, Chemring Nobel) and PBXN-9 (HMX 92%, HyTemp 2%, DOA 6%; shaped charges, Chemring Nobel). For shaped charge applications, formulations with HMX (82%) and GAP or PolyNimmo (18%) can also be used.

2.3 Propellant Charges

The oldest known propellant charge is blackpowder (or gunpowder) which is a mixture of 75% KNO_3, 10% sulfur and 15% charcoal dust. It is an easy to ignite powder with a burn rate of 600–800 m s^{-1}, which is still used today as a charge in military and civil *pyrotechnic* munitions. NC powder was discovered in 1846 by Schönbein and burns almost without any residue. Its burn rates at atmospheric pressure is 0.06–0.1 m s^{-1}, which is considerably slower than blackpowder, as long as no pressure is applied to the propellant charge on burning. In addition to the linear burn rate r (in m s^{-1}), the mass-weighted burning rate m (in g s^{-1}) is of interest as well (A: surface, in m^2; ρ: density, in g m^{-3}):

$$m = r A \rho$$

Single-base nitrocellulose powder is the oldest of all types of NC powders and is often referred to as "smokeless" powder. It is prepared by reacting cellulose with nitric acid.

Depending on the acid concentration used, NC is formed with different degrees of nitration (e.g. 11.5–14.0%) (Fig. 2.10). In addition to the nitrocellulose based, double- and triple-base propellants contain further substances such as nitroglycerine and nitroguanidine.

The dependence of the enthalpy of formation of nitrocellulose on the percentage nitrogen content of NC (degree of nitration) is shown in Fig. 2.10 [1g].

While single-base propellant charges (NC) are used in weapons from pistols to artillery weapons, the higher performance double-base propellant charges (NC + NG) are used predominantly in pistols and mortars. The disadvantage of the double-base powder is the strong erosion of the gun barrel (see below), resulting from the significantly higher combustion temperatures as well as the appearance of a muzzle flash because of the explosion of some of the combustion gases upon contact with air. To prevent erosion and the muzzle flash, a triple-base powder (NC + NG + NQ) with an NQ content of up to 50% is used, particularly in large calibre tank and NAVY weapons. However, the performance of triple-base powders is lower than that of double-base powders (Tab. 2.4). In a triple-base powder, particularly in large tank and NAVY cannons, NQ is replaced by RDX in order to increase the performance. However, the barrels suffer increased erosion problems again due to the significantly higher combustion temperatures.

Fig. 2.10: The dependence of the enthalpy of formation of NC on the percentage nitrogen content of the NC.

Tab. 2.4: Parameters for single, double and triple-base propellant charges.

propellant charge	ρ/g cm^{-3}	Ω/%	T_c/K	I_{sp}/s	N_2/CO
NC[a]	1.66	−30.2	2750	232	0.3
NC[a] / NG (50 : 50)	1.63	−13.3	3308	249	0.7
NC[a] / NG / NQ (25 : 25 : 50)	1.70	−22.0	2683	236	1.4
Hy-At / ADN (50 : 50)	1.68	−24.7	2653	254	6.3

[a] NC-13.3 (N content 13.3 %)

Fig. 2.10a: Gun propulsion.

The velocity v of a projectile on ejection from the gun barrel can roughly be estimated using the following equation:

$$v = \left(\frac{2m\,Q\eta}{M}\right)^{0.5}$$

$$Q = \Delta U_f^\circ \text{ (prop.charge)} - \Delta U_f^\circ \text{ (combustion products)}$$

Where m is the mass of the propellant charge (in g), M is the mass of the projectile (in g), Q is the heat of combustion (in J g^{-1}) and η is a constant, which is specific for the system. Here, the energy released in Q is once again of crucial importance.

Fig. 2.10a shows a schematic representation of a gun propulsion.

The erosion problems in guns are generally caused by the formation of iron carbide (Fe from the gun barrel, C from CO) at high temperatures. Modern research on propellant charges is therefore focused on the development of powders which burn at lowest possible temperatures, but show good performance (see Ch. 4.2.3). Moreover, the N_2/CO ratio, which lies at approximately 0.5 for conventional propellant charges, should be increased as much as possible. The formation of N_2 instead of CO should also strongly reduce the amount of erosion, since iron nitride has a higher melting point than iron carbide (Fe$_3$C) and furthermore, it forms a solid protective layer on the inside of the gun barrel. New studies have shown that the introduction of propellant charges which are based on nitrogen-rich compounds such as e.g. triaminoguanidinium azotetrazolate (TAGzT) result in a considerably better N_2/CO ratio and the life-span of high calibre NAVY cannons can be increased by up-to a factor of 4 [11]. The synthesis of azotetrazolate salts (e.g. Na$_2$ZT) occurs in a facile route from aminotetrazole by oxidation using KMnO$_4$ in alkaline conditions (NaOH). *Note:* the reduction of Na$_2$ZT in acidic conditions (HOAc) using Mg forms the neutral compound bistetrazolylhydrazine (BTH).

Also hydrazinium 5-aminotetrazolate (Hy-At) appears to show good potential in this context (Tab. 2.4) [12] (Fig. 2.10b).

For propellant charges the insensitivity is also playing an increasingly important role. Since approximately 1970 propellant charges have been developed and

Fig. 2.10b: Molecular structures of triaminoguanidinium azotetrazolate (TAGzT) and hydrazinium 5-aminotetrazolate (Hy-At).

used under the title **LOVA** (low-vulnerability ammunition). Under bullet impact, shaped charge impact or fire they respond with fire in the worst case scenario, but not in deflagration and definitely not in detonation. As energetic fillers mainly RDX or HMX were used. As energetic binders, energy and gas generating polymers such as e.g. GAP (see Fig. 1.4) are particularly suitable. Such composite propellant charges are considerably less sensitive than powders based on NC.

2.4 Rocket Propellants

Rocket propellants are similar to the propellant charge powders discussed above, because they combust in a controlled manner and do not detonate. However, propellant charge powders burn considerably quicker than rocket propellants, which results in a significantly higher pressure for gun propellants in comparison with rocket propellants. Typical pressures in the combustion chambers of rockets are 70 bar, which can be compared with up to 4000 bar for large artillery and NAVY cannons.

As is the case for propellant charges, the specific impulse for rocket propellants is also one of the most important performance parameters.

The **specific impulse** (I_{sp}) is the change in the impulse (impulse = mass × velocity or force × time) per mass unit of the propellant. It is an important performance parameter of rocket motors and shows the effective velocity of the combustion gases when leaving the nozzle, and is therefore a measure for the effectiveness of a propellant composition.

$$I_{sp} = \frac{F \cdot t_b}{m} = \frac{1}{m} \cdot \int_0^{t_b} F(t)dt$$

The force F is the time dependent **thrust** $F(t)$, or the average thrust $\bar{F}$; t_b is the combustion time (in s) and m is the mass of propellant (in kg). Therefore the specific impulse has the units N s kg^{-1} or m s^{-1}. Sometimes, predominantly in English speaking areas, the specific impulse is given based on the gravitation of the Earth g ($g = 9.81$ m s^{-2}), which is on the mass of the propellant, and therefore has the unit seconds (s):

$$I_{sp}^* = \frac{I_{sp}}{g}$$

The specific impulse can be formulated exactly as follows:

$$I_{sp} = \sqrt{\frac{2\, y\, R\, T_c}{(y-1)M}}$$

Whereby y is the ratio of the specific heat capacities of the gas mixture, R is the gas constant, T_c is the temperature (K) in the combustion chamber and M is the average molecular weight (kg mol^{-1}) of the formed combustion gases:

$$y = \frac{C_p}{C_v}$$

$$I_{sp}^* = \frac{1}{g}\sqrt{\frac{2\, y\, R\, T_c}{(y-1)M}}$$

The average thrust of a rocket $\bar{F}$ can in accordance with the equation above be given simply as:

$$F = I_{sp}\frac{\Delta m}{\Delta t}$$

Where I_{sp} is the specific impulse in (m s^{-1}), Δm is the mass of used propellant (in kg) and Δt is the duration of burning of the engine (in s). Therefore, the thrust has the units kg m s^{-2} or N and corresponds to a force.

In the following discussion we want to predominantly use the English term I_{sp}^*, which is smaller than the I_{sp} by a factor of approximately 10 and has the units s. Typical values for the specific impulse of solid boosters are 250 s, whereas for bipropellants they are found at approx. 450 s. For chemistry it is important that the specific impulse is proportional to the square root of the ratio of the combustion chamber's temperature T_c and the average molecular mass of the combustion product M:

$$I_{sp} \propto \sqrt{\frac{T_c}{M}}$$

It is important in the discussion of the specific impulse I_{sp}^* below, that as a rule of thumb an increase of the specific impulse by 20 s, causes the maximum possible carried payload (warheads, satellites) to be approximately doubled.

A more detailed description to derive the thrust or the specific impulse of a rocket is given below: In order to propel a rocket, a rocket engine ejects combustion gases with low molecular mass but high velocity z through the nozzle. The rocket has the mass M and initially moves with velocity u. If the rocket ejects combustion gases of the mass of Δm within the time Δt and with the velocity z, a decrease in mass $M - \Delta m$ and an increase in the velocity to $u + \Delta u$ results. Due to the conservation of the impulse, the following can be derived:

$$Mu = (M - \Delta m)(u + \Delta u) + \Delta m z$$
$$\cancel{Mu} = \cancel{Mu} + M\Delta u - \Delta m u - \Delta m \Delta u + \Delta m z$$
$$M\Delta u = \underbrace{\Delta m (u + \Delta u - z)}_{v_e}$$

$$M\frac{\Delta u}{\Delta t} = \frac{\Delta m}{\Delta t} v_e$$

$$M\frac{du}{dt} = \frac{dm}{dt} v_e$$

Therefore $v_e = u + \Delta u - z$ and is the velocity of the ejected combustion gases relative to the rocket. Here the force $(dm/dt)v_e$ corresponds to the **thrust** $F_{impulse}$, which originates from the impulse of the system:

$$F_{impulse} = \frac{dm}{dt} v_e$$

However, this is only equal to the total thrust F of the system when the pressure p_e at the end of the nozzle is the same as the pressure of its surroundings p_a. In general a correction term is needed which is called the pressure term $F_{pressure}$.

Figure 2.11 schematically shows the pressure profile which is one factor that influences rocket performance. The length of the arrows shows the contribution of the pressure from inside and outside the walls. While the atmospheric pressure outside is constant, the inside pressure of the combustion chamber is at its largest

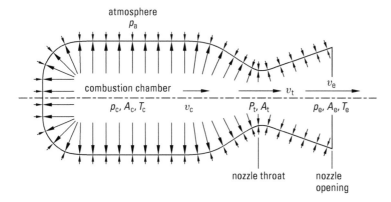

Fig. 2.11: Combustion chamber and nozzle.

and decreases in the direction of the nozzle end. The pressure term is proportional to the diameter A_e:

$$F = F_{\text{impulse}} + F_{\text{pressure}} = \frac{dm}{dt} v_e + (p_e - p_a) A_e$$

If the pressure at the end of the nozzle is smaller than the pressure of the surroundings (which occurs in the case of a so-called overexpanding nozzle), the pressure term then has a negative value and reduces the total thrust. Therefore, a pressure p_e which is the same or higher (underexpanding nozzle) than the air pressure is desirable.

Since the air pressure decreases with increasing flight altitude, at constant nozzle diameter, the total thrust increases with increasing flight altitude. This increase can correspond to approximately 10 to 30 % of the total thrust depending on the rocket. The maximum thrust is reached in vacuo. The so-called effective ejection velocity c_{eff} (of the combustion gases) is defined as the ratio between the thrust and the mass flux (dm/dt):

$$\frac{F}{\frac{dm}{dt}} = c_{\text{eff}} = v_e + \frac{(p_e - p_a) A_e}{\frac{dm}{dt}}$$

We can write:

$$F = \frac{dm}{dt} c_{\text{eff}}$$

The effective ejection velocity is an average. In reality the velocity distribution is not constant throughout the whole nozzle diameter. Assuming a constant ejection velocity allows a one-dimensional description of the problem.

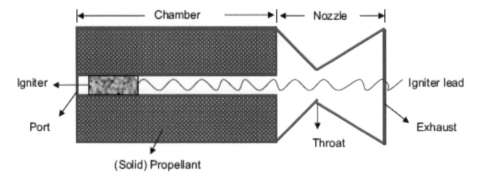

Fig. 2.11a: Schematic design of a solid rocket motor.

The so-called total impulse I_t is the integral of the total thrust integrated over the total combustion time t:

$$I_t = \int_0^t F\,dt$$

The actual design of a rocket motor is of crucial importance for its performance. A motor basically consists of a combustion chamber and a nozzle. The combustion chamber is usually a metallic tube sealed at one end and contains the solid rocket propellant. The rocket chamber is connected to the nozzle, which, in most cases, is a convergent-divergent (CD) nozzle, as shown in Fig. 2.11a. An igniter initiates the ignition of the entire propellant surface which results in the formation of gaseous products and high temperature and pressure in the combustion chamber. There is a great increase in the velocity of the gaseous products when they expand from the throat to the exit of the nozzle. As a consequence of Newton's third law of motion, the exiting gases propel the rocket into the opposite direction.

The total thrust with which a rocket is propelled has two components (see above). The first component of the thrust (F) is due to the thrust F_1 created due to the imbalance of the chamber pressure p_c and the exhaust gas pressure p_e acting on the throat, the area of which is A_t (Fig. 2.11). Therefore, F_1 can be written as:

$$F_1 = (p_c - p_e)\,A_t$$

Since the gases under high pressure in the combustion chamber are expanding after passing through the throat, p_c is always greater than p_e, therefore F_1 is always positive.

The second component of the thrust (F) is due to the thrust F_2 created due to the imbalance of the exhaust gas pressure (p_e) and the pressure p_a outside of the rocket acting on the exhaust (Fig. 2.11). Therefore, F_2 can be written as:

$$F_2 = (p_e - p_a) A_e$$

(Note: underexpanding nozzle, if: $p_e > p_a$; overexpanding nozzle, if: $p_e < p_a$; optimum expanding nozzle, if: $p_e = p_a$.)

The overall propulsive force (thrust F) a rocket experiences is the sum of F_1 and F_2:

$$F = F_1 + F_2 = (p_c - p_e) A_t + (p_e - p_a) A_e$$

The **specific impulse** is one of the most important values to characterize the performance of a rocket propulsion system. The higher this value is, the higher will the performance be. The specific impulse is defined as the total impulse per mass unit:

$$I_s^* = \frac{\int_0^t F dt}{g_0 \int_0^t \frac{dm}{dt} dt}$$

where $g_0 = 9.81$ m s^{-2} at sea level.

For constant thrust and mass flux, the equation above can be simplified to

$$I_{sp}^* = \frac{F}{\frac{dm}{dt} g_0}$$

Using the SI system, the specific impulse can be given in the units s^{-1}. However, the value g_0 is not always the same. Therefore, the specific impulse I_{sp} is often also given by the value

$$I_{sp} = \frac{F}{\frac{dm}{dt}}$$

This has the advantage that it is independent of the effect of gravity. It is given in the unit N s kg^{-1} or m s^{-1}. At sea level, both expressions are linked by a factor of approximately 10.

By inserting

$$F = \frac{dm}{dt} c_{eff}$$

in

$$I_{sp}^* = \frac{F}{\frac{dm}{dt} g_0}$$

the mass flux is shortened, and one obtains

$$I_{sp}^* = \frac{c_{eff}}{g_0}$$

The effective ejection velocity c_{eff} is different from the specific impulse I_{sp} only due to g_0 (see above).

The so-called **rocket equation** describes the fundamental equations of rocket propulsion. If we consider the simplest case, in which a monostage rocket accelerates in a gravity-free vacuum, i.e. a slow-down due to gravitation and friction is not taken into consideration. When the rocket has a velocity of zero at the start and ejects the propellant with a constant ejection velocity v_e, the velocity u of the rocket after time t corresponds to:

$$u(t) = v_e \ln\left(\frac{m(0)}{m(t)}\right)$$

Where $m(0)$ is the mass of the rocket at the start and $m(t)$ is the mass of the rocket at time t.

For rocket launches from earth, the formula must be changed to include the gravitation of Earth and corresponds to low altitudes with a constant earth gravitation g of 9.81 m s^{-2}:

$$u(t) = v_e \ln\left(\frac{m(0)}{m(t)}\right) - gt$$

Since g is dependent on the altitude, the following correction is needed:

$$u(t) = v_e \ln\left(\frac{m(0)}{m(t)}\right) - \int_0^t g(t')dt'$$

In addition to the gravitation of Earth a rocket launched from earth must also overcome the air resistance of the atmosphere, which means that the rocket equation for such cases is only an approximation. Planes, RAM and SCRAM jets which are propelled by jet engines, transport their fuel with them, but they also suck air in and use the oxygen from the air for the combustion of the fuel. They only carry the

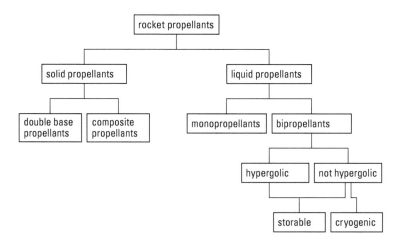

Fig. 2.12: Classification of rocket propellants.

fuel but not the oxidizer with them. The rocket equation is not valid for such vehicles, which are referred to as air-breathing engines.

	Propellant		
small arms propellant	mortar propellant	gun propellant	rocket propellant
– single-base – double-base	– single-base – double-base	– single-base – double-base – triple-base – nitramine-base – liquid	– double-base – composite – composite modified double-base (CMDB) – liquid

Generally, propellants can be classified based on either where they are going to be used as rocket or gun propellants or according to their chemical composition. A possible classification of propellants based on their use is shown in the diagram above.

Generally, one can categorize rocket propellants into those for solid propellants and those for liquid propellants (Fig. 2.12). The **solid propellants** can be categorized further into double-base (homogeneous) and composite propellants (heterogeneous). The homogeneous propellants are generally based on NC, whereas the heterogeneous propellants are generally based on AP. The double-base propellants are homogeneous NC/NG formulations. The composite propellants are mixtures of a crystalline oxidizer (e.g. ammonium perchlorate, AP) and a polymeric binder, which has been cured with isocyanate (e.g. hydroxy-terminated polybutadiene, HTPB/diisocyanate) and which contains the fuel (e.g. Al) (Fig. 2.13). The purpose of the polymer matrix (the binder) is to form a solid, elastic body of the propellant

Fig. 2.13: Homopolymeric R45HT (hydroxyl terminated polybutadiene, HTPB).

Tab. 2.5: Solid propellants.

type	composition	I_{sp}^* / s	T_c / K
double-base	NC (12.6 % N)[b] NG plasticizer other additives	200	2500
composite	AP Al HTPB additives	259	4273

[a] A typical double-base composition for minimum smoke missiles would be the M36 propellant with the following composition: NC (12.6 % N) 49.0 %, NG 40.6 %, Di-n-propyl adipate 3.0 %, 2-nitrodiphenylamine 2.0 %, ballistic modifier (e.g. lead resorcinate) 5.3 %, wax 0.1 %.

ingredients with sufficient mechanical properties. The binder is also used as a fuel because it contains mainly hydrogen and carbon. Two examples of double-base and composite propellants are given in Table 2.5.

The binder is a liquid, which is loaded with all solid components and acts as a fuel itself. It is the backbone of the polymeric system and determines the structural and mechanical properties of the final propellant.

The choice of suitable binders is limited. The most commonly used binders nowadays are epoxy-cured and polyurethane systems based on polybutadiene.

A typical basic missile design is shown in Fig. 2.13a (top), with a schematic drawing of the solid rocket motor shown in Fig. 2.13a (bottom).

PBAN
(Polybutadiene acrylonitrile)

HTPB
(Hydroxy-terminated polybutadiene)

GAP
(Glycidyl azide polymer)

Fig. 2.13a: Basic missile design (top) with a schematic drawing of the solid rocket motor (bottom) [reproduced with kind permission of Bayern Chemie, an MBDA company].

Presently, research is on-going into trying to find alternatives to AP/Al (see also Ch. 1.2.4). The problems with the AP/Al mixtures which contain HTPB as a binder, are two-fold. On the one hand AP is toxic and should be substituted for this reason alone (see Ch. 1.2.4). On the other hand, such formulations are also problematic in slow cook-off tests (SCO test, see Ch. 6.2). It appears to be the case that here the AP slowly decomposes during the formation of acidic side-products. These acidic side-products then react with the HTPB binder, which can result in the formation of cracks and cavities in the composite, which consequently negatively affects the performance and sensitivity. Possible alternatives for AP are ADN, HNF and TAGNF. However, they cause other problems, such as, for example, the low thermal stability (ADN melts at 93 °C and already decomposes at 135 °C) and the binder compatibility is not always guaranteed either. Further research work is absolutely necessary in order to find better oxidizers for solid propellants. In this context, the following requirements must be fulfilled:
1. density has to be as high as possible, if possible over 2 g cm^{-3},
2. oxygen balance has to be good, if possible better than that in AP,
3. melting point has to be above 150 °C,
4. vapor pressure has to be low,

5. decomposition temperature has to be above 200 °C,
6. synthesis has to be facile and economic,
7. compatible with binders (with HTPB),
8. sensitivity has to be as low as possible, definitely not more sensitive than PETN,
9. enthalpy of formation has to be as high as possible.

There are also advances in the area of propellants (Al). For example, attempts are being made to replace aluminum (which is normally used on the micro meter scale) with nano-Al. Thereby the combustion efficiency (quantitative combustion) would be increased, but the air (oxygen) sensitivity of the propellant would also increase significantly. ALEX is a nano material in the 20–60 nm region which is prepared by an electrical explosion (quick electrical resistance heating in vacuum or under an inert gas). The use of aluminiumhydrides, AlH_3, as replacements for pure Al are also being discussed currently, however these are even more sensitive to air (oxidation) when they are not stabilized through the use of wax.

Another possibility which is also currently being researched in order to increase performance is the replacement of the aluminum (Al) by aluminum hydride (AlH_3). Calculations have proven that the specific impulse (I_{sp}^*) can be increased by approximately 8 % (or 20 s) by replacing Al by AlH_3. A further, however somewhat smaller improvement can theoretically be reached by substituting AP by ADN:

AP / Al	(0.70 / 0.30)	$I_{sp}^* = 252$ s
AP / AlH_3	(0.75 / 0.25)	$I_{sp}^* = 272$ s
ADN / AlH_3	(0.70 / 0.30)	$I_{sp}^* = 287$ s

Silanes, and in particular the higher silanes such as e.g. *cyclo*-pentasilane, Si_5H_{10}, have also been discussed recently as possible propellants. These could possibly be suitable for RAM and SCRAM jets, since in addition to forming SiO_2, they could burn using nitrogen from the air to form Si_3N_4 and thereby use not only the oxygen, but also the nitrogen from the air as oxidizers (air-breathing engines).

Double-base propellants commonly consist of a binder polymer (nitrocellulose) which is plasticized with a nitric ester (nitroglycerine). In fact, Double-base propellants are considered as one of the oldest propellant families, and were developed as a consequence of the development of propulsion. At the end of World War I, gun propellants were essentially powders which were based on nitrocellulose. Later, it was found that the inclusion of nitroglycerine on nitrocellulose made it possible to increase the energy level, but due to the increase of the combustion temperature, its usage as a gun propellant has been limited. However, despite this, they have found use as rocket motor propellants.

Unfortunately, propellants containing NC continuously slowly decompose. The decomposition products released in the decomposition process increase the rate of decomposition, and a self-accelerating behavior is observed. To prevent this auto-

Fig. 2.13b: Currently used stabilizers for NC in single, double and triple based propellants.

catalysis, stabilizers are added to NC/NG-based propellants. The action of the stabilizers is to trap the nitrous decomposition products and form stable compounds, which prevent or delay further decomposition of the NC/NG-based propellant. The most common stabilizers are diphenylamine and its derivatives (Fig. 2.13b).

Since it was necessary to increase the energy level of the propellants in order to fulfill the special needs and requirements that are being continuously demanded for propellants, it was necessary to add some energetic additives, such as ammonium perchlorate (AP), Aluminum powder (Al), or some nitramines, such as RDX or HMX, to form so-called composite modified double-base propellants (CMDB) propellants. However, this resulted in a new set of problems such as the formation of toxic gases and an exhaust smoke plume due to using AP as an oxidizer additive or Al as a fuel in CMDB propellants respectively.

The generation of smoke has several detrimental effects such as: it may indicate the launch site of the missile and allow the missile to be located in flight, or it can result in the loss of control of the missile in cases involving optically guided missiles, and finally a negative environmental impact due to the emission of HCl gas in the exhaust plumes. The smoke produced from the combustion of composite solid rocket propellants can be classified into two general classes: primary smoke, which is a result of the gaseous products such as CO, CO_2, H_2, H_2O, Al_2O_3, and HCl, and secondary smoke which is caused by the combustion of the ammonium perchlorate-containing propellant producing HCl gas under certain atmospheric conditions (temperature and humidity). The combination with air or humidity results in the formation of a white cloud.

Composite propellants can be classified according to the composition of the smoke they produce into: high smoke propellants in which the composite propellant contains Al fuel and AP oxidizer as the main ingredients, and the exhaust plume which is produced contains high primary smoke (Al_2O_3) particles and HCl

Fig. 2.13c: Synthesis of three new oxidizers from trinitroethanol.

gas and reduced smoke propellants in which the composite propellant contains a small amount of Al or Mg and a reduced quantity of the AP oxidizer (by replacing AP with other materials), and the exhaust plume which is produced contains either a low amount of HCl gas or no HCl gas, as well as a low amount of primary smoke. Minimum Smoke propellants are those in which the propellant does not contain Al fuel or AP oxidizer. Instead, AP is replaced by a nitramine such as HMX, RDX or ammonium nitrate (AN) with nitramine and the exhaust plume is free of HCl gas.

Possible replacements for AP are high-energy density oxidizers (HEDOs) which have been discussed recently. These compounds consist of C, H, N, and O and have a high oxygen content. 2,2,2-Trinitroethanol (TNE) is one of the most suitable starting materials with an oxygen balance (Ω_{CO}) of 30.9 % and is easily synthesized through a Henry reaction. Several new oxidizers have been synthesized recently starting from TNE (Fig. 2.13c).

The energetic properties and the sensitivity towards different thermal and mechanical stimulus were determined. The experimental burning tests for the three formulations of the new propellants containing the three new oxidizer compounds (**1, 2,** and **3** from Fig. 2.13c) showed very good homogeneity between the fuel polymer (NC) indicated by homogeneous burning and observed using a high speed camera. Also, a high burning rate was observed in case of the second (NC/TNENC) and third (NC/TNEF) formulation of the propellants, while the first formulation (NC/BTNEO) showed a low burning rate and low flame in comparison with the other two compounds which produced a very bright flame on burning. This can be clearly observed in Fig. 2.13d. In addition, from the theoretical calculations, the high oxygen balance of the formulation containing NC/TNENC Ω_{CO_2} [%] ≈ 15 could be predicted as well as its high specific impulse.

Fig. 2.13d: Burning of three different new formulations of smokeless double-base propellants (from left to right: NC/BTNEO, NC/TNENC and NC/TNEF).

Fig. 2.13e: Effect of burning on the copper plate for the different three propellant formulations.

Furthermore, the effect of the temperature of the burning flame temperature on the copper plate (see Fig. 2.13e) shows that the first formulation (NC/BTNEO), which was based on oxalate, has a very low effect on the copper plate due to the low burning flame temperature, but the opposite is observed for the other two formulations, which are based on the carbamate and formate respectively.

The new high-performance, smokeless, double-base propellants based on the trinitroethyl unit have been prepared and studied on a laboratory scale. Three different formulations (NC/BTNEO, NC/TNENC and NC/TNEF) showed homogeneous burning with good burning rates and all were completely smokeless. The second formula with carbamate (NC/TNENC) showed the highest burning rate, followed by the third (NC/TNEF), with the lowest burning rate shown by the first formula containing oxalate (NC/BTNEO). Also, the lowest burning temperature (lowest effect on the copper plate during burning) and relatively small flame was the formula with oxalate (NC/BTNEO), while the highest burning temperature (highest effect on the copper plate during burning) and brightest flame was observed for the formula with the formate (NC/TNEF).

For **liquid rocket propellants**, there is a difference between mono and bipropellants. Monopropellants are endothermic liquids (e.g. hydrazine), which decompose exothermically – mainly catalytically (e.g. Shell-405; Ir/Al_2O_3) – in the absence of oxygen:

$$N_2H_4 \xrightarrow{\text{catalyst}} N_2 + 2\,H_2 \qquad \Delta H = -51 \text{ kJ mol}^{-1}$$

Monopropellants possess a relatively small energy content and a specific impulse and are therefore only used in small missiles and small satellites (for correcting orbits), where no large thrust is necessary. A summary of some monopropellants can be found in Table 2.6.

In a bipropellant system, the oxidizer and the fuel are transported in two tanks and are injected into the combustion chamber. The bipropellants can be separated into two different classes, either in cryogenic bipropellants, which can be handled only at very low temperatures and are therefore unsuitable for military applications (e.g. H_2/O_2) and storable bipropellants (e.g. monomethylhydrazine/HNO_3), or in accordance with their ignition behavior in hypergolic and non-hypergolic mixtures. **Hypergolic** describes rocket propellants whose components react spontaneously with one another (in less than 20 ms) when they are mixed or somehow contact each other. The components of hypergolic propellants are mostly oxidizers or reducing agents, which react immediately upon contact, and ignite partly explosively. Since the fuel reacts immediately by burning on contact with the oxidizer at the time of injection into the combustion chamber, it is not possible for too much fuel to accumulate in the combustion chamber prior to ignition which could lead to an explosion and damage of the rocket engine. A further important advantage is that the ignition definitely occurs, which is important for weapon systems such as intercontinental rockets, pulsed engines and upper stages of launch vehicles (re-ignition in space) for example. (*Note:* Hypergolic systems are also used in incendiary devices e.g. for mine clearance.) Hydrazine derivatives (hydrazine, MMH and UDMH) with HNO_3 or NTO (dinitrogen tetroxide) are practically the only hypergolic propellants used today. Table 2.7 shows a summary of different liquid bipropellant mixtures.

However, there are currently efforts to replace carcinogenic MMH and UDMH by toxicologically less damaging, but still hypergolic substances. One possible candidate is dimethylaminoethylazide (DMAZ) which has been produced by the company MACH I (PA):

DMAZ $>N-CH_2-CH_2-N_3$
2-Dimethylaminoethylazide

Tab. 2.6: Monopropellants.

monopropellant	formula	I_{sp}^* /s	T_c / K
hydrazine	H_2N-NH_2	186	1500
hydrogen peroxide	$HO-OH$	119	900
isopropylnitrate	$C_3H_7-O-NO_2$	157	1300
nitromethane	CH_3-NO_2	213	2400

Tab. 2.7: Bipropellants.

oxidizer	fuel	T_c / °C	T / K	I_{sp}^* / s
LOX	H_2	2740	3013	389
	H_2 / Be (49 : 51)	2557	2831	459
	CH_4	3260	3533	310
	C_2H_6	3320	3593	307
	B_2H_6	3484	3762	342
	N_2H_4	3132	3405	312
F_2	H_2	3689	3962	412
	MMH	4074	4347	348
	N_2H_4	4461	4734	365
OF_2	H_2	3311	3584	410
FLOX (30 / 70)	H_2	2954	3227	395
N_2F_4	CH_4	3707	3978	319
	N_2H_4	4214	4487	335
ClF_5	MMH	3577	3850	302
	N_2H_4	3894	4167	313
ClF_3	MMH	3407	3680	285
	N_2H_4	3650	3923	294
N_2O_4, NTO	MMH	3122	3395	289
MON-25 (25 % NO)	MMH	3153	3426	290
	N_2H_4	3023	3296	293
IRFNA (III-A)[b]	UDMH	2874	3147	272
IRFNA (IV HDA)[b]	MMH	2953	3226	280
	UDMH	2983	3256	277
H_2O_2	N_2H_4	2651	2924	287
	MMH	2718	2991	285

[a] IRFNA (III-A): 83.4 % HNO_3, 14 % NO_2, 2 % H_2O, 0.6 % HF
[b] IRFNA (IV HDA): 54.3 % HNO_3, 44 % NO_2, 1 % H_2O, 0.7 % HF

The work that has been reported so far indicates that DMAZ would make a logical alternative to MMH as a candidate for a gel fuel component in a hypergolic bipropellant design (GAP may be a suitable gellant). MACH I (along with Aerojet-Rocketdyne) has demonstrated that DMAZ-based fuel gels display several advantages such as a lower toxicity, higher density, and wider liquid range as a function of temperature. Furthermore, it has been shown that gels can be formulated with DMAZ to meet the rheological requirements of certain system designs. It has also been demonstrated that the issue of longer native ignition delay of DMAZ, compared to that of MMH, can be effectively addressed using additives (soluble Co or Mn salts).

Another interesting compound is tetramethyl tetrazene, which was suggested by CNRS, Airbus Safran Launchers. 1,1,4,4-tetramethyl-2-tetrazene (TMTZ) is also considered as a prospective replacement for toxic hydrazines used in liquid rocket propulsion. The heat of formation of TMTZ was computed and measured, giving values well above those of the hydrazines commonly used in propulsion. This led to a predicted maximum I_{sp} of 337 s for TMTZ/N_2O_4 mixtures, which is a value comparable to that of MMH. TMTZ is thus an attractive liquid propellant candidate, with a performance comparable to hydrazines but a lower vapor pressure and toxicity.

TMTZ

Out of the bipropellant mixtures shown in Table 2.7, only a few are used in practial applications. In particular, LOX / H_2 has proven useful in the cryogenic main engines of the civil Space Shuttle and Ariane V. The Aestus upper stage of the Ariane V relies on using NTO/MMH. The engines of the Delta (RS-27) and Atlas rockets work with LOX / HC (HC = hydrocarbons). Russian rockets in particular often use UDMH, which is very similar to MMH. As storable oxidizers, practially only NTO or RFNA (red fuming nitric acid) are used. Despite the very good specific impulse of fluorine containing (F_2 or FLOX) oxidizers and beryllium containing propellants, they have not found application due to technical (handling and pumping of liquid fluorine) and ecological (toxicity of HF and Be compounds) grounds.

At this point, the more recent development of so-called **gel propellants** should be mentioned. We have seen above that for military missiles, hypergolicity is a desired property. On the other hand, transportation of IRFNA or NTO in combination with MMH or UDMH on a ship or submarine for example also carries high risks and danger if there is an accident with unwanted contact of the oxidizer with the fuel. Gel propellants try to combine the desired and positive properties of a liquid propellant (flow behavior, pulse-mode) with the reliability of a solid propellant. The term geled fuel is used to describe a fuel which has an additional gelling agent (approx. 5–6%) and an additive, so that it behaves like a semi-solid and only begins to liquefy under pressure. Inorganic (e.g. Aerosil 200, SiO_2 or aluminium octanoate) or organic substances (e.g. cellulose or thixatrol = modified castor oil derivative) are suitable gelling agents. The gel becomes semi-solid as a result of the presence of strong π-π- or van-der-Waals interactions and therefore it not only has a low vapor pressure but often also a higher density in comparison to pure fuel, which makes it very safe to handle. The gel only liquidizes under pressure while the viscosity decreases with increasing shear stress. There are certain problems still exist today in the production of oxidizer gels as a result of the high reactivity of WFNA, RFNA or NTO. Hypergolic alternatives to the strongly corrosive nitric acid derivaties are also sought after.

Hybrid rocket motors use propellants in two different states of matter, usually a solid fuel and a liquid or gaseous oxidizer. Like liquid rockets and unlike solid rockets a hybrid engine can be turned off easily but due to the fact that fuel and oxidizer (different states of matter) would not mix in the case of an accident, hybrid rockets fail more benignly than liquid or solid motors.

The specific impulse is generally higher than that for solid motors. A hybrid engine usually consists of a pressurized tank (oxidizer) and a combustion chamber containing the solid propellant (fuel). When thrust is desired, the liquid oxidizer flows into the combustion chamber where it is vaporized and then reacted with the solid fuel at chamber pressures of about 7–10 bar.

Commonly used fuels are polymers, e.g. HTPB or polyethylene which can be metallized (aluminized) in order to increase the specific impulse. Common oxidizers include gaseous or liquid oxygen or nitrous oxide.

Some of the disadvantages of the fuels presently used are a relatively low regression rate and a low combustion efficiency. Such problems could be overcome in the future by using cryogenic solid fuels like frozen pentane (@ 77 K). Such fuels would form a melt layer during the combustion process and have a higher regression rate. Alternatively, paraffin wax (not cryogenic) would also form a melt layer during combustion and could easily be aluminized. Such systems have specific impulses of up to 360 s.

The most commonly used oxidizers have the disadvantage of either being cryogenic (LOX) or toxic (N_2O_4). Newer oxidizer systems which are presently under investigation include refrigerated (approx. –40 to –80 °C) mixtures of N_2O and O_2 ("Nytox", 80 % N_2O, 20 % O_2). The advantage of such a mixture would be a relatively high vapor pressure (compared to pure N_2O) and a high density due to the presence of O_2, but without being cryogenic. In combination with a metallized paraffin (wax) nytox as a fuel could help to overcome performance as well as toxicity issues.

Other developments go into the direction of adding up to approx. 10 % aluminum hydride (AlH_3) or $LiAlH_4$ to the HTPB fuel in order to increase the regression rate and deliver a specific impulse of up to 370 s.

2.4.1 Chemical Thermal Propulsion (CTP)

A further special area of propulsion systems is **Chemical Thermal Propulsion** (CTP). CTP is defined in contrast to STP (solar thermal propulsion) and NTP (nuclear thermal propulsion). In CTP, in a very exothermic chemical reaction in a closed system, heat but no pressure is generated since the products of the reaction are solid or liquid. The heat energy is then transferred to a liquid medium (the propel-

Tab. 2.8: Specific impulse I_{sp}^* of the CTP reaction Li/SF$_6$ occurring at 10 bar and 2500 K, depending on the propellant used.

propellant	M (propellant) / g mol^{-1}	I_{sp}^* / s
H$_2$	2	900
He	4	500
H$_2$O	18	320
N$_2$	28	230

lant) using a heat exchanger, which is responsible for the propulsion of for example, the torpedo. Suitable propellants are e.g. water (the torpedo can suck it in directly from its surroundsings) or H$_2$ or He, due to their very low molecular or atomic masses. The basic principles of CTP can also be used in special heat generators. A good example for a chemical reaction which is suitable for CTP is the reaction of (non-toxic) SF$_6$ (sulfur hexafluoride) with easily liquified lithium (m.p. 180 °C):

$$SF_6 + 8\,Li \rightarrow 6\,LiF + Li_2S \qquad \Delta H = -14727 \text{ kJ kg}^{-1} \text{ (mixture)}$$

In comparison to this, the reaction of MMH with NTO generates only 6515 kJ per kilogramm of a stoichiometric mixture. Table 2.8 clearly shows the influence of the average molecular mass of the "combustion products" (in this case the propellants) on the specific impulse.

Further chemical reactions which are principally suitable for CTP are outlined in the following equations:

8 Li + SF$_6$	→	Li$_2$S + 6 LiF	$\Delta_r H =$ −3520 kcal kg^{-1}
4 Li + OF$_2$	→	Li$_2$O + 2 LiF	−5415
6 Li + NF$_3$	→	Li$_3$N + 3 LiF	−3999
8 Li + SeF$_6$	→	Li$_2$Se + 6 LiF	−2974
6 Li + ClF$_5$	→	LiCl + 5 LiF	−4513
6 Li + BrF$_5$	→	LiBr + 5 LiF	−3212
6 Li + IF$_5$	→	LiI + 5 LiF	−2222
4 Li + ClF$_3$	→	3 LiF + LiCl	−4160
8 Li + ClO$_3$F	→	3 Li$_2$O + LiCl + LiF	−4869

It is obvious that the Li / SF$_6$ system (although it is energetically lower than others) is preferred over other systems because it is easy to handle and has low health and environmental hazards in comparison to other mixtures involving different oxidizers.

2.5 Pyrotechnics

2.5.1 Detonators, Initiators, Delay Compositions and Heat-Generating Pyrotechnics

A pyrotechnic composition is a substance or mixture of substances designed to produce an effect by heat, light, sound, gas or smoke or a combination of these, as a result of non-detonative self-sustaining exothermic chemical reactions. Pyrotechnic substances do not rely on oxygen from external sources to sustain the reaction.

The term pyrotechnic is derived from ancient Greek and translates as 'the art of fire' or 'the art of handling fire'. Similarly, as with explosives or propellants, the effects of pyrotechnics are also based on strongly exothermic reactions, whereby explosives show the largest and propellants relatively speaking the lowest speeds of reaction. Pyrotechnics are located somewhere in between. In contrast to explosives which release large quantities of gas during a reaction, pyrotechnics form mainly solid as well as gaseous products. Generally every pyrotechnic consists of an oxidizer as well as a reducing agent. Moreover, depending on the intended use, they can also contain a binder, a propellant charge and a colorant as well as smoke and noise generating additives. In contrast to explosives, which often contain the oxidizer and reducing moieties combined in one molecule (e.g. TNT, RDX), traditionally pyrotechnics are mixtures of different substances. Many pyrotechnic reactions are therefore solid state reactions, which result in a well-defined particle-size and highest possible homogeneity. The energy released in a pyrotechnic reaction usually results in a flame and some smoke, light and also gas formation. Pyrotechnics have many applications, some of which are summarized in Table 2.9.

Pyrotechnic mixtures which are used for ignition in blasting caps or detonators or which are used as first fires in propellant charges, are very easy to initiate. Here the initiator generates a flame if it is hit with a metal (percussion primer) or if it is initiated electrically using resistance heating (bridge wire). The first fire then initiates the propellant charge (percussion primer) or the first fire of an electrical match initiates a pyrotechnic mixture in a detonator, which then initiates a primary explosive which subsequently initiates a secondary explosive in the detonator. The shockwave of that detonator then initiates the main charge (not a component of the detonator). Figure 2.14 schematically shows the construction of a detonator which is initiated using an electrical match (initiator). For safety reasons, the initiator is introduced into the detonator only shortly before use. Typical examples for primary explosives used in detonators are lead azide and lead styphnate, and PETN for secondary explosives.

SINOXID (= without rust) is the trademark used for the traditional primer compositions of the company Dynamit Nobel. In contrast to the mixture of mercury fulminate and potassium chlorate which was used previously, SINOXID is made-

Tab. 2.9: Applications of pyrotechnics.

pyrotechnic	area of application	example
heat generating pyrotechnics	first fires, primers	
	pyrotechnic mixtures in detonators (blasting caps, detonators or primers)	see Fig. 2.14
	electric "matches" in initiators ("first fires")	Ox: KNO_3, BaO_2 Fuel: Mg, Ti, Si see Fig. 2.14
	incendiary devices percussion primers delay compositions (bombs, projectiles, grenades)	$KClO_3$ + TNT + PbO_2, Sb_2S_3 gassy: blackpowder gasless: metal oxide and –chromate + metal powder
	matches	head: Sb_2S_3 or S_8 + $KClO_3$ striking surface: P_{red} + glass
smoke-generating pyrotechnics	smoke munition	
light-generating pyrotechnics	signal flares	
	decoy munition (IR region) fireworks	MTV

Fig. 2.14: Schematic representation of a detonator with an electric initiator.

up of lead styphnate (Fig. 1.17), tetrazene (Fig. 2.2), barium nitrate, lead dioxide, antimony sulfide and calcium silicide. SINOXID compositions feature very good chemical stability and storage life, they are abrasion-, erosion- and corrosion-free and ignite propellants with precision. **SINTOX-** (non-toxic) is a recently developed primer composition of Dynamit Nobel. It was developed specifically for ambient air in indoor firing ranges, which should not to be polluted with combustion products containing lead, antimony or barium. SINTOX consist essentially of diazodinitro-

Fig. 2.14a: Schematic drawing of an electric detonator containing a delay element.

phenol (or strontium diazodinitroresorcinate) and tetrazene as initial explosives and zinc peroxide as the oxidizer.

In fireworks, blackpowder is still used as the first fire.

Delay compositions are divided into two types: gasless (e.g. metal oxides or metal chromates with an elemental fuel) and gassy (e.g. blackpowder). Gasless delays are used in conditions of confinement (e.g. bombs, grenades, projectiles) or at high altitudes, where it is important that there are no variations of normal, ambient pressure. The desired burn rate (mm ms^{-1} to mm s^{-1}) is dependent on the purpose of use. Delay compositions with high burning rates for example are used in projectiles and bombs which should explode on impact, whereas those with low burning rates are primarily used in ground chemical munitions such as smoke pots, smoke grenades and tear pots. The following mixtures are examples of gasless delay compositions:

1. Fuel: B, Si   oxidizer: $K_2Cr_2O_7$,
2. Fuel: W   oxidizer: $BaCrO_4$, $KClO_4$,
3. Fuel: Mn   oxidizer: $PbCrO_4$, $BaCrO_4$,
4. Fuel: Cr   oxidizer: $BaCrO_4$, $KClO_4$.

A typical delay composition would be: $BaCrO_4$ (56%), W (32%), $KClO_4$ (11%), VAAR (~1%).

Gasless (producing less than 10 mL g^{-1} of gas) delays (Fig. 2.14a) are used in a large number of items every year in a range of US military applications, for example in grenades of the fuse type M201A1. Up until now, they have often been based on thermite reactions such as that shown in the following equation (equation 1).

$$Pb_3O_4 + 2\,Si \rightarrow 2\,SiO_2 + 3\,Pb \qquad (1)$$

Different additives are used in order to increase the ignitability and combustion time. One of the most prominent mixtures with a burn rate (BR) between 0.6 and 150 mm s^{-1} is the W/$BaCrO_4$/$KClO_4$/diatomaceous earth (dilutant) composition.

A more complex reaction equation for this system is given in the following equation (equation 2).

$$W + {}^3/_8 KClO_4 + BaCrO_4 \rightarrow WO_3 + {}^3/_8 KCl + {}^1/_2 Cr_2O_3 + BaO + 508 \text{ kJ} \qquad (2)$$

Further examples of prominent gasless delay compositions are:
- barium chromate, potassium perchlorate and zirconium/nickel alloy (BR = 1.7–25 mm/s)
- boron and barium chromate (BR = 7–50 mm/s)
- potassium dichromate, boron and silicon (BR = 1.7–25 mm/s).

The time range of the delay element which is required varies strongly depending on the application. Very recently, progress on two condensed phase reactions e.g Ti/C (Ti + C → TiC, ΔH_R = –3079 J g^{-1}) and 3 Ni/Al (3 Ni + Al → Ni$_3$Al, ΔH_R = –753 J g^{-1}), as potential time delay compositions has been published.

Pyrotechnic time delays are reactive systems that burn for a desired period of time at a fixed rate, and are commonly used in military applications such as grenades and hand-held signals. More recently, there has been considerable advocacy for the development and use of environmentally friendly pyrotechnic ignition delays, by the removal of barium chromate (BaCrO$_4$). In a recent study by Groven et al., the viability of two possible replacements – strontium molybdate (SrMoO$_4$) and barium molybdate (BaMoO$_4$) – for the harmful component, BaCrO$_4$, in traditional delay compositions was evaluated.

The molybdates of strontium and barium have been shown to be viable options as replacements for BaCrO$_4$ in traditional ignition delay formulations.

In general, the current trend is to favor 'green' compositions which are targeted to replace compositions containing perchlorates, chromates, lead, and barium. Thermite-based reactions are still the most promising alternatives.

Table 2.9a summarizes the burn rates for slow and fast delay compositions and thermites.

Tab. 2.9a: Burn rates for slow and fast delay compositions and thermites (Focke, Tichapondwa et al., PEP 2019, 44, 55–93).

fuel/oxidant	tube material	burn rate/mm s^{-1}
Si / Pb$_3$O$_4$	cast metal	2.2–3.8
Si / CaSO$_4$	Al	7–30
Si / Sb$_2$O$_3$	Pb	7–14
Si / BaSO$_4$	Al	8.4–16
Si / PbCrO$_4$	Pb	10–30
Si / Bi$_2$O$_3$	Pb	15–155
Al / WO$_3$	–	0.1–4.1
Al / MoO$_3$	–	4–12
Al / MoO$_3$	acrylic	600–1000
Mn / Sb$_2$O$_3$	Pb	4–9
Mn / CuO	Al	5–10

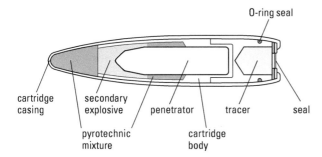

Fig. 2.15: Schematic representation of the construction of a 12.7 mm multipurpose projectile.

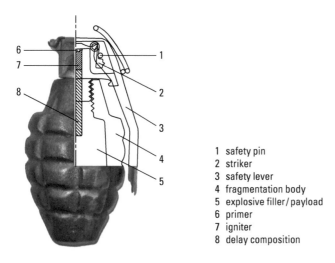

1 safety pin
2 striker
3 safety lever
4 fragmentation body
5 explosive filler / payload
6 primer
7 igniter
8 delay composition

Fig. 2.16: Schematic representation of the construction of a hand grenade with a pyrotechnical delay device.

Figure 2.15 shows a schematic representation of the construction of a 12.7 mm multipurpose projectile with a pyrotechnic mixture (incendiary) and secondary explosive. Figure 2.16 schematically shows the construction of a hand grenade with a pyrotechnic delay composition. Bombs such as, for example, the MK 80 series, where approx. 45% of the total mass is the explosive, usually contain pyrotechnic ("nose" or "tail") fuses.

2.5.2 Light-Generating Pyrotechnics

Pyrotechnic formulations which emit electromagnetic radiation in the visible region are used as parachutes or hand-held signal flares for large-area illumination or to mark specific positions (targets, landing areas for planes or paratroopers etc.).

Tab. 2.10: Spectra of the most important atoms and molecules for pyrotechnics.

element	emitter	wavelength / nm	color
Lithium	atomic Li	670.8	red
		610.4	red
		460	blue
		413	violet
		497	blue/green
		427	violet
Sodium	atomic Na	589.0, 589.6	yellow
Copper	CuCl	420–460	blue/violet
		510–550	green
Strontium	SrCl	661.4, 662.0, 674.5, 675.6	red
	SrCl	623.9, 636.2, 648.5	orange
	SrCl	393.7, 396.1, 400.9	violet
	SrOH	605.0, 646.0	orange
		659.0, 667.5, 682.0	red
	atomic Sr	460.7	blue
Barium	BaCl	507, 513.8, 516.2, 524.1, 532.1,	green
		649	red
	BaOH	487,	blue/green
		512	geen
	BaO	604, 610, 617, 622, 629	orange
	atomic Ba	553.5,	green
		660	red

[13–16] For example, red is often used by aircraft personnel, to indicate their position in the event of an emergency landing or accident. Green and yellow are used by ground troops to mark their positions and white (Mg + Ba(NO$_3$)$_2$ + KNO$_3$; Al, Ti and Zr are also used as fuels for producing white lights) allows large-area illumination at night. The intensity and wavelength (color of the emitted light) depend on the components in the burning pyrotechnic mixture. Pyrotechnics which burn at approximately 2200 °C, usually contain perchlorates as the oxidizer as well as organic propellants (e.g. nitrocellulose). In order to increase the flame temperature to 2500–3000 °C, metal powders such as magnesium are added. The emission of specific spectral colors occurs as a result of the addition of other metals and other metal salts. Sodium is therefore used for yellow, strontium for red and barium is usually used for green. The main emitters of the corresponding spectral bands are atomic Na for yellow, Sr$^{(I)}$OH and Sr$^{(I)}$Cl for red as well as Ba$^{(I)}$Cl and Ba$^{(I)}$OH for green (Tab. 2.10). BaO is also a strong emitter in Barium illuminants (Tab. 2.10). The hydrogen necessary for the formation of the monohydroxides SrOH and BaOH comes from the decomposition products of the binder or the PVC (Tab. 2.10), the Cl atoms come from the perchlorate oxidizing agent but also from the PVC. Table 2.11 shows the composition of some important illuminants in the visible range.

Tab. 2.11: Composition of selected illuminants in the visible region (%).

ingredient	Red Navy	Red Highway	Green Navy	Yellow Navy	White Navy
Mg	24.4		21.0	30.3	30
$KClO_4$	20.5	6.0	32.5	21.0	
$Sr(NO_3)_2$	34.7	74.0			varied
$Ba(NO_3)_2$			22.5	20.0	53
PVC	11.4		12.0		
$Na_2(COO)_2$				19.8	
Cu powder			7.0		
Bitumen	9.0			3.9	
Binder (VAAR)		10.0	5.0	5.0	5
S_8		10.0			

The emission in pyrotechnic compositions generally originates from one of the four following sources:
 - Atomic (elemental) emission (e.g. Na) → line spectrum,
 - Molecular emission (e.g. BaCl, SrOH) → band spectrum,
 - Incandescence → continuous spectrum,
 - Luminescence → band spectrum.

Colored flames are usually produced by the excitation of species which emit metal atoms (atomic emission). Other metal atoms may combine with other radicals in the flame, forming a color emitting molecule (molecular emission). The vapors of the emitting species are then excited by the thermal energy of the flame. The atoms or molecules in an excited state then relax to the ground state with the emission of visible light. For example, for the color red, strontium chloride is the main emitting species (Fig. 2.16a). The sequence of emission phenomena occurs in rapid succession. Figure 2.16a demonstrates the entire sequence of emission phenomena for strontium chloride.

In order to design colored illuminants, the properties of the human eye have to be considered (sample rate about 20 Hz). An eye can distinguish in the wavelength range of approximately 400 to 700 nm at an accuracy of about ± 2.5 nm, although an eye measures over very wide spectral bands in only three channels (Fig 2.16b). The high resolution is achieved by a highly advanced, real time signal processing of overlapping bands, each of which has a different response over the entire VIS range (visual light range). The red color cells also respond to blue light, which enables the perception of purple hues. Without this property, an eye would not be able to perceive a large sector of the chromaticity diagram, because no single wavelength can represent a purple color (Fig 2.18).

All three channels in an eye can record light over an even wider range than 400 to 700 nm (Fig. 2.16b), but the sensitivity outside of this region is very low. Hence, it is possible to see even near infrared radiation, if the source is strong

Fig. 2.16a: Sequence of emission phenomena for the SrCl molecule.

Fig. 2.16b: CIE 1931 tristimulus functions (sensitivity curves of each color cell type in a human eye). x-axis: wavelength (nm), y-axis: sensitivity.

enough, such as a laser. Since such sources are rare, the range of visible wavelengths in the standard observer response functions has been cut to include the above VIS range only. These cut-off gain functions are called the color matching functions. The response of these functions was measured for the first time in 1928 and published in 1931, which is why these functions are called CIE 1931 standard observer. A revised version is called CIE 1964. In this work, the CIE 1931 color matching functions with a white point at the coordinates (1/3,1/3) were used, since most emission measurements in the literature are based on these. In addition, conversion functions are available to convert from one color space to another. The white point was set into the center of the chromaticity diagram (cf. Fig. 2.18), because the measurements were not convolved with another light source. The broad bands of the tristimulus functions (Fig. 2.16b) enable creating relatively broad band

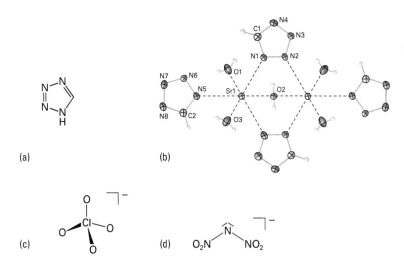

Fig. 2.17: Molecular structure of 1-H-tetrazole (a), the strontium compound as colorant in the environmentally friendly red pyrotechnic mixture (b), the perchlorate anion (c) and the dinitramide (d).

light sources with high color saturation. It is, however, important to keep all emissions within a narrow enough band to excite only one of the detectors in the eye. For red light, this means that the emissions should be kept above 600 nm wavelength, to make sure the green color detector in the eye is not appreciably excited. From the ecological and toxicological point of view the relative high perchlorate content of all of the illuminating compositions (Tab. 2.11) as well as the use of the heavy metal barium in green illuminants is problematic; during the production and burning of the light-generating compositions, toxic perchlorate and soluble barium salts can get into ground water. The ClO_4^- anion is a possible teratogen and has shown negative effects on the thyroid function. The up-take of iodine into the thyroid is suppressed by chronic perchlorate up-take which in consequence causes thyroid subfunction. Moreover, the binders and other ingredients (PVC, Bitumen, sulfur) often result in burning with strong smoke. Current research is therefore concerned with the development of:
- Perchlorate-free red and green colored illuminants,
- Barium in the form of sparingly soluble Ba salts (not soluble $Ba(NO_3)_2$),
- Barium (heavy metal) free green illuminants,
- weakly / smoke-free red, green and yellow illuminants.

Many of the newly developed pyrotechnics are based on metal complexes with very nitrogen-rich ligands, mainly tetrazole derivatives (Fig. 2.17). Unlike conventional energetic substances they do not derive their energy not through the oxidation of a carbon back-bone but through their high (positive) heat of formation. These compounds are suited for application as propellants, reducing agents or colorants – preferably with less toxic metal salts such as Cu^{II} instead of Ba^{II}.

Since environmentally compatible pyrotechnics should not contain perchlorates or heavy metals in the first instance, new red and green formulations are being developed which contain Cu instead of Ba and which contain other oxidizers (e.g. nitrate or dinitramide, Fig. 2.17) instead of perchlorates. The high nitrogen content also guarantees a significant reduction of the smoke and particulate matter released and consequently results in a considerably higher color brilliance.

Furthermore, high-N red and green formulations show an increase in burn time, an increase in luminous intensity and an increase in spectral purity. Two examples of new red and green formulations developed by ARDEC/LMU are:

red:	$Sr(NO_3)_2$	39%
	Mg	29%
	PVC	15%
	high-N Sr-tetrazolate salt	10%
	binder	7%
green:	$Ba(NO_3)_2$	46%
	Mg	22%
	PVC	15%
	high-N Ba-tetrazolate salt	10%
	binder	7%

In the above mentioned red and green formulations it is possible to replace the high-N metal tetrazolate salt with more metal nitrate and free aminotetrazole. This causes longer burn times but lower intensities.

The strontium formulation shows a reasonably high red color purity as can be seen from the chromaticity diagram in Fig. 2.18. The color purity (p_e) of a visible flare with the chromaticity (x, y) is its difference from the illuminant's white point relative to the furthest point on the chromaticity diagram with the same hue (dominant wavelength for monochromatic sources): where (x_n, y_n) is the chromaticity of the white point and (x_I, y_I) is the point on the perimeter whose line segment to the white point contains the chromaticity of the stimulus.

$$p_e \sqrt{\frac{(x - x_n)^2 + (y - y_n)^2}{(x_I - x_n)^2 + (y_I - y_n)^2}}$$

The color purity is the relative distance from the white point. Contours of constant purity can be found by shrinking the spectral locus about the white point. The points along the line segment have the same hue, with p_e increasing from 0 to 1 between the white point and position on the spectral locus (position of the color

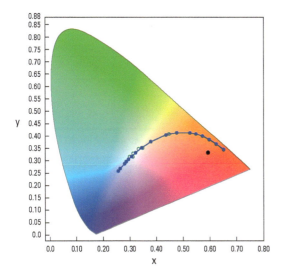

Fig. 2.18: Chromaticity diagram for the high-N strontium-based flare.

Tab. 2.12: Effect of boron carbide in pyrotechnics.

fuel (B : B$_4$C) in formulation[b]	burn time [s]	luminous intensity [cd]	dominant wave-length [nm]	spectral purity [%]
control[b]	8.15	1357	562	61
100:0	2.29	1707	559	55
50:50	5.89	2545	563	54
40:60	6.45	2169	563	54
30:70	8.67	1914	562	53
20:80	8.10	1819	563	53
10:90	8.92	1458	562	52
0:100	9.69	1403	563	52

[a] Formulation: KNO$_3$ 83%, fuel 10%, epoxy binder 7%
[b] Ba(NO$_3$)$_2$ 46%, Mg 33%, PVC 16%, Laminac 4116/Lupersol binder 5%

on the horseshoe shape in the diagram) or (as at the saturated end of the line shown in the diagram) position on the line of purples.

A breakthrough discovery was reported by Sabatini (ARDEC) in the area of green illuminants. Formulations without any heavy metal can be based on boron carbide (B$_4$C, fuel) with a suitable oxidizer (e.g. KNO$_3$). Table 2.12 shows the effect of boron carbide (B$_4$C) in pyrotechnical compositions. It can be seen that flares with 100% boron carbide as fuel show longer burn times and higher luminous intensity than the control barium nitrate based flare (M125 A1) while the spectral (color) purity is slightly lower.

Tab. 2.12a: Shimizus' chlorine-containing blue-light-emitting formulation (reference formulation).

	KClO$_4$[wt%]	Cu[wt%]	PVC[wt%]	Starch [wt%]
Control	68	15	17	5

Tab. 2.12b: Formulations A and B.

	Cu(IO$_3$)$_2$[wt%]	Guanidine nitrate[wt%]	Mg[wt%]	Urea [wt%]	Cu[wt%]	Epon 828/Epikure 3140[wt%]
A	20	50	10	–	15	5
B	30	35	9	21	–	5

Due to issues such as the ratio of ingredients employed, the molecular behavior of light-emitting species, and the combustion temperatures typically associated with the generation of colored pyrotechnic flames, the production of blue-light-emitting pyrotechnic illuminants is believed to be a most challenging field for chemists. Typically, blue flames are generated by the use of copper or copper-containing compounds in conjunction with a chlorine source (Tab. 2.12a). Chlorine sources used in blue-light-emitting pyrotechnics include ammonium perchlorate, potassium perchlorate, or chlorine donors such as poly(vinyl) chloride. When copper combines with a chlorine source at elevated temperature, copper(I) chloride (CuCl) is formed, which has always been believed to be the lone molecular emitter toward achieving high quality blue flames.

However, the use of perchlorates in pyrotechnic formulations is discouraged due to the supposed toxicity of this chemical which has been linked to causing thyroid disorders. Furthermore, the combustion of polychlorinated organic materials such as poly(vinyl) chloride has been shown to produce polychlorinated biphenyls (PCBs), polychlorinated dibenzo-p-dioxins (PCDDs), and polychlorinated dibenzofurans (PCDFs). These polychlorinated chemicals are highly toxic and are potent carcinogens. Therefore, the removal of perchlorates and chlorinated organic materials in pyrotechnic formulations would eliminate the formation of these aforementioned pollutants.

In a joint research effort between LMU Munich (Klapötke and Rusan) and ARDEC (Sabatini), the development of the first known perchlorate-free and chlorine-free blue-light-emitting pyrotechnic formulation has been developed. The optimal formulation (Tab. 2.12b) relies on the generation of copper(I) iodide (CuI), which serves as a strong blue-light-emitting species (Tab. 2.12c, Fig. 2.18a). In addition to the environmentally friendly goal of eliminating perchlorates, the formation of polyiodated biphenyls – which are likely to form during the combustion process – is not perceived to be an environmental hazard. Polyiodated biphenyls are used as contrast agents for radiological purposes in medicinal applications.

Tab. 2.12c: Performance and sensitivity data of formulations A and B.

	burn time [s]	Dw [nm]	Sp [%]	LI [cd]	LE [cdsg^{-1}]	IS [J]	FS [N]	grain size[μm]	T_{dec} [°C]
Control	4	475 (552)	61	54	360	8	324	307	307
A	6	477 (555)	64	80	1067	> 40	> 360	< 100	198
B	6	476 (525)	63	78	780	> 40	> 360	< 100	180

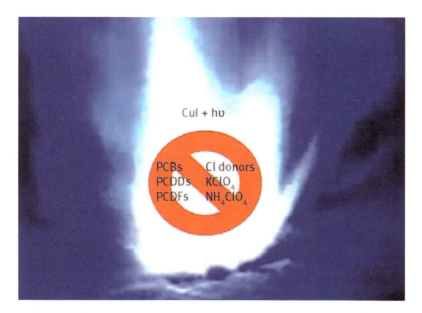

Fig. 2.18a: Blue flame originating from formulation **A**.

The US military has embarked on many initiatives over the years to mitigate environmental concerns associated with their munitions. Since the US military has an interest in blue-light-emitting signaling flares, the new formulations may provide a significant benefit to the Department of Defense. In addition, civilian firework companies have been under increasing scrutiny from the Environmental Protection Agency (EPA) to "green" their firework displays. The removal of perchlorates and chlorinated organic compounds while generating blue light represents an important step toward that goal, since blue is believed to be the most difficult of all colors to generate in an environmentally friendly fashion.

Extremely important for the safe handling of such pyrotechnic mixtures are, in addition to the chemical stability (compatability with the binder and other additional compounds) particular the thermal stability and the lowest possible impact, friction and electrostatic sensitivities. Figure 2.18b shows an example of a DSC ther-

Fig. 2.18b: DSC thermogram of the new environmentally friendly red mixture.

"Sustain the Mission – Secure the Future"

Fig. 2.19: Triple bottom line plus.

mogramm (Differential Scanning Calorimetry), which shows that the new environmentally friendly red mixture (colorant components from Figure 2.17) is thermally stable up to approx. 260 °C. But the sensitivity of the mixture towards electrostatic discharge (ESD) was lowered to 1 J (cf. typical values for the human body are within the range 0.005–0.02 J).

The protection of the environment through the use of sustainable technologies is one of the biggest and most important tasks of our time. Within the concept "triple bottom line plus" (mission, environment, community + economy) (Fig. 2.19), the US Army in co-operation with leading scientists world-wide, is trying to implement the sustainability of their operations and missions. A good example of this is the research into new, less environmentally hazardous pyrotechnics, which contribute to a successful operation as a result of their properties, but don't damage the environment due to their "green" characteristics. Furthermore, they are of general economic interest (new innovative technology, reduce the cost of purifying ground water) and of particular interest for local communities (reduced exposure to particulate matter, heavy metals and perchlorate in the vicinity of troop training camps).

2.5.3 Decoy Flares

The terms decoy flare or countermeasure munition are used to denote a system, which imitates the IR signal of a plane and therefore ground-air, water-air or air-

Fig. 2.20: The AIM-9 Sidewinder (AIM = Aerial Intercept Missile) is a heat-seeking guided missile for short-distance combat, which can be carried in most western fighter planes. In its 50 year history, more planes have been shot down using the Sidewinder than with any other device.

air rockets, so-called heat-seeking missiles, get lured away from their targets. One of the first and best-known IR seekers – "Sidewinder" – was developed in China Lake (Fig. 2.20).

The Sidewinder has a high explosive warhead and an IR-based heat-seeking guidance system. Its other components are an optical target detector, the fins and the rocket engine. The guidance-system directs the missile straight into the hot engine of the enemy's plane (or whatever the missile considers this to be). An infrared unit costs less than any other guidance-system. After it has been launched, the missile does not need anymore support from the launching platform and steers itself towards the target.

In order to be protected against such heat-seeking guided missiles, pyrotechnic decoy flares, which are based on blackbody radiation (or better greybody radiation), have been and are still being developed since the late 1950's. The specific radiant emittance of a radiating body (W in W cm^{-2} µm^{-1}) can be described using Planck's rule:

$$W_\lambda = \frac{2\pi h c^2}{\lambda^5} \frac{1}{e^{hcl/\lambda kT} - 1}$$

where λ (in µm) is the wavelength, h is Planck's constant (6.626 10^{-34} W s^2), T is the absolute temperature of the radiator (in K), c is the speed of light (2.998 10^{10} cm s^{-1}) and k is the Boltzmann constant (1.38 10^{-23} W s K^{-1}).

In accordance with the rule of Wien, the maximum wavelength of the blackbody radiation λ_{max} (µm) shifts towards shorter wavelengths (higher energy) with increasing temperature:

$$\lambda_{max} = 2897.756 \text{ µm K } T^{-1}$$

Since in reality there is no actual blackbody radiator (for which the Planck law applies, W), but a more realistic so-called greybody radiator (W'), the value ε has been defined as the quotient W'/W, where ε can be given values between 0 and 1 (1 = genuine blackbody). For example, soot ($\varepsilon > 0.95$) behaves almost like a blackbody radiator, whereas MgO behaves more like a greybody radiator.

Heat-seeking missiles mainly work with photo-conductive IR detectors in the so-called α band (PbS, 2–3 µm) or in the β band (PbSe, 3–5 µm). While the hot tail pipes of a plane turbine typically emit a greybody radiation with λ_{max} between 2 and 2.5 µm, this is superimposed by the selective radiation of the hot exhaust gases (CO_2, CO, H_2O) in the region of λ_{max} = 3–5 µm. Figure 2.21 shows the typical IR signiature of a plane in the α and β bands. In order to imitate this signiature, initially Al/WO_3 thermites in graphite-balls were used (diameter 10 cm, thickness 2 mm), since graphite in the region < 2.8 µm emits a very good blackbody radiation with ε = 0.95. Nowadays, most of the pyrotechnic mixtures used are mixtures of Mg and perfluorinated polymers (Mg/Teflon/Viton, MTV), since the carbon soot ($\varepsilon \approx 0.85$) which is produced at very high temperatures (approx. 2200 K) is a good IR emitter. The schematic construction of a IR decoy flare is shown in Figure 2.22. The main reaction between Mg and Teflon can be formulated as follows, where m ≥ 2:

$$m \text{ Mg} + -(C_2F_4)- \rightarrow 2 \text{ MgF}_2(l) + m\text{-}2 \text{ Mg}(g) + 2 \text{ C} + h\nu$$

The highly exothermic reaction for the formation of MgF_2 heats up the carbon soot formed to approx. 2200 K, which then emits the IR radiation. Moreover, in Mg rich formulations (m ≥ 2), the evaporating Mg is oxidized in the gas phase (3100 K). In addition, the carbon which is formed from the reductive elimination of fluorine from Teflon can be oxidized further to CO or CO_2 by atmospheric oxygen:

$$m \text{ Mg} + -(C_2F_4)- + O_2 \rightarrow 2 \text{ MgF}_2(l) + m\text{-}2 \text{ MgO}(s) + 2 \text{ CO}_2 + h\nu$$

Therefore, during the radiation of a MTV decoy flare, the greybody radiation of the carbon soot (intensive) is superimposed with the less intensive selective radiation of the CO or CO_2, respectively. In addition to the typical wavelength dependent intensity of the radiation of a plane, Figure 2.21 also shows that of a MTV decoy flare.

Typical pyrotechnic compositions for the generation of blackbody radiation contain magnesium as the fuel and a fluorine-containing oxidizer such as e.g. polytetrafluoroethylene, PTFE, $(C_2F_4)_n$. During the reaction, (1) large quantities of carbon

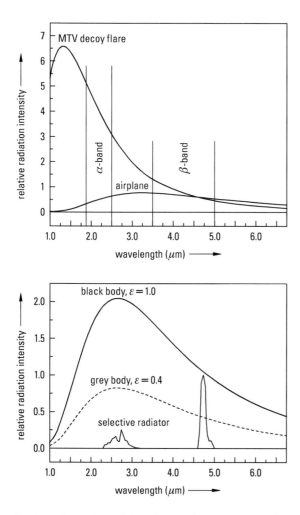

Fig. 2.21: Comparison of the relative radiation intensity of a plane and a MTV decoy flare.

Fig. 2.22: Schematic construction of an IR decoy flare.

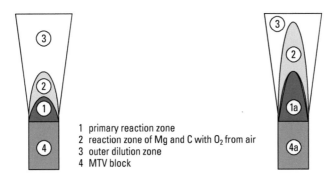

Fig. 2.22a: Typical reaction zones of a conventional decoy flare (left) and a high-nitrogen flare based on perfluoroalkylated tetrazoles (right).

form (which is a strong emitter in the IR region) and become thermally excited to radiation as a result of the heat of reaction. Decoy compositions are always rich in magnesium (50–70 weight-% Mg) and use the oxygen in the air as a complimentary oxidizer (2). This will now be described in greater detail by using a MTV countermeasure as an example, consisting of 63 weight-% Mg and 37 weight-% PTFE:

$$2.592\ Mg\,(s) + 0.3699\ (C_2F_4)_n \rightarrow 0.7398\ MgF_2(s)\ 0.7398 + 1.8522\ Mg\,(s)$$
$$+ 0.7398\ C(gr) + 532\ kJ \quad (1)$$

$$1.8522\ Mg\,(s) + 0.7398\ C(gr) + xs.\ O_2 \rightarrow 1.8522\ MgO(s) + 0.7398\ CO_2 + 1403\ kJ \quad (2)$$

The primary reaction (1) contributes only about a third of the total heat of the reaction. Despite this, the primary reaction significantly influences the mass flow and the radiation characteristics. Therefore, it is sensible to increase the exothermicity of the primary reaction, in order to reach the highest possible spectral efficiency, E, [J g^{-1} sr^{-1}]. For this reason, it has been suggested to use oxidizers which would result in a higher primary enthalpies of reaction than PTFE. Suitable substances with a higher molar enthalpy of reaction with magnesium than PTFE are graphite fluoride, $(CF_x)_n$, perfluorinated carbocyclene $(CF)_n$ or difluoroamine substituted compounds $\{C(NF_2)\}_n$.

The flame of a MTV composition consists of an inner zone of higher emission intensity, which contains a high concentration of carbon particles, which are formed through a primary reaction (1) (Fig. 2.22a). The primary reaction zone is surrounded by a second layer of lower emission intensity in which the oxidation of the magnesium and the carbon with oxygen from the air occurs (eq. 2). This intermediate zone is finally surrounded by an outer zone in which the cooling of the reaction products CO_2 and MgO occurs through radiation transfer and conduction of heat to the surrounding atmosphere. The high combustion temperature of the second layer heats up the inner zone even further and thereby influences the radiation characteristics considerably. Furthermore, the emission is also affected by the spatial expansion of the inner zone as well as the concentration of the carbon particles in it. Often, the concentration of the carbon particles in the inner zone is so high, that they cannot emit directly to the surroundings and therefore do not contribute to the radiation intensity. By dilution of the inner zone using an optically transparent gas such as e.g. nitrogen, the inner zone can be expanded and the concentration can also be reduced, so that a larger proportion of carbon particles can emit to the surroundings. Therefore, a radiation optimized decoy composition should deliver a sufficiently high proportion of nitrogen during combustion (Fig. 2.22a).

Recently it has been discovered that mixtures of magnesium and perfluoralkylated tetrazolates show a higher radiation intensity than compositions which are currently state-of-the-art and based on magnesium and e.g. PTFE, although the molar enthalpies of reaction for the reactions analogous to equation (1) are lower than e.g. MTV.

It is obvious that the intensity of the radiant intensity of MTV flares increases with the mass flow (kg s^{-1}, larger flares) and the burning rate. Current research is amongst other problems concerned with finding additives (e.g. 10 % Zr powder), which raise the burning rate by a factor of up to 1.5. The largest influence on the burning rate r is the pressure p (see Ch. 1.3), while the coefficient β describes the influence of the temperature ($\beta = f(T)$) on the linear burn rate and the index α, the pressure dependence:

$$r = \beta p^\alpha$$

To increase the temperature of reaction, components which have less negative heats of formation than PTFE can be used as well. Therefore, in some mixtures graphite fluoride or polycarbonmonofluoride (PMF) are used instead of PTFE. This achieves higher combustion enthalpies. Furthermore, PMF shows higher thermal stability than PTFE resulting in a better (longer) storability. A typical PMF mixture has the following approximate composition:
- 55–65 %   Mg   (fuel),
- 30–40 %   PMF   (oxidizer),
-     5 %   Viton (binder).

As a result of the pressure dependence, the linear burn rate also decreases considerably with increasing altitude (Fig. 2.23).

The biggest problem and therefore also the biggest challenge for research is however, the intensity ratio Θ of the relative intensities in the α (I_α) and β bands (I_β). Whereas the values for this relationship lie approximately between 1.3–1.4 for MTV mixtures Θ (MTV) = (I_α) / (I_β), planes show Θ (plane) = (I_α) / (I_β) values of between approx. 0.5 to 0.8 (see Fig. 2.24). This means that modern heat-seeking guided missiles can differentiate between a real target (plane) and a MTV decoy flare.

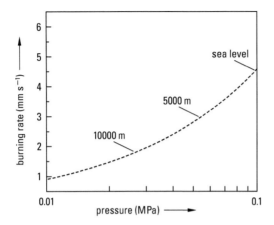

Fig. 2.23: Pressure dependence of the linear burning rate of a MTV mixture (50 % Mg, 45 % PTFE, 5 % Viton).

Fig. 2.24: Color ratio $\Theta_{\alpha/\beta}$ for a blackbody as a function of the temperature.

New decoy flares which show a strong emission (in β band) as a result of CO_2 or HBO_2 could solve this problem. One possibility would be to add organic substances (organic fuels) or boron compounds (formation of HBO_2) as well as additional oxidizers (AP) to the decoy flare compositions, but this often results in a too low intensity of the radiation, which is problematic. This is a current area of research and cannot be answered conclusively at this time.

Figure 2.21 (see above) displays the relative radiant intensity for classical MTV flares and a target signature. It can be seen that the radiant intensity distribution is totally different for MTV compared to the actual target signature. While MTV yields ratios $\Theta_{\alpha/\beta} = 1.33$, true targets exhibit ratios between $0.5 < \Theta_{\alpha/\beta} < 0.8$ depending on the type of engine. Figure 2.24 shows the intensity ratio for both α and β

bands as a function of the temperature. It is obvious that only a body with a temperature of less than 900 K will provide a color ratio $\Theta_{\alpha/\beta}$ smaller than 0.8. The lower graph of Fig. 2.21 shows the total flame temperature of a plane taking into account the large amount of selective CO_2 radiation. In practice, this could mean the use of relatively cool continuous radiators, such as e.g. pyrophoric metal foils, or even the combustion of hydrocarbons to CO_2 and H_2O.

2.5.4 Smoke Munitions

Smoke-generating munition is used predominantly as obscurants for self-protection (camouflage in visible and IR regions), in smoke signals and for blinding of the enemy (Fig. 2.25). An obscurant smoke is an aerosol cloud brought into the line of sight between an observer and a target (Fig. 2.26). When considering an obscurant aerosol one has to distinguish between hygroscopic (P_4O_{10}) and non-hygroscopic aerosols. Almost all aerosols which are currently used are based on red phosphorus and are therefore hygroscopic, meaning that their effectiveness is also dependent on the relative humidity (Fig. 2.27).

In order to increase the combustion efficiency and the burning rate, formulations containing sodium nitrate or potassium nitrate, an organic binder as well as

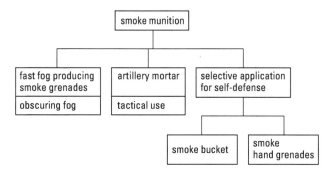

Fig. 2.25: Use of smoke munition.

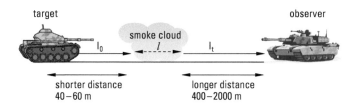

Fig. 2.26: Typical obscurant scenario (I_0 = intensity of the incoming light; I_t = intensity of the transmitted light; l = length of the fog cloud).

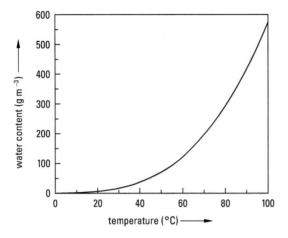

Fig. 2.27: Maximum water content of the atmosphere as a function of temperature.

an aerosol agent (NH$_4$Cl) are used in addition to red phosphorus. Typical, simple smoke generating compositions, which are based on red phosphorus and contain approx. 10–75 % P are:

Composition 1	red phosphorus	smoke generator	10 %
	KNO$_3$	oxidizer	30 %
	NH$_4$Cl	aerosol additive	60 %
Composition 2	red phosphorus	smoke generator	50 %
	CaSO$_4$	oxidizer	37 %
	boron	fuel & burn rate modifier	10 %
	Viton	binder	3 %
Composition 3	red phosphorus	smoke generator	75 %
	CuO	oxidizer	10 %
	Mg	fuel & burn rate modifier	11 %
	various binders	binder	4 %

Increasingly, a shielding in the mid and far IR region in addition to the simple obscurent effects in the visible region is desirable, since detection by the enemy is possible using electro-optical heat sensors. Although smoke generating formulations which are based on red phosphorus provide excellent shielding in the visible (0.36–0.76 μm) and near IR regions (0.76–1.3 μm), they screen only moderately in the mid (1.5–5.5 μm) and far IR (5.5–1000 μm) regions; they are almost exclusively used for this region because of their unsurpassed properties (aerosol formation,

visible region, low toxicity etc.). Therefore, the only way to guarantee a good shielding in the mid and far IR region as well is by achieving higher mass flow rates (per time) of red phosphorus. This can be achieved using, amongst others, the following strategies:
1. higher density of the formulations by pressing → better thermal conductivity,
2. higher combustion pressure (see also Ch. 1.3),
3. higher thermal conductivity of composite materials (metallic fuels or carbon fibers),
4. more strongly exothermic chemical reactions → higher temperatures,
5. larger surface areas (no spherical particles which have a low surface to volume ratio),
6. better ignition process (detonative ignition).

Particularly interesting is aspect (4) which is concerned with more strongly exothermic reactions. This makes the presence of an energetic additive necessary. Suitable additives are, for example:
a) Metals / Metal nitrates
b) Metals / fluorocarbons (see also Ch. 2.5.3)
c) Metals / Oxides
d) Other homogeneous highly energetic materials.

Examples of such "advanced" smoke-generating compositions are the following two formulations:

Composition 4	Mg	7–12%
	red phosphorus	65–70%
	PTFE	17%
	polychloroprene binder	6%
Composition 5	Mg	12%
	red phosphorus	66%
	KNO_3	16%
	polyvinylalcohol (PVA)	6%

One fundamental problem with formulations containing red phosphorus and magnesium, in particular in formulations which contain a substoichiometric quantity of oxidizer is that a thermodynamically favorable side-reaction between the phosphorus and the magnesium can occur:

$$2P + 3Mg \rightarrow Mg_3P_2$$

The magnesium phosphide that is subsequently formed can react with atmospheric moisture to form phosphine, PH_3 which is toxic:

$$Mg_3P_2 + 6\,H_2O \rightarrow 3\,Mg(OH)_2 + 2\,PH_3$$

This unwanted reaction can also occur very slowly if smoke munition is stored for several years. This can result in a considerable contamination of the munition depot (mainly bunkers or tunnel-shaped caves) with gaseous PH_3. For this reason, in modern formulations Mg is often replaced by Ti or Zr. These metals could in principal also form the corresponding metal phosphides, however, these metal phosphides do not react with water under normal conditions to form PH_3. Moreover, boron and/or silicon additives increase the shielding in the mid and far IR regions as a result of the formation of B_2O_3 (4–5 µm) or SiO_2 emmtters (9–10 µm). A typical Mg free smoke formulation, which also shields good in the mid and far IR regions is:

Composition 6	red phosphorus	smoke generator	58.5 %
	KNO_3	oxidizer	21.1 %
	Zr	energetic fuel	4.7 %
	Si	energetic fuel	4.7 %
	B	energetic fuel	4.7 %
	chloroprene	binder	6.3 %.

The potential of **phosphorus(V) nitride, P_3N_5,** as a replacement for red phosphorus (PR) in pyrotechnic obscurants has been theoretically and experimentally investigated by Koch and Cudzilo [Angew. Chem. Int. Ed. **2016**, 55, 15439–15422]. P_3N_5 can be safely mixed with KNO_3, and even $KClO_3$ and $KClO_4$. The corresponding formulations are surprisingly insensitive to friction and only mildly impact-sensitive. P_3N_5 / KNO_3 pyrolants with 20–80 wt % P_3N_5 burn 200 times faster than the corresponding mixtures based on PR and generate a dense smoke. An oxygen-balanced formulation would have the following approximate composition: 35 % P_3N_5, 65 % KNO_3. Furthermore, unlike PR (which slowly degrades in moist air, forming phosphoric acids and phosphine (PH_3)), P_3N_5 is stable under these conditions and does not produce any acids or PH_3.

The size of the smoke cloud and its duration are of course strongly dependent on the amounts of smoke munition used (small hand grenades or 18 kg floating smoke pot) as well as the weather. While no good fog is formed (see Fig. 2.27) in extremely dry conditions, the duration of the cloud is significantly shorter in strong winds than in windfree conditions. Typical values e.g. for a 155 mm artillary smoke greande are summarized in Table 2.13.

Previously, ZnO / Al / hexachlorethane (HC) mixtures were used both as smoke- and fog-generating munitions as well. In those mixtures, the intermediate $ZnCl_2$ (and C, graphite) is formed, which then quickly hydrolyses forming HCl and ZnO. Such nonhygroscopic formulations are only used in handheld flares where the formation of acidic products (H_3PO_4) can't be tolerated. However, some of the side-

Tab. 2.13: Typical values using a 155 mm artillery smoke grenade.

	visible region (0.4–0.7 µm)	mid IR region (3–5 µm)	far IR region (8–14 µm)
duration of smoke	260 s	180 s	180 s
effective length	400 m	180 m	180 m
effective height	12–16 m	8–12 m	8–12 m
effective depth	ca. 40 m	ca. 40 m	ca. 40 m

products of the reactions have been found to be toxicologically and ecologically critical (e.g. hexachlorbenzol, hexachlorbutadiene, chlorinated dibenzofurane, dibenzodioxine), therefore research is underway to find suitable replacements for HC. Other munitions, which have to the knowledge of the author of this book, never been fielded by NATO countries contain TiO_2/Al/HC and also Ti/ZnO/HC.

$$C_2Cl_6 + 2\,Al \rightarrow 2\,C + 2\,AlCl_3$$

$$2\,AlCl_3 + 3\,ZnO \rightarrow Al_2O_3 + 3\,ZnCl_2$$

$$3\,ZnCl_2 + 3\,H_2O \rightarrow 3\,ZnO + 6\,HCl$$

In addition to phosphorus-based obscurant smokes for self-protection there are also colored smoke compositions for signaling purposes, often in a military context. It can be produced by smoke grenades, or by various other pyrotechnical devices.

Smokes, in general, are cooler burning pyrotechnic mixtures. The reason is two-fold. 1. A cooler burning smoke will rise gradually from the ground when deployed, which will serve as an effective screening/obscuration tool for military personnel. A smoke mixture which is too hot in temperature will result in a smoke rising too rapidly. When this happens, the screening/obscuration purposes of the smoke is lost. 2. A smoke which reaches high temperatures will result in combustion of the smoke dye (*i.e.* oxidation to carbon monoxide and carbon dioxide). Smoke dyes are added into smoke formulations for sublimation purposes. Not only does the sublimation event keep smokes cool (*i.e.* sublimation is an endothermic process), but it is the sublimation of the smoke dye which gives the smoke its distinct color (*i.e.* blue, red, yellow, green, black, etc.).

As has been commonly taught in smokes for years, the temperatures at which oxidizers and fuels melt is largely responsible for the temperature which a smoke will ultimately reach. Potassium chlorate has been identified as the "only oxidizer" to be used in smoke mixes because it has an exothermic decomposition, thus contributing energy to the pyrotechnic system, but also has a relatively low melting point. Scientists at ARDEC are currently looking at ways to replace potassium chlorate oxidizer in smoke mixes with other oxidizing agents. The EPA is looking at

Fig. 2.28: Ingredients for colored smoke.

potassium chlorate now, and now environmentalists are questioning whether potassium chlorate is also bad like potassium perchlorate. Another problem with potassium chlorate is its sensitivity. Potassium chlorate used to be used in many of the early fireworks, but it was replaced by potassium perchlorate because potassium perchlorate was found to be inherently safer to work with.

As pointed out above, $KClO_3$ is much less stable than perchlorate and therefore more hazardous (especially mixtures with sulfur and phosphorus). On the other hand, it shows high burning rates and easy ignition. $KClO_3$ is slightly more hygroscopic than potassium nitrate and produces smoke of KCl. Furthermore, it can act as a chlorine donor.

The mixture used for producing colored smoke is usually a cooler-burning formula based on potassium chlorate oxidizer, $KClO_3$ (ca. 35%) sugar as a fuel (ca. 20%), and one or more dyes, with about 40–50% content of the dye. About 2% sodium bicarbonate may be added as a coolant, to lower the burning temperature. Its coolant properties arise because 2 moles of sodium bicarbonate will endothermically decompose into one mole each of sodium carbonate, carbon dioxide and water. In addition to reducing flame temperatures of smokes, coolants are also used to reduce the sensitivity of smoke mixes, given the large amount of potassium chlorate used in these mixes. The presence of the coolants minimizes undesired acid formation and helps to prolong storage of the smoke mixes.

Typical colorants for smokes are:

Red: Disperse Red 9 (1-methylamino anthraquinone, a)
Orange: Sudan I, 1-phenylazonaphth-2-ol (b)
Yellow: Vat Yellow 4 (c) together with Benzanthrone (BZA)
Violet: Disperse Red 9 (a) with 1,4-Diamino-2,3-dihydroanthraquinone (1,4-Diaminoanthracene-9,10-dione)
Green: Vat Yellow 4 (c) together with with benzanthrone and Solvent Green 3
Black: naphthalene and anthracene

In the US, Nation Ford Chemical is the main supplier to the US military of the complete line of domestically produced dyes for smoke signaling and cover: http://www.nationfordchem.com/products/smoke-dyes/.

Benzanthrone

Vat yellow 4

Solvent yellow 33

Fig. 2.28a: Chemical structures of yellow dyes.

Tab. 2.13a: In-service and new formulations for yellow smoke.

In-service M194 Ingredients / wt%	New formulation Ingredients / wt%
Vat Yellow 4 / 13	Solvent Yellow 33 / 37
KClO$_3$ / 35	KClO$_3$ / 34.5
Sucrose / 20	Sucrose / 21.5
NaHCO$_3$ / 3	Mg(CO$_3$)$_4$(OH)$_2 \cdot$ 4 H$_2$O / 5.5
Benzanthrone / 28	stearic acid / 1
VAAR / 1	Fumed silica / 0.5

Sabatini, Moretti et al. from ARDEC located at Picatinny Arsenal, N.J. have now developed a prototype of a hand-held signal flare with an environmentally benign yellow smoke formulation to replace the environmentally and toxicologically hazardous mixture currently specified for the U. S. Army's M194 yellow smoke handheld signal (*ACS Sustainable Chem. Eng.* **2013**, *1*, 673–678). This new formulation meets the burn time parameters outlined in military specifications and is composed entirely of dry, powdered, solid ingredients without the need of solvent-based binders. In addition, this formulation was found to have relatively low sensitivity towards impact, friction and electric discharge.

The main goal was to replace toxic dyes such as Benzanthrone and Vat Yellow 4 – which are presently used in the in-service M194 formulation – with environmentally benign dyes such as Solvent Yellow 33 (Fig. 2.28a, Tab. 2.13a).

The newly developed formulation contains: the environmentally and toxicologically benign dye Solvent Yellow 33 as the smoke sublimating agent, hydrated basic magnesium carbonate instead of sodium bicarbonate as the endothermic coolant (eq. 1 and 2) and stearic acid as the lubricant and processing aid. Fumed silica was introduced in the absence of a binder to promote homogeneity

$$2\,\text{NaHCO}_3\,(s) \rightarrow \text{Na}_2\text{CO}_3\,(s) + \text{H}_2\text{O}\,(g) + \text{CO}_2\,(g) \quad \Delta H_r = +0.81 \text{ kJ/g} \quad (1)$$

$$\text{Mg}_5(\text{CO}_3)_4(\text{OH})_2 \cdot 4\,\text{H}_2\text{O} \rightarrow 5\,\text{MgO}\,(s) + 5\,\text{H}_2\text{O}\,(g) + 4\,\text{CO}_2\,(g) \quad \Delta H_r = +0.80 \text{ kJ/g} \quad (2)$$

Tab. 2.13b

	Military requirement for M-194	New formulation
Burn time / s	9–18	15
Burn rate / g s^{-1}	3.89–7.78	4.64
IS / J		17
FS / N		> 360
ESD / J		> 0.25

Despite the fact that both reactions (1) and (2) are endothermic and have comparable per gram enthalpies, reaction (1) occurs completely within the cool-burning temperature range of sucrose/KClO$_3$ smoke compositions, whereas reaction (2) is incomplete. Sodium bicarbonate is therefore a more aggressive coolant than hydromagnesite, as it effectively removes more energy from the smoke composition in the temperature range in which they burn. The use of NaHCO$_3$ in place of Mg$_5$(CO$_3$)$_4$(OH)$_2 \cdot$ 4H$_2$O results in longer burn times, i.e. slower burn rates. Therefore, it is essential to employ Mg$_5$(CO$_3$)$_4$(OH)$_2 \cdot$ 4 H$_2$O in order to achieving a yellow smoke formulation that meets the short burn time requirements of M194.

Table 2.13b summarizes the performance and sensitivity data of the new smoke formulation in comparison with the military requirements and Fig. 2.28b shows an image of a smoke fountain generated by the new yellow smoke formulation.

It should also be noted that white smokes are also of interest as well. There as several ways to produce a white smoke. In the old days, sublimation of sulfur did the trick (again, a low melting fuel), but the formation of sulfur dioxide always was seen as a problem. The hexachloroethane (HC) smokes was around for years (see above), which consisted of hexachloroethane, zinc oxide and aluminum. The trick in using the HC smoke was that it led to the formation of the Lewis acids zinc chloride and aluminum trichloride, and these smokes readily reacted with moisture in the air to produce aluminum hydroxide, zinc hydroxide, and most importantly from a smoke perspective, large amounts of hydrochloric acid. It was the hot HCl droplets that formed which produced a thick smoke upon cooling and pulling additional moisture from the air. Metal and metal oxide particles also contributed to the smoke production due to these finer particle scattering over a rather wide area. However, HC is now deemed toxic, and in fact, HC smokes are no longer being produced by the military in the USA. The military has been complaining about the lack of an HC replacement. Right now, the candidate to replace it is the terephthalic acid (TA) smoke mix. But this mix is not as good as HC in terms of efficiency (40–50 % of the mix is simply residue). It takes 2–3 TA grenades to equal the obscuration power of one HC grenade, and the TA smoke does not burn nearly as long. The smoke is less efficient for the TA smoke because instead of pulling moisture from the air and producing smoke with acid droplets, one is relying on sublimation and condensation of TA to produce the smoke. TA does not pull moisture from the

Fig. 2.28b: Image of a smoke fountain generated by the new yellow smoke formulation.

air, which hampers its ability as a smoke obscuration agent. Nonetheless, efforts are underway at ARDEC to find a quality replacement for the HC smoke mix. Another method of producing white smoke would be the volatilization of oil, but this has not been proven to work to any appreciable extent, even after much research.

For white (camouflage) smokes, the yield factor Y, is defined as the mass of aerosol (m_a) divided by the mass of the pyrotechnic payload (m_c).

Yield factor = mass of aerosol / mass of pyrotechnic payload:

$$Y = \frac{m_a}{m_c}$$

The figure of merit (FM_m) is a quality criteria which allows different compositions to be compared with each other. It is defined as the product of the yield factor multiplied with the extinction coefficient:

$$FM_m = \alpha_\lambda \cdot Y = \left(\frac{-V \cdot \ln T}{m_a \cdot T}\right) \cdot \left(\frac{m_a}{m_c}\right) = \frac{-V \cdot \ln T}{m_c \cdot L}$$

The extinction coefficient α_λ is defined according to the Lambert-Beer law,

$$\alpha_\lambda = -\frac{\ln T(obsc)}{c\,L} \qquad T(obsc) = \frac{I}{I(0)}$$

with:

α_λ = extinction coefficient
T = degree of transmission
$I(0)$ = intensity without smoke
c = conc. of aerosol in obscurant cloud [kg/m^3]
L = path length [m]
V = chamber volume [m^3].

It should also be mentioned that stearic acid or graphite is some times added in small amounts as an additive to smoke compositions. These materials serve as a lubricant, which really helps reduce sensitivities (especially to friction and impact) of smoke mixes. As for pyrotechnic binders in smokes, vinyl alcohol acetate resin (VAAR) has been used in the past. The most commonly used binder system now in smokes is polyvinyl alcohol (PVA), and it is applied wet.

The development of sugar-free multi-colored smoke formulations – so-called fuel mixes – has recently been reported (Küblböck et al., N. J. Chem.: DOI: 10.1039/c8nj01786g). These simple four ingredient-based mixtures of dye, potassium chlorate, 5-amino-1H-tetrazole, and a magnesium carbonate derivative are able to produce a variety of colors by applying the same pyrotechnical system (Fig. 2.28c). All components except the dye are pre-mixed; the dye is added in the final step. Based on previous results which indicated an overall higher smoke performance in terms of efficiency and persistence by applying 5-amino-1H-tetrazole as the fuel in smoke

Fig. 2.28c: The concept of fuel mixes.

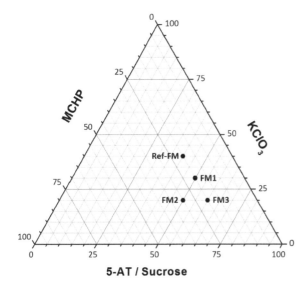

Fig. 2.28d: Developed fuel mixes. FM1 = 5-AT (50 wt%), $KClO_3$ (30 wt%), MCHP (20 wt%); FM2 = 5-AT (50 wt%), $KClO_3$ (20 wt%), MCHP (30 wt%); FM3 = 5-AT (60 wt%), $KClO_3$ (20 wt%), MCHP (20 wt%); Ref-FM = sucrose (40 wt%), $KClO_3$ (40 wt%), MCHP (20 wt%).

formulations, new colored smoke formulations were developed. For large producers as well as consumers, the development of fuel mixes is an effective way to reduce costs and provide a higher degree of safety. The main focus was on dyes applied in the U.S. M18 colored smoke grenades.

The most promising fuel mixes are illustrated within the ternary diagram shown in Fig. 2.28d. The first fuel mix, FM1, consisted of 50 wt% 5-AT, 30 wt% $KClO_3$, and 20 wt% MCPH and therefore, contained the highest amount of oxidizer. FM2 had a composition of 50 wt% 5-AT, 20 wt% $KClO_3$, and 30 wt% MCPH and finally fuel mix FM3, contained the highest amount of fuel with 60 wt% 5-AT and equal wt.% of $KClO_3$ and MCPH (20 wt%). The reference fuel mix Ref-FM, contained 40 wt% sucrose, 40 wt% $KClO_3$, and 20 wt% MCPH.

It was therefore successfully demonstrated that 5-AT-based fuel mixes can be used to produce green, yellow, red and violet smoke. The formulations that have been developed revealed that non-traditional high-nitrogen fuels can produce smoke of high color quality. More precisely, whereas FM1-based mixtures result in rapid, strong smoke generation in a short time, FM3-based compositions are characterized by a slow, continuous smoke generation over a much longer period of time. 5-AT is only one many potential candidates which will be investigated in the future. All the formulations are insensitive towards friction and in addition, all the 5-AT-based colored smoke formulations are completely insensitive towards impact. The fuel mixes were more sensitive towards mechanical stimuli. A comparison of the collected aerosol revealed similar yields for FM1-based and sugar-based formulations. The superiority of the sugar-based reference formulations was illustrated by the measured transfer rates. Future investigations will no doubt focus on providing even more colored smoke formulations, e.g. blue and black, by applying the same fuel mixes. However, in order to achieve a proper evaluation, new strategies to characterize the aerosol produced by a mixture of two dyes (e.g. green dye mix) have to be established.

2.5.5 Near-Infrared (NIR) Compositions

Since the mid-20th century, electro-optical systems such as night vision goggles and near infrared vision devices have become available. In general, these systems enable vision at night and make use of either non-visible photons and/or intensify the scarcely visible ones. The NIR range is considered to be between 700 nm and 2000 nm. However, for detection purposes, the range between 700–1000 nm is used extensively by several common detector materials in the UV, VIS and NIR range. For clandestine signalling and illuminating purposes in this range (700–1000 nm) pyrotechnic compositions based on either potassium (K) and/or cesium (Cs) compounds that are based on the intense emission lines of K and Cs in the infrared, are used. These flare compositions contain silicon as a high-energy non-luminous fuel, which is known to burn in the condensed phase. A typical composition is: KNO_3 70 %, hexamine 16 %, Si 10 %, binder 4 %. Furthermore, these compositions contain nitrogen-rich compounds like e.g. hexamine or azodicarbonamide as blowing agents to peel off the condensed reaction products. While hexamine is certainly a historically used fuel in pyrotechnic IR applications, especially given its relatively low light output due to its high-nitrogen content, one of the drawbacks of hexamine (at least in the USA), is a) the cost of the material is rising and b) hexamine is getting harder and harder to obtain. In this context, lactose monohydrate has potential to serve as a hexamine replacement. This would solve the supply issue and reduce cost if it can be used in IR applications, as lactose monohydrate is very cheap, and is a fraction of the cost of hexamine. It has a higher heat of combustion than hexamine, but still gives off a substantial amount of gas (i.e. lots of CO_2 and some CO). While materials with higher heats of combustion are likely to increase visible light output, this can be minimized or eliminated altogether by adjusting oxidizer/fuel ratios in the formulations. Presently research is focussing on compounds with an even higher nitrogen content such as potassium or cesium azotetrazolate or potassium or cesium bis(tetrazolyl)amine. Typical oxidizers are the peroxides or barium, strontium and zinc and stannic oxide. Besides NIR tracers, NIR illuminants comprise parachute signal flares as well as illuminating rounds for mortars. For clandestine aerial reconnaissance, NIR illumination rockets can be used.

From Fig. 2.29 (top) it is clear that NIR flares based on pure cesium formulations ($CsNO_3$) are superior emitters in the NIR region, whereas cesium/potassium

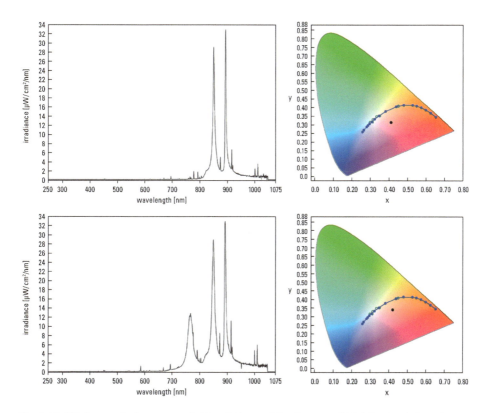

Fig. 2.29: NIR flare emission spectra (left) and chromaticity diagrams (right) for $CsNO_3$ based flares (top) and $KNO_3/CsNO_3$ based formulations (bottom).

based formulations ($CsNO_3$/KNO_3) (Fig. 2.29 bottom) show a NIR color impurity with tailing into the visible red region.

The radiant intensity (I) is a measure of radiometric power per unit solid angle Ω ($\Omega = A/r^2$), expressed in watts per steradian (W sr^{-1}). The value of the solid angle Ω is numerically equal to the size of that area divided by the square of the radius of the sphere. For hand-held NIR flares one wants a reasonable burn time (ca. 45 s), a high concealment index X which is defined as the ratio of the radiant intensity in the near IR region (I_{NIR}, 700–1000 nm) over the radiant intensity in the visible region (I_{VIS}, 400–700 nm) and high radiant intensities in the NIR regions I_1(600–900 nm) and I_2(695–1050):

$$X = I_{NIR}/I_{VIS} > 25; I_1 > 25 \text{ W sr}^{-1}; I_2 > 30 \text{ W sr}^{-1}; \text{luminous intensity (visible)} < 350 \text{ cd.}$$

Check your knowledge: See Chapter 14 for Study Questions.

3 Detonation, Detonation Velocity and Detonation Pressure

In Chapter 1.3 we defined a detonation as the propagation of a chemical reaction through an energetic material under the influence of a shock-wave at speeds faster than the speed of sound in the material. The velocity at which the energetic material decomposes is therefore only dependent on the velocity of the shock-wave. It is not determined by a heat-transfer process as it is the case for deflagration or combustion.

A detonation can result from either a continuously accelerating combustion (DDT, Deflagration-to-Detonation Transition) or from a shock (use of a primary explosive in a detonator to initiate a secondary explosive).

In the first case, we can assume that the linear burning rate increases proportionally to the pressure on the surface of the explosive:

$$r = \beta p^\alpha$$

Where β is the temperature dependent coefficient ($\beta = f(T)$) and α is the index of the burning rate, which describes the pressure dependence. For deflagrations $\alpha < 1$, however, this value increases to $\alpha > 1$ for detonations. The DDT transition can occur when an explosive is ignited in a confined tube, where the gases formed cannot fully escape. This results in a sharp increase of the pressure and reaction velocity. Therefore, in a detonating explosive, the reaction velocity can increase above the speed of sound, turning the deflagration into a detonation.

In the second case we assume that an explosive is subjected to a shock-wave. This method is used in detonators (see Ch. 2.5.1); the shock-wave of a detonating primary explosive initiates a secondary explosive. This shock-wave compresses the secondary explosive enough to cause the temperature to increase to higher than the decomposition temperature through adiabatic heating and the explosive material that is directly behind the shock-wave front reacts. During this reaction the shock-wave accelerates as a result of the strongly exothermic reaction of the secondary explosive. As a result of the influence of the shock-wave, the density of the explosive shortly before the reaction zone increases to 1.3–1.5 times the maximum density in the crystal (TMD), while in the thin (up to ca. 0.2 mm) chemical reaction zone directly behind the shockwave front, temperatures of up to and over 3000 K and at its end pressures of over 330 kbar can result. If the propagation of the shock-wave through the explosive occurs at speeds faster than the speed of sound, the process is described as a detonation. The shock-wave moves through the explosive under constant acceleration until it reaches a stationary state. The **stationary state** is reached, when the **free energy released** through this exothermic chemical reaction **is the same** as the **energy which is released to the surroundings as heat + the energy which is necessary to compress and move the crystal**. This means that when these chemical reactions occur with the release of heat at constant pressure and temperature, the propagation of the shock-wave will become a self-sustaining process.

A critical characteristic of energetic materials is the Chapman-Jouguet (C-J) state. This describes the chemical equilibrium of the products at the end of the reaction zone of the detonation wave before the isentropic expansion. In the classical Zel'dovich-Neumann-Döring (ZND) detonation model, the detonation wave propagates at constant velocity. This velocity is the same as at the C-J point which characterizes the state of reaction products in which the local speed of sound decreases to the detonation velocity as the product gases expand.

According to the classical ZND model, the thermodynamic quantities of a material in the unshocked state and the final shocked state are related by the conservation equations of mass, momentum and energy across the shock front according to:

$$\rho_0 D = \rho (D-u)$$

$$p - p_0 = \rho_0 u D$$

$$e + pv + \frac{1}{2}(D-u)^2 = e_0 + p_0 v_0 + \frac{1}{2}D^2$$

Where D is the constant velocity of the unsupported detonation wave, ρ is the density, u is the velocity of the products behind the detonation wave, p is the pressure, e is the specific energy, and v is the specific volume. The term "specific" refers to the quantity per unit mass, while the subscript "0" refers to the quantity in the initial unshocked state.

Under the influence of the dynamic behavior of the shock-wave, a thin layer of still unreacted explosive gets compressed alongside the schock adiabat of the corresponding explosive (or Hugoniot adiabat) from the original specific volume V_0 ($V_0 = 1/\rho_0$) to volume V_1 (Fig. 3.1). As a consequence of the dynamic compression, the pressure increases from p_0 to p_1, which in turn causes an increase in the temperature in the thin compressed layer of the explosive (Fig. 3.1), which causes an initiation of the chemical reaction results. At the end of the chemical reaction, the specific volume and pressure have the values V_2 and p_2. This state corresponds to the point of the detonation product on the shock adiabat (Fig. 3.1). At this point it is important to emphasize again that in a deflagration, the propagation of the reaction occurs through thermal processes, while for the considerably quicker occurring detonation a shockwave mechanism occurs.

The desirable situation for the occurrence and linear propagation of a detonation for homogeneous explosives exists in confined tubes or cylindrical shaped explosives, in which the system should not fall below the critical diameter (characteristic for every individual explosive) as it otherwise causes the wave front to be

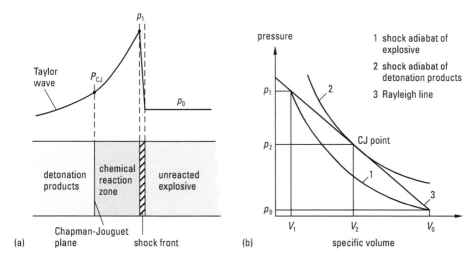

Fig. 3.1: Schematic representation of the detonation process and the shock-wave structure (a) as well as the shock adiabat for an explosive and the detonation products (detonation in a stationary state) (b).

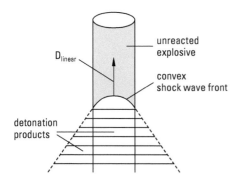

Fig. 3.2: Schematic propagation of a shock wave in a cylindrical shaped explosive.

disturbed ("loss" of energy to outside) and therefore the detonation velocity will be reduced. While for many secondary explosives one inch is a "good" tube diameter, for primary explosives often considerably smaller diameters of 5 mm are sufficient. Figure 3.2 schematically shows the propagation of a shockwave in a cylindrical shaped explosive. Since the wave front is convex and not level, the linear detonation velocity is highest in the centre of the cylindrical explosive and decreases towards the surface. For large diameters this effect plays a negligible role, but for very small diameters the surface effects can become dominant, which causes the wave front to be unstable. This so-called critical diameter must be taken into account for measurements of the detonation velocity, so that one stays above the critical diameter.

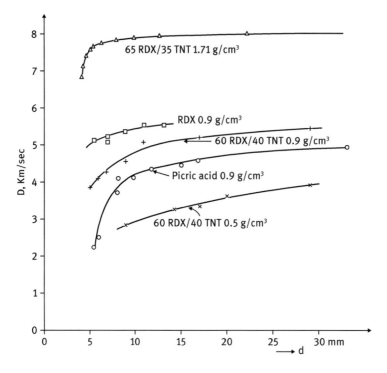

Fig. 3.2a: The dependence of the detonation velocity on the charge diameter of various high (secondary) explosives at low densities.

The critical diameter is defined as the diameter of a cylindrical charge, below which stable detonation cannot propagate, no matter how high the velocity of the initiating shock wave is. Primary explosives have usually very small critical diameters (sometimes in the µm region), whereas secondary explosives typically have larger critical diameters (in the mm to cm range). The critical diameter is also usually larger in cast than in pressed charges. Moreover, the critical diameter is of importance in considering the safety of proposed designs for detonators containing detonatable materials.

Above the critical diameter, the dependence of the detonation velocity (D) on the charge diameter (d) is characteristic for many explosives, particularly at lower densities and for explosives which exhibit non-ideal behavior. The reason for this is due to the fact that radial loses of energy are higher at lower charge diameters. Fig. 3.2a shows the dependence of the detonation velocity on the charge diameter of various high (secondary) explosives at low densities.

If the measured detonation velocity (D) is plotted against the reciprocal diameter of the cylindrical charge used (Fig. 3.2b) the detonation velocity for an infinitely large diameter (D_∞) can be extrapolated according to the following equation, where A_L is a constant characteristic for the explosive used:

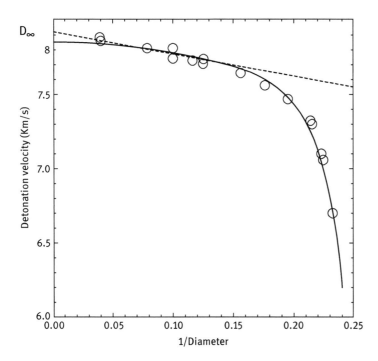

Fig. 3.2b: Extrapolation of the detonation velocity D for an infinitely large diameter of a cylindrical charge.

$$D = D_\infty \left(1 - A_L \frac{1}{d}\right)$$

The general detonation behavior of an explosive is schematically shown in Fig 3.2c.

As previously stated, this discussion is valid for homogeneous explosives, such as the ones used in the military, since their reactions are predominantly intramolecular. Such explosives are often referred to as "ideal" explosives, in particular when they can be described using the steady state model of Chapman and Jouguet. In heterogeneous explosives (non-ideal), which are currently used in civil applications, intermolecular (diffusion controlled) mechanisms are predominant for the air bubbles, cavities or cracks (etc.). As a general rule detonation velocities increase proportional to the diameter.

A novel method has been introduced which allows the simple estimation of the critical diameter of CHNO pure and composite explosives (M. H. Keshavarz et al., J. Energ. Mat., in press). It uses the number of oxygen and nitrogen atoms as well as an additional value for C_{Shock} which is based on shock sensitivity and impact sensitivity for desired explosives under certain situations. Equation (2) enables good predicted values to be achieved for a wide range of CHNO pure and composite high explosives containing nitro, nitramine and nitrate ester functional groups.

Fig. 3.2c: General detonation behavior of an explosive.

$$d_{c,core} = -11.97\, n_N + 14.11\, n_O \tag{1}$$

$$d_c = -3.19\, n_N + 5.38\, n_O + 21.80\, C_{Shock} \tag{2}$$

Different values of C_{Shock} are given as:
a) For pure explosives, the following statements have to be taken into account:
   (i) The value of C_{Shock} equals -0.4 if $P_{98\,\%TMD} > 38$ kbar and $E_{IS} > 28$ J.
   (ii) For $E_{IS} < 22$ J there are not sufficient experimental values to determine values for C_{Shock}. Thus, d_c is taken to be equal to 2.5.
b) For composite explosives, the value for C_{Shock} can be estimated as follows:
   (iii) For composite explosives containing PETN, the value of C_{Shock} is -0.3.
   (iv) The value of C_{Shock} equals -0.1 if $P_{98\,\%TMD} < 24.4$ kbar.
c) For $d_{c,core} < 0$: The value of C_{Shock} is 0.8.

Equation (2) provides the easiest pathway for predicting critical diameters using the conditions of $P_{98\,\%TMD}$ of pure and composite explosives as well as the calculated E_{IS} and the negative values of $d_{c,core}$.

In accordance with the detonation model for the stationary state, the points (V_0, p_0), (V_1, p_1) and (V_2, p_2) lie on one line (Fig. 1.22) which is called the **Rayleigh line**. The gradient of the Rayleigh line is determined by the detonation velocity of the explosive. In agreement with the postulate of Chapman and Jouguet, this Rayleigh line is also a tangent to the shock adiabat of the detonation products at exactly the point which corresponds to the end of the chemical reaction (V_2, p_2). This point is therefore also referred to as the **Chapman-Jouguet Point** (C-J point).

Tab. 3.1: Comparison of the predicted critical diameter (mm) using equations (1) and (2) with the experimental data (the values of $P_{98\%\ TMD}$ in kbar and E_{IS} in J were calculated).

Name	Density (g.cm^{-3})	Formula	$P_{98\%\ TMD}$/ kbar	E_{IS}/J	Exp./mm	Eq. (1)/ mm	Dev/mm	C_{Shock}	Eq. (2)/ mm	Dev/mm
PBX-9502 (TATB/Kel-F 800 95/5)	1.895	$C_{2.30}H_{2.23}N_{2.21}O_{2.21}Cl_{0.038}F_{0.13}$	28.11		9.0	4.74	-4.26	0	4.84	-4.16
PBX-9404 (HMX/NC/CEF 94/3/3)	1.846	$C_{1.40}H_{2.75}N_{2.57}O_{2.69}Cl_{0.03}P_{0.01}$	21.47		1.18	7.20	6.02	-0.1	4.09	2.91
XTX 8003 (PETN/Silicon rubber 80/20)	1.53	$C_{1.80}H_{3.64}N_{1.01}O_{3.31}Si_{0.27}$	26.13		0.36	34.62	34.26	-0.3	8.04	7.68
XTX 8004 (RDX/Sylgard 80/20)	1.53	$C_{1.62}H_{3.78}N_{2.16}O_{2.43}Si_{0.27}$	23.67		1.4	8.44	7.04	-0.1	4.00	2.60
DAAF (3,3'-Diamino-4,4'-azoxyfurazan)	1.60	$C_4H_4N_8O_3$	32.08	26.95	1.25	-53.40	-54.65	0.8	8.05	6.80
BPTAP (2,4,8,10-Tetranitro-benzopyrido-1,3a,6,6a-tetraazapentalene)		$C_{11}H_3N_9O_8$	35.40	23.03	3.0 (<3)	5.18	2.18	0	2.50	-0.50
RDX/Wax 95/5	1.05	$C_{1.64}H_{3.28}N_{2.57}O_{2.57}$	27.03		4.5 (4.0–5.0)	5.50	1.00	-0.1	3.44	-1.06
RDX/Wax 90/10	1.1	$C_{1.93}H_{3.86}N_{2.43}O_{2.43}$	28.61		4.5 (4.0–5.0)	5.21	0.71	-0.1	3.14	-1.36
RDX/Wax 80/20	1.25	$C_{2.51}H_{5.02}N_{2.16}O_{2.16}$	31.77		4.4 (3.8–5.0)	4.63	0.23	0	4.73	0.33
RDX/Wax 72/28	1.39	$C_{2.97}H_{5.95}N_{1.95}O_{1.95}$	34.29		4.4 (3.8–5.0)	4.17	-0.23	0	4.26	-0.14

Tab. 3.1 (continued)

Name	Density (g.cm^{-3})	Formula	$P_{98\% TMD}$/ kbar	E_{IS}/J	Exp./mm	Eq. (1)/ mm	Dev/mm	C_{Shock}	Eq. (2)/ mm	Dev/mm
B 2214 (NTO/HMX/HTPB 72/12/16)		$C_{2.44}H_{3.23}N_{2.54}O_{1.99}$	25.85		65	−2.32	−67.32	1	24.40	−40.60
B 2248 (NTO/HMX/HTPB 46/42/12)		$C_{2.15}H_{3.19}N_{2.55}O_{2.20}$	24.70		11	0.53	−8.97	0	3.70	−5.80
NTO (3-nitro-1,2,4-triazol-5-One)		$C_2H_2N_4O_3$	25.45	21.56	25.4	−5.54	−30.94	0.8	20.82	−4.58
RMSD							21.77 (All data except $d_c = 2.5$ in Eq.(2))			3.96 (All data except $d_c = 2.5$ in Eq. (2))

Tab. 3.2: The coefficients of volume expansion.

CL-20	$1.5 \cdot 10^{-4}$ K^{-1}
HMX *	$1.6 \cdot 10^{-4}$ K^{-1}
TNT	$2.26 \cdot 10^{-4}$ K^{-1}
BTF	$3.57 \cdot 10^{-4}$ K^{-1}
TKX50	$1.10 \cdot 10^{-4}$ K^{-1}

* C. Xu, et al., *Propellants, Explos. Pyrotech.* 2010, 35, 333–338

Generally, for homogeneous explosives, the **detonation velocity** increases proportional to the density of the energetic formulation. This means that reaching the maximum loading density is essential in order to obtain good performance. Different technical processes can be used in order to achieve the highest possible density including pressing, melt casting or extruding processes. For pure substances, the limiting, theoretical maximum density (TMD) is that of the single crystal (obtained from X-ray data at room temperature) (Tab. 3.1).

Usually the experimental densities are determined by single crystal X-ray diffraction at low temperature. Therefore it is necessary to convert the low-temperature densities into room-temperature values, in order to obtain the theoretical maximum density (TMD) (Tab. 3.2):

$$d_{298\,K} = \frac{d_T}{1 + \alpha_V (298 - T_0)} \rightarrow \alpha = \frac{d_T}{d_{298}} \div (298 - T) - 1 \div (298 - T)$$

As a good approximation the following formula can also be used:

$$d_{298\,K} = \frac{d_T}{1 + \alpha_V (298 - T_0)} \quad \text{with} \quad \alpha_V = 1.5 \cdot 10^{-4} \, K^{-1}$$

Kamlet and Jacobs suggested an empirical relationship between the detonation velocity and the **detonation pressure**. In this, the detonation velocity D is linear and the detonation pressure p_{C-J} to the power of two dependent on the loading density ρ_0 (in g cm^{-3}) [17–19]:

$$p_{C-J}\,[\text{kbar}] = K \rho_0^2 \Phi$$

$$D\,[\text{mm}\,\mu s^{-1}] = A\,\Phi^{0.5}(1 + B\,\rho_0)$$

The constants K, A and B are defined as follows:

K = 15.88
A = 1.01
B = 1.30

The value Φ is therefore

$$\Phi = N(M)^{0.5}(Q)^{0.5},$$

where N is the number of moles of gas released per gram of explosive, M is the mass of gas in gram per mole of gas and Q is the heat of explosion in cal per gram.

The Kamlet-Jacobs equations for the Chapman-Jouguet pressure (p_{C-J}) and the detonation velocity D (VoD), show that detonation velocity and detonation pressure are strongly dependent on (a) the loading density and (b) the parameter φ. In the past, the importance of the role of the density was greatly emphasized, since it appears to a higher power than φ in the Kamlet-Jacobs equations. This resulted in the density being referred to as "the primary physical parameter in detonation performance" (P. Politzer, J. S. Murray, *High Performance, Low Sensitivity: Conflicting or Compatible?*, PEP, **2016**, *41*, 414–425). In subsequent years, however, it became more and more clear, that not only is the parameter φ as important in determining the detonation velocity as the density is, it is also needed in order to reliably predict the detonation pressure.

The Kamlet-Jacobs parameter φ reflects the specific inherent properties of an individual explosive compound. The loading density, on the other hand, depends not only on the crystal structure of the explosive compound, but also on external factors such as the method of preparation used to prepare the explosive. As a result of this difference, Politzer and Murray suggested that φ should be considered to be a measure of the intrinsic detonation potential of the explosive compound (P. Politzer, J. Murray, *PEP*, **2019**, *44*, in press: DOI: 10.1002/prep.201900002). The intrinsic detonation potential is highly interesting, since it can provide useful insights into factors which promote a high detonation performance. An explosive compound that has a high φ value – even if coupled with a low density which will result in an unsatisfactory detonation performance – may draw attention to desirable molecular features that the compound possesses. It is worthwhile to point out that nitramines and nitrates usually show the highest intrinsic detonation potentials, with all having φ ≥ 6.48. In contrast, compounds which contain C—NO$_2$ linkages usually have φ < 5.2. One exception to the aforementioned is FOX-7, which has a φ value of 6.10. Generally, it can be therefore be concluded that explosives containing nitramine (N—NO$_2$) and nitrate (O—NO$_2$) linkages will generally show a much higher level of detonation performance than for explosive compounds containing only the C—NO$_2$ linkage. Interestingly, even attaching four nitro groups to a single carbon atom is not sufficient to achieve a high intrinsic detonation potential, since tetranitromethane (TNM) has the lowest φ of 4.02. Unfortunately, it has been found that the desirable detonation behavior of nitramines and nitrates is usually accompanied by a high level of sensitivity (impact and friction) which is undesirable. At this point it is worthwhile to point out that RDX and HMX have exactly the same φ value, which means that in the case of these two compounds,

Tab. 3.3: Loading density dependency of the detonation velocity D and the detonation pressure p_{C-J} (experimental data) [20].

explosive	density, ρ / g cm^{-3}	detonation velocity, D / m s^{-1}	detonation pressure, p_{C-J} / kbar
TNT	1.64	6950	210
	1.53	6810	171
	1.00	5000	67
RDX	1.80	8750	347
	1.66	8240	293
	1.20	6770	152
	1.00	6100	107
PETN	1.76	8270	315
	1.60	7750	266
	1.50	7480	240
	1.26	6590	160

the additional $-CH_2-N(NO_2)$ group in HMX does not increase the intrinsic detonation potential. The considerably better detonation performance of HMX can be entirely and exclusively attributed to its higher density.

Conceptually, the intrinsic detonation potential can be considered to be similar to the property known as the explosive power, which was discussed in chapter 4.1.

It has been demonstrated that if the intrinsic detonation potential of explosives is known, it can provide useful insight with respect to the detonation performances of explosives. Even if the density of an explosive compound is low – which will result in an unsatisfactory detonation performance for that explosive – if the compound shows a high value of φ, it may reveal desirable molecular features which could then be target features to include in hitherto unknown explosive compounds. The elucidation of a high φ value for an explosive compound may also act as a motivational force to increase efforts aimed at modifying the crystallization procedure to increase the density of the explosive in order to take advantage of the high φ.

Similarly, empirical relationships have also been published to correlate the specific impulse (I_{sp} / N s g^{-1}) with the detonation velocity (D / km s^{-1}) and the detonation pressure (p_{C-J} / kbar) [see also: M. H. Keshavarz, *Chem. of Heterocyclic Comp.*, **2017**, in press]:

$$D \text{ [km s}^{-1}\text{]} = 1.453 \, I_{sp} \text{ [N s g}^{-1}\text{]} \, \rho_0 \text{ [g cm}^{-3}\text{]} + 1.98$$

$$p_{C-J} \text{ [kbar]} = 44.4 \, I_{sp} \text{ [N s g}^{-1}\text{]} \, \rho_0^2 \text{ [g}^2 \text{ cm}^{-6}\text{]} - 21$$

Table 3.3 shows how the detonation velocity and the detonation pressure are dependent on the density for selected explosives.

Non-ideal explosives have Chapman-Jouguet (C-J) detonation parameters (pressure/velocity) significantly different from that expected from a thermochemical / hydrodynamic code for equilibrium and steady state calculations. Therefore, common computer codes such as Cheetah, EXPLO5 and others often do not correctly predict the detonation parameters of non-ideal explosives such as ANFO or metalized explosives.

For a wide range of CHNOFClAl(AN) explosives a very useful and simple method for the calculation of the detonation parameters of CHNOFCl, aluminized and ammonium nitrate explosives was recently suggested by M H Keshavarz et al.:

$$D = 5.468\, \alpha^{0.5}\, (\overline{Mw}_g\, Q_d)^{0.25}\, \rho_0 + 2.045$$

D, Q_d and ρ_0 are detonation velocity (in km s^{-1}), heat of detonation (in kJ g^{-1}), and initial density (in g cm^{-3}), respectively. α is the number of moles of gaseous products per gram of explosive (α) and $\overline{Mw}_g$ the average molecular weight of gaseous products.

Quite recently, Keshavarz also provided a general and simple method for predicting the detonation pressure (DP) of different kinds of ideal and non-ideal explosives containing aluminum (Al) and ammonium nitrate (AN). The new model can be applied for CHNO and CHNOFCl explosives in pure form or in mixtures, as well as non-ideal mixed explosives containing Al and AN. It can also be used for different plastic bonded explosives (PBXs). The detonation pressure can be calculated according to the following equation:

$$DP = p_{C\text{-}J} = 24.436\, \alpha\, (\underline{MW}_g\, Q_d)^{0.5}\, \rho_0^2 - 0.874$$

The parameters are:

α moles of gaseous products per gram of explosive (mol g^{-1}),
$\underline{MW}_g$ average molecular mass of gaseous products (g mol^{-1}),
Q_d heat of detonation (kJ g^{-1}),
ρ_0 density (g cm^{-3}).

Check your knowledge: See Chapter 14 for Study Questions.

4 Thermodynamics

4.1 Theoretical Basis

As we have already discussed above, the main performance criteria for secondary explosives are:
1. heat of explosion Q (in kJ kg^{-1}),
2. detonation velocity D (m s^{-1}),
3. detonation pressure p (in kbar),

and, slightly less importantly,
4. explosion temperature T (K) and
5. volume of gas released V per kg explosive (in l kg^{-1}).

To calculate the detonation velocity and the detonation pressure we require (see Ch. 3) thermodynamic values such as e.g. the heat of detonation, from which the detonation temperature can be obtained.

Before we are concerned with detailed thermodynamic calculations, it is important not to forget that for the good design of a secondary explosive, a balanced oxygen content (**oxygen balance Ω**) is desirable. Generally, oxygen balance describes the relative amount of oxygen excess or defect (with negative sign "–"), to achieve a balanced ratio between the oxidizer and the combustable components (fuel). A mixture of substances with a balanced oxygen balance ($\Omega = 0$) can be converted into fully oxidized products by heating them in a closed container, without any added external oxygen and without any excess oxidizer or fuel. In accordance with this definition, a CHNO-containing explosive is completely converted to CO_2, H_2O and N_2. For a compound with the general formula $C_aH_bN_cO_d$, the oxygen balance Ω (in %) is defined as follows:

$$\Omega = \Omega_{CO_2} = \frac{\left[d - (2a) - \left(\frac{b}{2}\right)\right] \times 1600}{M} \; ; \quad \Omega_{CO} = \frac{\left[d - a - \left(\frac{b}{2}\right)\right] \times 1600}{M}$$

while M is the molecular mass of the explosive. For example, TNT has a molecular mass of $M(\text{TNT}) = 227$ g mol^{-1} and a composition with the formula $C_7H_5N_3O_6$. Therefore the oxygen balance for TNT is $\Omega(\text{TNT}) = -74\,\%$:

$$C_7H_5N_3O_6 \rightarrow 7\ CO_2 + 2.5\ H_2O + 1.5\ N_2 - 10.5\ O$$

Table 4.1 shows a summary of the oxygen balances of important secondary explosives.

Tab. 4.1: Oxygen balances of some secondary explosives.

secondary explosive	formula	oxygen balance, Ω_{CO_2} / %
Ammonium nitrate, AN	NH_4NO_3	+20.0
Nitroglycerine, NG	$C_3H_5N_3O_9$	+ 3.5
Nitropenta, PETN	$C_5H_8N_4O_{12}$	−10.1
RDX	$C_3H_6N_6O_6$	−21.6
HMX	$C_4H_8N_8O_8$	−21.6
Nitroguanidine, NQ	$CH_4N_4O_2$	−30.7
Picric acid, PA	$C_6H_3N_3O_7$	−45.4
Hexanitrostilbene, HNS	$C_{14}H_6N_6O_{12}$	−67.6
Trinitrotoluene, TNT	$C_7H_5N_3O_6$	−74.0

Fig. 4.1: Influence of the oxygen balance Ω on the heat of explosion Q.

Figure 4.1 shows the influence of the oxygen balance on conventional CHNO-explosives. Usually (this is the case for very nitrogen-rich compounds as well) a good oxygen balance results in a greater (more negative) heat of explosion and therefore leads to a better performance of the explosive.

The oxygen balance of an explosive alone does not help, when the thermodynamics of the reaction are to be estimated. In this case, it is (particularly for compounds with too-low oxygen balances) necessary to estimate how much of the oxygen gets converted into CO, CO_2 and H_2O. Additionally, since all explosions occur at high temperatures (approx. 3000 K), even with oxygen balances of $\Omega = 0$, not only CO_2 but also CO will form as stated in the Boudouard equilibrium. An approximate, but very simple scheme for the estimation of the detonation products is given by the so-called modified **Springall-Roberts rules**. These are given below and must be applied in order from 1 to 6:

Tab. 4.2: Determination of the detonation products of TNT ($C_7H_5N_3O_6$) according to the Springall-Roberts-Rules.

C atoms are converted into CO	$6\,C \rightarrow 6\,CO$
if O atoms remain, they oxidize hydrogen to H_2O	all oxygen is already used-up
if there are remaining O atoms, they oxidize the already formed CO to CO_2	all oxygen is already used-up
all nitrogen is converted to N_2	$3\,N \rightarrow 1.5\,N_2$
one third of the CO formed is converted into C and CO_2	$2\,CO \rightarrow C + CO_2$
one sixth of the CO formed originally is converted into C and water	$CO + H_2 \rightarrow C + H_2O$
total reaction	$C_7H_5N_3O_6 \rightarrow 3\,CO + CO_2 + 3\,C + 1.5\,H_2$ $+ H_2O + 1.5\,N_2$

1. C atoms are converted into CO
2. if O atoms remain, they oxidize hydrogen to H_2O
3. if there are remaining O atoms, they oxidize the already formed CO to CO_2
4. all nitrogen is converted to N_2
5. one third of the CO formed is converted into C and CO_2
6. one sixth of the CO formed originally is converted into C and water.

As an example, in Table 4.2, the Springall-Roberts-Rules have been applied to work out the detonation products of TNT.

Therefore, the enthalpy of explosion (ΔH_{ex}) for TNT corresponds to the reaction enthalpy for the following reaction:

$$C_7H_5N_3O_6(s) \rightarrow 3\,CO(g) + CO_2(g) + 3\,C(s) + 1.5\,H_2(g) + H_2O(g) + 1.5\,N_2(g) \quad \Delta H_{ex}(TNT)$$

In contrast, the **enthalpy of combustion** for TNT (ΔH_{comb}) corresponds to the change in enthalpy for the following reaction and is more negative (more exothermic) than the explosion:

$$C_7H_5N_3O_6(s) + 5.25\,O_2(g) \rightarrow 7\,CO_2(g) + 2.5\,H_2O(g) + 1.5\,N_2(g) \quad \Delta H_{comb}(TNT)$$

For the definition of the enthalpy of combustion (ΔH_{comb}), it is important to check if there are values are given for $H_2O(g)$ or $H_2O(l)$.

Since the exact standard enthalpies of formation for all possible detonation products are known (see e.g. http://webbook.nist.gov/chemistry/), the enthalpy of detonation can easily be calculated, as long as only the enthalpy of combustion is exactly known from the experimental data, which unfortunately is not always the case.

According to the first fundamental law of thermodynamics, energy is conserved in any process involving a thermodynamic system and its surroundings. It is convenient to focus on changes in the assumed internal energy (U) and to regard them as due to a combination of heat (Q) added to the system and work done by the system (W). Taking ΔU as an incremental change in internal energy, we can write:

$$\Delta U = W + Q$$

Since work, W is defined as:

$$W = -\int_{V_1}^{V_2} p\, dV = -p\Delta V$$

for $\quad V = \text{const.}: \Delta U = Q_v$
and for $\ p = \text{const.}: \Delta U = Q_p - p\Delta V$

For propellant charges, which burn in a gun barrel with constant pressure and for secondary explosives, a good approximation is $V = \text{const.}$ (isochoric) and therefore $\Delta U = Q_v$, while for rocket fuels (free expansion of the gases in the atmosphere) $p = \text{const.}$ and therefore $\Delta U = Q_p - p\Delta V$ is a good approximation.

Since the equation of state H (enthalpy) is defined as $H = U + pV$, the following generalization can be made:

$$\Delta H = \Delta U + p\Delta V + V\Delta p$$

Therefore, for $p = \text{const.}: \quad \Delta H = Q_p$.
This means that we can generalize as follows:

$$Q_v = \sum \Delta_f U^\circ_{\text{(detonation products)}} - \sum \Delta_f U^\circ_{\text{f (explosive)}}$$

and

$$Q_p = \sum \Delta_f H^\circ_{\text{(detonation products)}} - \sum \Delta_f H^\circ_{\text{f (explosive)}}$$

In a first approximation, the energy of an explosion can be taken to be equal to that of the Helmholtz Free Energy:

$$\int p\, dV = -\Delta A = -\Delta U + T\Delta S$$

The molar energies of formation U and enthalpies H for a given explosive with the composition $C_a H_b N_c O_d$ can be connected as follows:

$$H = U + \Delta n R T$$

Tab. 4.3: Calculated volumes of gas released in an explosion (under STP).

explosive	V_0 / l kg^{-1}
NG	740
PETN	780
RDX	908
HMX	908
NQ	1077
PA	831
HNS	747
TNT	740

here, n is the chance of moles of the *gaseous* substances and R is the general gas constant. If we consider the hypothetical "equation of formation (from the elements)" for TNT for example, the change of moles is negative and Δn corresponds to –7:

$$7\,C(s) + 2.5\,H_2(g) + 1.5\,N_2(g) + 3\,O_2(g) \rightarrow C_7H_5N_3O_6(s)$$

In practice, in a good approximation H and U can often be taken to be equal.

The liberated volume of detonation gases (V_0, usually back-calculated for standard conditions of 273 K and 1 bar) is also an important parameter when assessing the performance of an explosive. V_0 can easily be calculated using the reaction equation derived from the Springall-Roberts rules and applying the ideal gas law and it is usually given in l kg^{-1}.

Table 4.3 shows a summary of typical values for the volumes of gas released during an explosion (under STP).

The **Explosive Power** has been defined as the product of the volume of gas released V_0 (STP) and the heat of explosion Q. Traditionally Q is given in kJ kg^{-1}, and V_0 in l kg^{-1} (Tab. 4.3). The value of the explosive power is usually compared with the explosive power of a standard explosive (picric acid) resulting in power indices, as shown below (Tab. 4.4).

$$\text{Explosive power} = Q\,(\text{kJ kg}^{-1}) \times V_0\,(\text{l kg}^{-1}) \times 10^{-4}\,(\text{kg}^2\,\text{kJ}^{-1}\,\text{l}^{-1})$$

$$\text{Power index} = \frac{Q \times V_0}{Q_{PA} \times V_{PA}} \times 100$$

The detonation or explosion temperature T_{ex} is the theoretical temperature of the detonation products, assuming that the explosion occurs in a confined and undestroyable adiabatic environment (adiabatic conditions). The detonation temperature

Tab. 4.4: Explosive power and power index.

explosive	$-Q$ / kJ kg^{-1}	V_0 / l kg^{-1}	explosive power	power index
Pb(N$_3$)$_2$	1610	218	35	13
NG	6195	740	458	170
PETN	5794	780	452	167
RDX	5036	908	457	169
HMX	5010	908	455	169
NQ	2471	1077	266	99
PA	3249	831	270	100
HNS	3942	747	294	109
TNT	4247	740	314	116

Tab. 4.5: Average heat capacities C_V (in J K^{-1} mol^{-1}).

T_{ex} / K	CO$_2$	CO	H$_2$O	H$_2$	N$_2$
2000	45.371	25.037	34.459	22.782	24.698
2100	45.744	25.204	34.945	22.966	24.866
2200	46.087	25.359	35.413	23.146	25.025
2300	46.409	25.506	35.865	23.322	25.175
2400	46.710	25.640	36.292	23.493	25.317
2500	46.991	25.769	36.706	23.665	25.451
2600	47.258	25.895	37.104	23.832	25.581
2700	47.509	26.012	37.485	23.995	25.703
2800	47.744	26.121	37.849	24.154	25.820
2900	47.965	26.221	38.200	24.309	25.928
3000	48.175	26.317	38.535	24.260	26.029
3100	48.375	26.409	38.861	24.606	26.129
3200	48.568	26.502	39.171	24.748	26.225
3300	48.748	26.589	39.472	24.886	26.317
3400	48.924	26.669	39.761	25.025	26.401
3500	49.091	26.744	40.037	25.158	26.481
3600	49.250	26.819	40.305	25.248	26.560
3700	49.401	26.891	40.560	25.405	26.635
3800	49.546	26.962	40.974	25.527	26.707
3900	49.690	27.029	41.045	25.644	26.778
4000	49.823	27.091	41.271	25.757	26.845
4500	50.430	27.372	42.300	26.296	27.154
5000	50.949	27.623	43.137	26.769	27.397

can be calculated with the assumption that the heat content of the detonation products must be the same as the calculated heat of detonation (Q). We can assume, that the detonation products of the initial temperature T_i (normally 298 K) will be brought up to a temperature of T_{ex} due to the heat of detonation, i. e. T_{ex} is dependent on Q. The connection between Q and T is given in the following equation, whereby C_V is the molar heat capacity of the detonation products:

$$Q = \Sigma \int_{T_i}^{T_{ex}} C_V dT \quad \text{with:} \quad C_V = \left(\frac{\partial Q}{\partial T}\right)_V = \left(\frac{\partial U}{\partial T}\right)_V$$

$$\text{and} \quad C_P = \left(\frac{\partial Q}{\partial T}\right)_P = \left(\frac{\partial H}{\partial T}\right)_P$$

Therefore, the detonation temperature T_{ex} can be estimated as follows:

$$T_{ex} = \frac{Q}{\Sigma C_V} + T_i$$

If a value which is slightly too low and a value which is slightly too high are used for the detonation temperature and the value for Q is calculated using the values given in Table 4.5, the "correct" detonation temperature T_{ex} can be estimated iteratively.

4.2 Computational Methods

4.2.1 Thermodynamics

The thermodynamic data as well as the detonation parameters can nowadays be very reliably obtained by using quantum-mechanical computer calculations. On the one hand it is important to check experimental results, and on the other hand – and even more importantly – it is important to predict the properties of potential new energetic materials without any prior experimental parameters, for example during the planning of synthetic work. Moreover, such computational methods are ideal for the estimation of the detonation parameters of newly synthesized compounds, which have not been obtained in the 50–100 g quantities which are necessary for the experimental determination of such detonation parameters (e.g. detonation velocity).

In order to be able to calculate the detonation parameters of a particular neutral or ionic compound, it is advisable to calculate the enthalpy (H) and free energy (G) quantum chemically with very exact methods (e.g. G2MP2, G3 or CBS-4M). To achieve this, the Gaussian (G03W or G09W) programme is suitable. In the following section, we will concentrate on the CBS-4M method developed by Petersson and co-workers. In the CBS method, (complete basis set) the asymptotic convergence behavior of natural orbitals is used, in order to extrapolate the energy limit for an infinitely large basis set. The CBS method begins with a HF/3–21G(d) calculation in order to optimize the structure and for the calculations of the zero-point energy. Then, using a larger basis set, the so-called base-energy is calculated. A MP2/6 – 31 + G calculation with a CBS extrapolation gives the perturbation-theory correct-

Tab. 4.6: CBS-4M values for NG, the ions NH_4^+ and $N(NO_2)_2^-$ and the relevant atoms H, C, N and O.

	$-H^{298}$ / a. u.	$-G^{298}$ / a. u.
NG,	957.089607	957.149231
NH_4^+	56.796608	56.817694
$N(NO_2)_2^-$	464.499549	464.536783
H	0.500991	0.514005
C	37.786156	37.803062
N	54.522462	54.539858
O	74.991202	75.008515

Tab. 4.7: Literature values for $\Delta_f H^°_{(atoms)}$ (in kcal mol^{-1}).

	Ref. [21]	NIST [24]
H	52.6	52.1
C	170.2	171.3
N	113.5	113.0
O	60.0	59.6

ed energy, which takes the electron correlation into account. A MP4(SDQ)/6–31+(d,p) calculation is used, to estimate the correlation contributions of a higher order. The most widely used CBS-4M version today is a re-parametrization of the original CBS-4 version, which contains additional empirical correction terms (M stands for "minimal population localization" here).

The enthalpies of the gaseous species M can now be calculated using the method of the atomisization energies [21–23]:

$$\Delta_f H^°(g, m) = H^°_{(molecule)} - \sum H^°_{(atoms)} + \sum \Delta_f H^°_{(atoms)}$$

Two examples should be considered for further discussion: the covalent nitroglycerine (NG) and the ionic ammonium dinitramide (ADN). The calculated enthalpies (H) and free energies (G) for NG, the ions NH_4^+ and $N(NO_2)_2^-$ as well as the relevant atoms H, C, N and O using the CBS-4M method, are summarized in Table 4.6.

Therefore in Table 4.6 we already have the $H^°_{(molecules)}$ and $H^°_{(atom)}$ values (given in a. u. = atomic units; 1 a. u. = 1 H = 627.089 kcal mol^{-1}). The values for $\Delta_f H^°_{(atoms)}$ are easily obtained from the literature and are summarized in Table 4.7.

According to the equation given above, we can now easily calculate the standard enthalpies of formation $\Delta_f H^°(g)$ for the gas phase species NG, NH_4^+ and $N(NO_2)_2^-$ (Tab. 4.8).

In order to be able to convert the standard enthalpies of formation $\Delta_f H^°(g)$ for the gas-phase into values for the condensed phase, for covalent molecules (NG) we additionally require the enthalpy of sublimation $\Delta H_{sub.}$ (for solids) or the enthalpy

Tab. 4.8: Calculated standard enthalpies of formation $\Delta_f H°$ (g) for the gas-phase.

gas-phase species	formula	$\Delta_f H°(g)$ / kcal mol^{-1}	$\Delta_f H°(g)$ / kJ mol^{-1}
NG	$C_3H_5N_3O_9$	− 67.2	−281.1
ammonium ion	NH_4^+	+151.9	+635.5
dinitramide ion	$N(NO_2)_2^-$	− 29.6	−123.8

Tab. 4.9: The values for the constants α and β for calculating the lattice energy by using the Jenkin's method.

salt type	α / kJ mol^{-1}	β / kJ mol^{-1}
AB	117.3	51.9
AB_2	133.5	60.9
A_2B	165.3	−29.8

of vaporization $\Delta H_{vap.}$ (for liquids). Both values can be estimated using the Trouton's rule, in which T_m is the melting point of the solid and T_b is the boiling point of the liquid [25]:

$$\Delta H_{sub.} \, [\text{J mol}^{-1}] = 188 \, T_m \, [\text{K}]$$
$$\Delta H_{vap.} \, [\text{J mol}^{-1}] = 90 \, T_b \, [\text{K}]$$

NG is a liquid and its extrapolated boiling point is 302 °C (= 575 K). Therefore, the enthalpy of vaporization is calculated to be $\Delta H_{vap.}(NG)$ = 51.8 kJ mol^{-1} (12.3 kcal mol^{-1}).

In the case of ionic solids of the type AB, AB_2 or A_2B, the lattice energy (ΔU_L) and lattice enthalpy (ΔH_L) can be calculated by using the Jenkin's method proposed [26–29]. Only the molecular volumes of the ions are required. These can be most easily obtained from single crystal X-ray diffraction data:

$$\Delta U_L = |z_+| \, |z_-| \, v \left[\frac{a}{\sqrt[3]{V_M}} + \beta \right]$$

Here $|z_+|$ and $|z_-|$ are the dimensionless charges of the cations and anions and v is the number of ions per 'molecule' (2 for ADN, 3 for Ba(DN)$_2$). V_M is the volume in nm^3 of a formula unit (e.g. V_M(ADN) = V_M(NH$_4^+$) + V_M(DN$^-$); V_M(Ba(DN)$_2$) = V_M(Ba^{2+}) + 2 V_M(DN$^-$)). The constants α and β are dependent on the composition of the salt, and are summarized in Table 4.9.

The lattice energy ΔU_L can easily be converted into the corresponding lattice enthalpy ΔH_L:

$$\Delta H_\text{L}(A_pB_q) = \Delta U_\text{L} + \left[p\left(\frac{n_A}{2} - 2\right) + q\left(\frac{n_B}{2} - 2\right)\right]RT$$

$n_A, n_B:$ = 3 for monoatomic ions
5 for linear, polyatomic ions
6 for non-linear, polyatomic ions.

From X-ray diffraction data it is known that the molecular volume of ADN 110 Å³ is 0.110 nm³ [30]. If this value was not known, it could have been calculated using the literature values for the volumes of $V_M(NH_4^+) = 0.021$ nm³ and $V_M(DN^-) = 0.089$ nm³ as being $V_M(ADN) = 0.110$ nm³. Therefore, the following values for ADN can be written:

$$\Delta U_\text{L}(ADN) = 593.4 \text{ kJ mol}^{-1}$$
$$\Delta H_\text{L}(ADN) = 598.4 \text{ kJ mol}^{-1}$$

Since the enthalpy of vapourization for NG and the lattice enthalpy for ADN are known, the gas-phase enthalpy values can easily be converted into the standard enthalpies of formation for the condensed phase:

$\Delta_f H°(NG)$ = −332.9 kJ mol⁻¹ = −80 kcal mol⁻¹
$\Delta_f H°(ADN, s)$ = −86.7 kJ mol⁻¹ = −21 kcal mol⁻¹.

A comparison with experimental values from literature shows that the calculation outlined here gives a good result, without having to rely on experimental data (with the exception of the density and the molecular volume):

$\Delta_f H°(NG)$ = −88 kcal mol⁻¹
$\Delta_f H°(ADN)$ = −36 kcal mol⁻¹

As we have already shown in Chapter 1.4, the molar energies of formation U and enthalpy H for a given explosive with the composition $C_aH_bN_cO_d$ are related as follows:

$$H = U + \Delta n\, RT$$

Here n is the change in the number of moles of gaseous substances and R is the ideal gas constant. Therefore, for NG ($\Delta n = -8.5$) and ADN ($\Delta n = -6$) the following values can be written:

$\Delta_f U°(NG)$ = −311.8 kJ mol⁻¹ = −1373.0 kJ kg⁻¹
$\Delta_f U°(ADN)$ = −71.8 kJ mol⁻¹ = −579.0 kJ kg⁻¹

4.2.2 Detonation Parameters

There now are various different codes available for computing detonation parameters (e.g. TIGER, CHEETAH, EXPLO5 etc.). In this discussion, we will concentrate on the application of one such code, namely the program EXPLO5. This program is based on the chemical equilibrium, a steady-state model of detonation. It uses the Becker-Kistiakowsky-Wilson's equation of state (BKW EOS) for gaseous detonation products and Cowan-Fickett's equation of state for solid carbon [31–34]. The calculation of the equilibrium composition of the detonation products is done by applying the modified White, Johnson and Dantzig's free energy minimization technique. The program is designed to enable the calculation of detonation parameters at the C-J point (see Ch. 3).

The ideal gas law of the form

$$pV = n\,RT$$

with the ideal gas constant R enables us to calculate an ideal gas the pressure p for a certain temperature T and also the volume V for a known molar amount n.

To calculate the detonation pressure, this equation is however not good enough, because there are strong deviations from the ideal gas behavior. The Becker-Kistiakowsky-Wilson equation of state (BKW-EOS)

$$\frac{pV}{RT} = 1 + x\,e^{\beta x} \qquad x = \frac{k}{VT^a}$$

which contains the covolume constant k takes into account the residual volume of the molecules of the gaseous components. The α and β parameters are obtained by comparisons with the experimental data (fitted empirically). For extremely low temperatures, however, the pressure becomes infinitely large. Cowan and Fickett developed the Becker-Kistiakowsky-Wilson equation of state to:

$$\frac{pV}{RT} = 1 + x\,e^{\beta x} \qquad x = \frac{\kappa \Sigma X_i k_i}{V(T+\theta)^a}$$

whereby also $(\delta p / \delta T)_V$ remains positive. The covolume constant is replaced by the weighted sum of the products of the mole fractions X_i and the geometric covolumes k_i of the gaseous components i. The Becker-Kistiakowsky-Wilson-Neumann parameters α, β, κ and θ were obtained empirically by adapting the computational results to the experimental data.

However, this equation can not be used for solids. For such cases Cowan and Fickett suggest an equation of the following form with the factors $p_1(V)$, $a(V)$ and $b(V)$ as polynomial functions of the compression of the material η relative to the crystal density of the solid in the standard state, which is known as the Cowan-Fickett equation of state for solids:

$$p = p_1(V) + a(V)T + b(V)T^2$$

$$\eta = \frac{V°(T°)}{V} = \frac{\rho}{\rho°}$$

The $p(V)$ diagram shows the shock adiabat of the explosive, which is also called the Hugoniot curve or Hugoniot adiabat (see Fig. 1.22). The shock adiabat can be calculated for both the non-reacted explosive as well as for the reaction products (see Fig. 1.22). The C-J point, which represents the point where the C-J conditions are fulfilled, is therefore the point where the shock adiabat of the reaction products touches the Raleigh line (tangent), which is described by the following equation:

$$p - p_0 = \rho_0^2 \, U^2 \, (V_0 - V)$$

Here, ρ_0 is the material density of the unreacted explosive, u is the impact velocity and V and V_0 are the specific volumes. The Raleigh line is a straight line with the slope D^2/V^2 connecting (V_0, p_0), (V_1, p_1) and (V_2, p_2) in the $p(V)$ diagram.

At the point where the Raleigh line touches the schock adiabate of the reaction products the slope of both functions is the same, and the following relationship, where U is the velocity of the products, is valid:

$$\frac{\partial p}{\partial V} = \frac{D^2}{V^2} = \frac{U^2}{V^2}$$

Under the assumption of a steady-state model of detonation, the EXPLO5 program allows the calculation of the detonation parameters and also the chemical composition of the system at the C-J point. For the calculations, the BKW equation of state is applied for gases, where X_i is the mole fraction of the i-th gaseous component and k_i is the molar covolume of the i-th gaseous detonation products:

$$\frac{pV}{RT} = 1 + x \, e^{\beta x} \qquad x = \frac{\kappa \, \Sigma \, X_i \, k_i}{V(T + \theta)^\alpha}$$

$\alpha = 0.5, \quad \beta = 0.176, \quad \kappa = 14.71, \quad \theta = 6620$ (EXPLO5 V5.03)
$\alpha = 0.5, \quad \beta = 0.096, \quad \kappa = 17.56, \quad \theta = 4950$ (EXPLO5 V5.04)

For solid carbon, the Cowan-Fickett equation of state is applied in the following form, with the presumption that carbon is present in the graphite modification:

$$p = p_1(V) + a(V)T' + b(V)T'^2$$

where:
$T' = T/11605.6$ K,
$p_1(V) = -2.467 + 6.769\,\eta - 6.956\,\eta^2 + 3.040\,\eta^3 - 0.3869\,\eta^4$,
$a(V) = -0.2267 + 0.2712\,\eta$,
$b(V) = 0.08316 - 0.07804\,\eta^{-1} + 0.03068\,\eta^{-2}$,

$$\text{compression of the material } \eta = \frac{V^\circ(T^\circ)}{V} = \frac{\rho}{\rho^\circ}$$

The minimization of the free energy in accordance with White-Johnson-Dantzig allows the determination of the composition of the detonation products in equilibrium state. The thermodynamic parameters (enthalpy, entropy, free enthalpy, free energy) of the detonation products can be calculated using functions based on the enthalpy $H_T^\circ - H_0^\circ$. The coefficients $c_0, \ldots c_4$ can be found in literature [J. Chem. Phys **1958**, *28*, 751–755; Prop., Expl. Pyrotech. **1985**, *10*, 47–52]:

$$(H_T^\circ - H_0^\circ) = c_0 + c_1 T + c_2 T^2 + c_3 T^3 + c_4 T^4$$

The EXPLO5 program calculates the parameters of state of the products along the shock adiabat, starting from a density of a given explosive (ρ_0) and then increasing it in an arbitrary chosen step up to the density of about $1.5 \cdot \rho_0$. Then, it determines the C-J point as a point on the shock adiabat at which the detonation velocity has a minimum value (this minimum, D_{min}, is determined by the minimum of the first derivative of the Hugoniot adiabat, see Fig. 4.2). Once the C-J point is determined the detonation parameters can be calculated applying the well-known relationships between them.

Fig. 4.2: Change of detonation velocity with specific volume of detonation products.

The hydrodynamic detonation theory allows to correlate the detonation parameters on the basis of the laws of mass, impulse and energy conservation and to calculate them independently of p and V:

$$\frac{D}{V_0} = \frac{D-U}{V}$$

$$\frac{D^2}{V_0} + p_0 = \frac{(D-U)^2}{V} + p$$

$$U_0 + D^2 + p_0 V_0 = U + 1/2\,(D-U)^2 + pV$$

From this, for the detonation velocity and the velocity of the shock front the following equations can be written:

$$D = V_0 \sqrt{\frac{p-p_0}{V_0 - V}}$$

$$U = (V_0 - V)\sqrt{\frac{p-p_0}{V_0 - V}}$$

The detonation parameters for NG, TNT and RDX calculated using the EXPLO5-program are summarized in Table 4.10, as well as the experimental data for comparison.

Sometimes, the terms "heat of explosion" and "heat of detonation" get mixed-up and are often (wrongly!) used as if they were equivalent. The heat of explosion and the heat of detonation are both heats of reaction but are obtained under differ-

Tab. 4.10: Calculated and experimentally determined detonation parameters.

explosive	density	method	D/m s^{-1}	p_{C-J}/kbar	T_{ex}/K	Q_{C-J}/kJ kg^{-1}
NG	1.60	experimental	7700	253	4260	
		EXPLO5	7819	242	4727	−6229
TNT	1.64	experimental	6950	210		
		EXPLO5	7150	202	3744	−5087
RDX	1.80	experimental	8750	347		
		EXPLO5	8920	345	4354	6033
HNS	1.65	experimental	7030	215		
		EXPLO5	7230	212	4079	−5239
PETN	1.76	experimental	8270	315		
		EXPLO5	8660	311	4349	5889

ent conditions (by applying Hess' law). In the detonation run, the heat of reaction corresponds to the "heat of reaction at the C-J point", and is called heat of detonation. The heat of reaction obtained under "constant volume condition" combustion is usually called the "heat of combustion" or "heat of explosion". There is one additional term which is worth mentioning here and that is the "heat of complete combustion". This term is usually used to denote constant volume combustion in an oxygen atmosphere.

4.2.3 Combustion Parameters

Using the calculated thermodynamic data outlined in 4.2.1, not only the detonation parameters (4.2.2) but also the most important performance parameters for rocket propellants can be calculated. For this there are various programs which are available as well, from which we are going to refer back to and concentrate again on the EXPLO5 code.

The combustion of an energetic material is an irreversible process in which mainly gaseous and to a lesser extent solid combustion products are formed. One can differentiate between two borderline cases of combustion:
1. isobaric
2. isochoric

In an **isobaric combustion**, the combustion occurs without any loss of heat to the surroundings (adiabatic) and at a constant pressure (isobar), while the combustion products are found in a chemical equilibrium (e.g. rocket propellants).

In an **isochoric combustion** it is also assumed that the combustion occurs without any loss of heat to the surroundings (adiabatic) but at a constant volume (isochoric) and the combustion products are found in chemical equilibrium (e.g. gun propellants).

For **rocket propellants**, as we have already seen in Chapter 4.1, the assumption of free expansion of the gases in the atmosphere or in space, p = const. (and thereby $\Delta U = Q_p - p\Delta V$) is a good approximation and therefore the process is best described as **isobaric**. For the theoretical calculation of the performance of rocket propellants, the following assumptions are made:
1. the pressure in the combustion chamber and at the nozzle throat is constant,
2. energy and momentum conservation equations are applicable,
3. the velocity of the products of combustion in the combustion chamber is equal to zero,
4. there is no temperature and velocity lag between condensed and gaseous species,
5. the expansion in the nozzle (nozzle, Fig. 4.3) occurs isentropically (a change of state is isentropic, when the entropy S remains constant, i.e. S = const. or $dS = 0$).

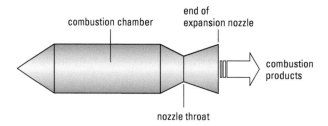

Fig. 4.3: Schematic representation of a rocket combustion chamber with expansion nozzle.

The important performance parameters for the performance of a rocket propellant are the thrust F and the specific impulse I_{sp}^*.

As we have already seen above (Ch. 2.4), the average thrust and the specific impulse I_{sp} are related to each other as follows ($I_{sp}^* = I_{sp}/g$):

$$F = I_{sp} \frac{\Delta m}{\Delta t}$$

For the thrust we can finally write:

$$F = \frac{dm}{dt} v_e + (p_e - p_a) A_e$$

Here, v_e is the velocity of the combustion gases at the end of the expansion nozzle (see Fig. 2.20), p_e and p_a are the pressure at the end of the expansion nozzle and atmospheric pressure respectively and A_e is the cross section at the end of the expansion nozzle.

Therefore, we can write the following for the specific impulse:

$$I_{sp} = \frac{F}{\frac{dm}{dt}} = v_e + \frac{(p_e - p_a) A_e}{\frac{dm}{dt}}$$

Using the EXPLO5 programme, we can calculate the following performance parameters (amongst others) for rocket propellants under isobaric conditions for different pressures in the combustion chamber:
- isobaric heat of combustion Q_p (kJ kg^{-1})
- isobaric combustion temperature T_c (K)
- composition of the products of combustion
- temperature and pressure at the nozzle throat
- flow velocity at the nozzle throat
- temperature at the end of the expansion nozzle ($p_e = 1$ bar)
- specific impulse

Fig. 4.4: Synthesis of hydrazinium aminotetrazolate (**2**, Hy-At).

Tab. 4.11: Calculated combustion parameters for hydrazinium aminotetrazolate – Hy-At /ADN formulations with p = 45 bar.

oxidizer ADN[b]	fuel Hy-At	ρ / g cm^{-3}	Ω / %	T_c / K	I_{sp} / s
10	90	1.573	−65.0	1863	227
20	80	1.599	−55.0	1922	229
30	70	1.625	−44.9	2110	236
40	60	1.651	−34.8	2377	246
50	50	1.678	−24.7	2653	254
60	40	1.704	−14.6	2916	260
70	30	1.730	− 4.5	3091	261
80	20	1.756	+ 5.6	2954	250
90	10	1.782	+15.7	2570	229
AP[b]	Al[c]				
70	30	2.178	− 2.9	4273	255

[a] ADN, ammonium dinitramide; [b] ammonium perchlorate, [c] aluminum

For isochoric combustion processes, the following parameters (amongst others) can be calculated:
- isochoric heat of combustion Q_v (kJ kg^{-1})
- total pressure in closed systems (bar)
- composition of the combustion products
- specific energy: $F = n\, RT_c$ (J kg^{-1}),
  T_c = isochoric combustion temperature.

Currently, a new propellant is being researched. This propellant is hydrazinium aminotetrazolate (**2**), which is obtained in a facile route from hydrazine and aminotetrazole (**1**) (Fig. 4.4) [35a].

The programme EXPLO5 was used to calculate the rocket propellant parameters for this compound in combination with ADN as the oxidizer while presuming a combustion chamber pressure of 45 bar (Tab. 4.11).

We can see that the specific impulse of a 70:30 mixture of Hy-At is approximately 5 s higher than that of a stoichiometric mixture of AP and Al. Therefore, the new fuel Hy-At in combination with the environmentally friendly ADN (in contrast to AP) could potentially be of interest.

Tab. 4.11a: Propellant charge compositions.

propellant charge formulation	ingredient	amount / % (weight)
M1	NC (13.25)	86
	2,4-DNT	10
	dibutyl phthalate (DBP)	3
	diphenyl amine (DPA)	1
EX-99	RDX	76
	cellulose acetate	12
	BDNPA/F[a]	8
	NC (13.25)	4
High-N-1	RDX	56
	TAGzT	20
	cellulose acetate	12
	BDNPA/F[a]	8
	NC (13.25)	4
High-N-2 (HNP)	RDX	40
	TAGzT	20
	FOX-12/GUDN	16
	cellulose acetate	12
	BDNPA/F[a]	8
	NC (13.25)	4
TAG$_2$-BT-High-N-2 (NICO)	RDX	40
	TAG$_2$-BT[b]	20
	FOX-12/GUDN	16
	cellulose acetate	12
	BDNPA/F[a]	8
	NC (13.25)	4

[a] Bis(dinitropropyl)acetal (BDNPA) : Bis(dinitropropyl)formal (BDNPF) = 50 : 50
[b] TAG$_2$-BT = (CN$_6$H$_9$)$_2$-5,5′-Bistetrazolate

For **gun propellants** we can assume isochoric combustion with the specific energy f_E or force or impetus ($f_E = n\,RT$), the combustion temperature T_c (K), the co-volume b_E (cm^3 g^{-1}) and the pressure p (bar; 3000–4000 bar) being the most important parameters. Moreover, a large N$_2$/CO ratio is desirable in order to avoid erosion problems. The loading densities are by far not as important as for high explosives. Comparing M1 and EX-99 with the newly developed high-nitrogen propellant (HNP) which is based triaminoguanidinium azotetrazolate (TAGzT) and with NICO (based on TAG$_2$BT) (Tab. 4.11a), very similar performance values for M1, HNP and NICO, but a greatly reduced errosion coefficient (N$_2$/CO) can be found (Tab. 4.11b, Fig. 4.4a). It is also worth mentioning that the HNP and NICO are similar in their performance to NILE (Navy Insensitive Low Erosion Propellant: 40 % RDX, 32 % GUDN, 7 % acetyl triethyl citrate, 14 % cellulose acetate butyrate, 5 % hydroxyl propylcellulose, 2 % plasticizer).

Tab. 4.11b: Calculated performance of various gun propellants[b].

	T_c / K	$p_{max.}$ / bar	f_E / kJ g^{-1}	b_E / cm^3 g^{-1}	N_2/CO (w/w)
M1	2834	2591	1.005	1.125	0.23
EX-99	3406	3249	1.257	1.129	0.71
High-N-1	2922	3042	1.161	1.185	0.95
High-N-2 (HNP)	2735	2848	1.088	1.181	1.05
TAG$_2$-BT-High-N-2 (NICO)	2756	2896	1.105	1.185	1.03

[a] loading density = 0.2 g cm^{-3}; virial equations of state (real gas)

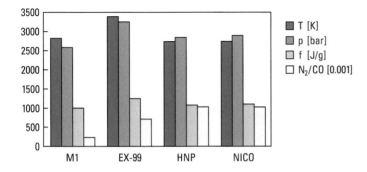

Fig. 4.4a: Calculated performance of various gun propellants (HNP = High-N-2, NICO = TAG$_2$-BT-High-N-2).

The term gun erosion refers to the loss of steel in the bore (inside) of a gun barrel. Consequently, if gun erosion occurs, it results in the wear of the rifle and/or an increase in the bore diameter.

Since as far back as even World War II, it has been common knowledge that the erosion of gun barrels leads to two main types of problems:
(i) financial: due to barrel replacement costs over the lifespan of the weapon system

and
(ii) reduced operational effectiveness: due to inconsistent gun performance and availability.

These problems are consequences of the effects of repeated firings on gun barrels. Even under normal firing conditions and usage, damage to the bore surface and therefore a continuous and progressive increase in the bore diameter result as consequences. Financial costs ensue not only due to the necessity of replacing the barrel itself, but also for the transportation of the gun barrel from its production facility to the location of the gun system. The latter is a logistical cost which depends on the distance and fuel prices, but is not trivial in its sum.

Three main approaches are employed to attempt to reduce the amount of gun barrel erosion which occurs, namely: the development of propellants which cause less erosion, the use of coatings and the production of treated barrel materials and liners, as well as the employment of additives and lubricants which reduce erosion. Out of these three approaches, the use of protective coatings, or treating the surface of the gun barrel bore have found the most wide-spread use. One of the most commonly employed methods used to achieve this includes the deposition of a thin layer of hard chromium inside the gun barrel or by nitriding (hardening) the inside of the barrel. Such coatings/treatments are designed to combat all of the mechanisms of erosion which are discussed below. Although gun barrel erosion is complex and no one single mechanism is usually exclusively responsible, it can be very approximately divided into three categories:
(i) mechanical action,
(ii) heat transfer effects and
(iii) changes in chemical composition.

These three categories can be classified as mechanical, thermal and chemical erosion respectively, and are highly interdependent in that they act in combination to erode the barrels. Mechanical erosion results from friction due to a projectile (or its driving band) moving through the gun barrel. Thermal erosion, on the other hand, is due to morphological changes of the gun barrel steel, (*i.e.* austenite and martensite phase transformation of steel) as well as melting of the barrel material. That the latter occurs is unsurprising when it is considered that depending on the propellant which is used and the rate of fire of the gun system, the steel temperature of the barrel can reach 1800 K in a matter of only milliseconds, which is high enough to cause partial melting of the gun steel, since typical gun steel melts at 1723 K. Such temperatures are, however, only achieved if very hot propellants are used and usually the temperature of the gun barrel will be significantly lower. A good illustrative example here is the barrel surface temperature for a Navy 5"/54 gun which has been evaluated at 1450 K for a certain propellant which has a flame temperature of 2654 K.

Chemical erosion is the term used to describe the effect of chemical reactions that occur between the gun steel and the combustion gases generated by the propellant. The main combustion gases generated by a gun propellant are carbon monoxide (CO), carbon dioxide (CO_2), water (H_2O), hydrogen (H_2) and nitrogen (N_2). At elevated temperatures, each of the above-mentioned gases can react with gun steel to form various compounds. Some of these reactions are shown in equations 1 to 4.

$$Fe + CO_2 \rightarrow FeO + CO \qquad (1)$$

$$3Fe + 2CO \rightarrow Fe_3C + CO_2 \qquad (2)$$

$$4Fe + CO \rightarrow FeO + Fe_3C \tag{3}$$

$$5Fe + CO_2 \rightarrow 2\,FeO + Fe_3C \tag{4}$$

Iron carbide, Fe_3C is formed due to the presence of CO and CO_2 in the gun barrel. In order to minimize erosion of the gun barrel, the formation of iron carbide (with its low melting point of approximately 1420 K) and other low melting point ferrous compounds should ideally be avoided in order to minimize erosion. It is interesting and worthwhile to note that there is currently no consensus on which combustion gas is the most erosive out of the combustion gases that are produced. Despite this, the following order has been proposed which lists the most (CO_2) to least (N_2) chemically erosive combustion gas. The four compounds to the left of "0" correspond to gases which contribute to gun erosion, whereas N_2 to the right of "0" with a chemical erosion value of less than zero, contributes to the stability of a gun barrel due its formation of iron nitride on reaction with the barrel.

$$CO_2 > CO > H_2O > H_2 > 0 > N_2$$

The formation of iron nitride as a consequence of nitrogen gas being present in the combustion gases has been established, and it is generally accepted that in contrast to CO_2, CO, H_2O and H_2, nitrogen gas a lowering effect on gun barrel wear. Consequently, it has been proposed that combustion gases which possess a high-nitrogen content could, in fact, contribute to the re-nitridation of the gun barrel and thereby increase the service lifetime by a significant amount (up to a factor of four).

The interdependence between the propellant and the thermal and chemical erosion can be more easily understood since combustion of the propellant is responsible for generating the heat in the gun and the composition of the combustion gases depend not only on the molecular composition of the propellant, but also on its flame temperature. And of course, the reactions of the steel and the combustion gases are not only dependent on what combustion gases are present, but also on the temperature of the gases.

Therefore, it is not surprising and also to be expected, that changes to either the flame temperature and/or the composition of the combustion gases will have a direct impact on the erosion capability of a propellant. The amount of CO that is produced can be decreased by increasing the oxygen balance of the propellant, resulting in combustion gases with a higher CO_2 and H_2O content, but without affecting the nitrogen content. Since nitrogen is a combustion gas which is not considered to be erosive, the generation of nitrogen should be prioritized. Importantly, since the N_2/CO ratio of combustion gases can be predicted using thermochemical calculations which negate the use of experiments, this ratio can be used to estimate how erosive a particular propellant will be.

If all of the above-mentioned points are taken into consideration, then it can be concluded that in designing a gun propellant which shows low erosion properties, it is of particular importance to achieve a lowering of the flame temperature, as well as to increase the nitrogen content of the combustion gases. Consequently, nitrogen-rich materials are particular promising since being nitrogen-rich they show very high (often >80 % by weight) nitrogen-contents. In addition, the advantageous effect of nitrogen-rich materials on the burning rates of propellants has also been shown. In fact, it has been demonstrated that for certain nitrogen-rich propellants, even although the flame temperature is lower, the experimentally determined maximum pressure in a closed vessel was the same as those of formulations that did not contain nitrogen-rich components, which indicates that the same level of performance is achievable.

The most common equation of state for interior ballistics is that of Nobel-Abel:

$$p(v - b_E) = n\,RT$$

(with: b_E = co-volume, n = mol number, R = gas constant).

The co-volume is a parameter which takes the physical size of the molecules and any intermolecular forces created by their procimity to one another into account.

If experimental test results (manometric pressure bomb) for the maximum pressure using different loading densities are available, the force f_E and the co-volumes b_E can be calculated as follows:

$$f_E = \frac{p_2}{d_2} \times \frac{p_1}{d_1} \times \frac{d_2 - d_1}{p_2 - p_1}$$

$$b_E = \frac{\frac{p_2}{d_2} - \frac{p_1}{d_1}}{p_2\,p_1}$$

where:
f_E force,
b_E co-volume,
p_1 maximum pressure at the lower density d_1,
p_2 maximum pressure at the higher density d_2,
d_1 lower loading density,
d_2 higher loading density.

Assuming that the powder combustion occurs at a temperature equal to the explosion temperature T_{ex} and under consideration of the fact, that after combustion, the specific volume V and the charge density (d) are equal, the following equation applies:

$$\frac{p_{max}}{d} = b_E\,p_{max} + f_E$$

Fig. 4.4b: Solid grain design (top two figures) and actual grains of manufactured HNP (see Tab. 4.11a) with the following parameters: l = 0.055 inches, outer diam = 0.06 inches, inner diam = 0.03 inches, web = 0.016 inches.

The EXPLO5 code can also be used for the calculation of the thermodynamic properties of propellant gases and shows an error of usually less than 5% (loading densities ca. 0.2 g cm^{-3}) when the maximum pressure, the specific energy and the co-volume are considered.

Once a suitable propellant formulation has been identified, the design of the solid grain becomes an important issue (and goes beyond the scope of this book). Fig. 4.4b shows some typical solid grain designs for regressive, neutral, and progressive burning.

The pressure-time profile inside the gun chamber and the actual value of the peak pressure are very important. The propellant grain configuration should, therefore, be designed so that the required peak pressure is achieved within a few milliseconds to propel the projectile with the desired velocity. If the surface area of the grain starts decreasing as the burning proceeds, it is called regressive burning (e.g. a solid cylindrical grain) (Fig. 4.4.b). If the surface area of the grain remains constant during burning, we talk about neutral burning (e.g. tubular grains) (Fig. 4.4.b). If a multi-perforated grain is used, e.g. a hepta-tubular grain, the rate of increase of the surface area originating from the seven holes overcompensates the decrease of the surface area due to burning from the periphery. As a result, the chamber pressure increases during burning (progressive burning) (Fig. 4.4b). The multi-tubular geometry of the grain is common among gun propellants for high performance guns such as tank guns.

4.2.4 Example: Theoretical Evaluation of New Solid Rocket Propellants

Solid propellants of essentially all solid rocket boosters are based on a mixture of aluminum (Al, fuel) and ammonium perchlorate (AP, oxidizer).

Ammonium perchlorate (AP) has applications in munitions, primarily as an oxidizer for solid rocket and missile propellants. It is also used as an air-bag inflator in the automotive industry, in fireworks, and is a contaminant in agricultural fertilizers. Because of these uses and ammonium perchlorate's high solubility, chemical stability, and persistence, it has become distributed widely in surface and ground water systems. There is little information about the effects of perchlorate in these systems or on the aquatic life that inhabits them. However, it is known that perchlorate is an endocrine disrupting chemical that interferes with normal thyroid function and that thyroid dysfunction impacts both growth and development in vertebrates. Because perchlorate competes for iodine binding sites in the thyroid, the addition of iodine to culture water has been examined in order to determine if perchlorate effects can be mitigated. Finally, perchlorate is known to affect normal pigmentation of amphibian embryos. In the US alone the cost for remediation is estimated to be several billion dollars, money that is deeply needed in other defense areas.

In the course of the emerging global interest in high-energetic, dense materials (HEDM) the LMU group is currently developing new energetic materials, which preferably have a positive oxygen balance value (see Ch. 4.1) [35b].

The objectives of the ongoing work is to explore the chemical synthesis of possible replacements for AP as an oxidizer in tactical missile rocket motors. The synthesis, sensitivities, thermal stability, binder compatibility and decomposition pathways of these new high-oxygen materials are currently being researched. In the following example, we theoretically want to evaluate, the suitability of nitrosyl (NO^+) and nitronium (NO_2^+) oxalate as a potential ingredient for solid rocket propellants.

Fig. 4.5: Optimized molecular structures of (ON)O$_2$C–CO$_2$(NO) (a) and (O$_2$N)O$_2$C–CO$_2$(NO$_2$) (b).

Tab. 4.12: Solid state energies of formation ($\Delta_f U°$).

	$\Delta_f H°(s)$ / kcal mol^{-1}	Δn	$\Delta_f U°(s)$ / kcal mol^{-1}	M / g mol^{-1}	$\Delta_f U°(s)$ / kJ kg^{-1}
[NO$_2$]$_2$[O$_2$C–CO$_2$]	– 86.6	–5	– 83.6	180.0	–1943.2
[NO]$_2$[O$_2$C–CO$_2$]	–107.0	–4	–104.6	148.0	–2957.1
O$_2$N–O$_2$C–CO$_2$–NO$_2$	–113.5	–5	–110.5	180.0	–2568.5
ON–O$_2$C–CO$_2$–NO	– 96.5	–4	– 94.1	148.0	–2660.2

The molecular structures of neutral (ON)O$_2$C–CO$_2$(NO) and (O$_2$N)O$_2$C–CO$_2$(NO$_2$) were fully optimized without symmetry constraints in both cases to C_i symmetry (Fig. 4.5).

Table 4.12 presents the calculated energies of formation for the solid neutral species and salts based on the CBS-4M method (see Ch. 4.2.1). Furthermore we see from table 4.12 that for the nitronium ([NO$_2$]$^+$) species the covalently bound form is favored over the ionic salt by 26.9 kcal mol^{-1} while for the nitrosonium species ([NO]$^+$) the salt is favored over the covalent isomer by 10.5 kcal mol^{-1}. This change from the preferred covalent form of $-$NO$_2$ compound (actually a nitrato ester) to the ionic nitronium salt can be attributed almost exclusively to the increased lattice enthalpy of the (smaller!) NO$^+$ species (ΔH_L(NO$^+$ – NO$_2^+$ salt) = 31.4 kcal mol^{-1}) (N.B. The difference in the ionization potentials of NO (215 kcal mol^{-1}) and NO$_2$ (221 kcal mol^{-1}) is only marginal).

For a solid rocket propellant free expansion of the combustion products into space (or atmosphere) with p = const. can be assumed and therefore the below equation is a good approximation, i.e. the combustion process as isobaric can be considered:

$$\Delta U = Q_p - p\,\Delta V$$

In this study we assume firing the rocket motor against an ambient atmosphere (p = 1 bar) as it is commonly the case for tactical missiles.

The following combustion calculations were carried out under isobaric conditions, based on the assumption that the combustion of fuel proceeds without any heat loss to the surroundings (i.e. adiabatically) and that the state of chemical

equilibrium establishes in the combustion products. The calculation of the theoretical rocket performances was based on the following assumptions:
1. the pressure in the combustion chamber and the chamber cross-section area are constant
2. the energy and momentum conservation equations are applicable
3. the velocity of the combustion products at the combustion chamber is equal to zero
4. there is no temperature and velocity lag between condensed and gaseous species
5. the expansion in the nozzle is isentropic (*N.B.* In thermodynamics, an isentropic process or isoentropic process is one during which the entropy of the system remains constant.)

The theoretical characteristics of the rocket motor propellant may be derived from the analysis of the expansion of the combustion products through the nozzle. The first step in the calculation of the theoretical rocket performance is to calculate the parameters in the combustion chamber, and the next step is to calculate the expansion through the nozzle (see Fig. 4.3). The expansion through the nozzle is assumed to be isentropic ($\Delta S = 0$). The EXPLO5 and ICT program code provide the following options:
- *frozen* flow (composition of combustion products remains unchanged, frozen, during the expansion through the nozzle, i.e. equal to composition in the combustion chamber),
- *equilibrium* flow (composition of combustion products at any station in the nozzle is defined by chemical equilibrium).

The frozen performance is based on the assumption that the composition of combustion products remains constant ("frozen"), while equilibrium performance is based on the assumption of instantaneous chemical equilibrium during the expansion in the nozzle.

The specific impulse I_{sp} is the change of the impulse (impulse = mass × velocity, or force × time) per propellant mass unit. The specific impulse is an important parameter for the characterization of rocket propellants and can be interpreted as the effective exhaust velocity of the combustion gases when exiting the expansion nozzle:

$$I_{sp} = \frac{F \times t_b}{m} = \frac{1}{m} \int_0^{t_b} F(t) \, dt$$

The force F is the time-dependent thrust, $F(t)$, or the average thrust, F, t_b is the burning time of the motor and m the mass of the propellant. Therefore, the unit of the specific impulse I_{sp} is N s kg^{-1} or m s^{-1}.

Tab. 4.13: Combustion properties (solid rocket motor) of Al formulations with essentially zero oxygen balance (frozen expansion).

	$O_2N-O_2C-CO_2-NO_2$: Al = 0.70 : 0.30	$[NO]_2[O_2C-CO_2]$: Al = 80 : 20	AP : Al = 0.70 : 0.30
condition	isobaric	isobaric	isobaric
p / bar	70	70	70
ρ / g cm^{-3}	1.93	1.82	2.18
Ω / %	−2.0	−0.6	−2.8
Q_p / kJ kg^{-1}	−6473	−5347	−6787
$T_{comb.}$ / K	4642	4039	4290
I_{sp}^* / s	223	220	243

It is conventional to divide the specific impulse I_{sp} by g_0 (standard gravity, g_0 = 9.81 m s^{-2}) so that the resulting specific impulse I_{sp}^* has the unit s (seconds):

$$I_{sp}^* = \frac{I_{sp}}{g_0}$$

The specific impulse I_{sp}^* can also be defined according to the following equation with $y = C_p / C_v$.

$$I_{sp}^* = \frac{1}{g}\sqrt{\frac{1\, y\, RT_C}{(y-1)M}} \qquad y = \frac{C_p}{C_v}$$

Table 4.13 summarizes the calculated propulsion parameters for aluminized formulations in which the Al content has been varied in order to achieve an oxygen balance that is close to zero (with respect to CO_2, see eq. 2). Table 4.13 contains the corresponding values for a AP/Al formulation for comparison as well. Finally, Table 4.14 shows the calculated specific impulses for equilibrium expansion for the three optimized formulations (covalent $O_2N-O_2C-CO_2-NO_2$/Al, ionic $[NO]_2[O_2C-CO_2]$/Al and AP/Al). The results of Table 4.14 are graphically summarized in Figure 4.6.

From Table 4.13 and Figure 4.6 we can conclude that generally the agreement between the EXPLO5 and ICT calculated equilibrium specific impulses is reasonably good with the ICT code always predicting a slightly better performance. It is further apparent that the formulation with covalent $O_2N-O_2C-CO_2-NO_2$ and Al results in better performance than the formulation using ionic $[NO]_2[O_2C-CO_2]$ and Al. The specific impulse of the chlorine and perchlorate-free formulation with covalent $O_2N-O_2C-CO_2-NO_2$ and Al (**I**) is just slightly lower than that of the AP/Al (**III**) formulation. Therefore it can be concluded that bis(nitronium) oxalate (or oxalic acid dinitrato ester), $O_2N-O_2C-CO_2-NO_2$, may be a promising new perchlorate-free and environmentally benign oxidizer for formulations to be used in solid rocket motors.

Tab. 4.14: Specific impulses (solid rocket motor) of Al formulations calculated using different codes (EXPLO5 and ICT).

formulation	$O_2N-O_2C-CO_2-NO_2$: Al = 0.70 : 0.30		$[NO]_2[O_2C-CO_2]$: Al = 80 : 20		AP : Al = 0.70 : 0.30	
	I		II		III	
	EXPLO5	ICT	EXPLO5	ICT	EXPLO5	ICT
condition	isobaric		isobaric		isobaric	
p / bar	70		70		70	
ρ / g cm^{-3}	1.82		1.74		2.13	
Ω / %	−1.8		−0.5		−2.85	
I_{sp}^* / s	223	206	220	215	243	229
frozen	226	245	225	230	247	257
equilibrium						

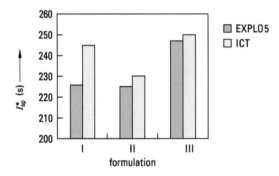

Fig. 4.6: Graphical representation of the calculated equilibrium specific impulses for the optimized formulations **I**: covalent $O_2N-O_2C-CO_2-NO_2$ / Al, **II**: ionic $[NO]_2[O_2C-CO_2]$ / Al and **III**: AP / Al (see Tab. 4.14).

The above discussion clearly revealed that the covalently bound oxalic acid dinitrate ester is the most promising candidate as a high oxidizer and potential replacement for AP in this series. In order to evaluate its thermodynamic and kinetic stability we calculated the decomposition into CO_2 and NO_2. The reaction enthalpy of $\Delta H = -56.5$ kcal mol^{-1} clearly indicates that oxalic acid dinitrate ester is (as expected) thermodynamically unstable with respect to its decomposition into CO_2 and NO_2.

$$O_2N-O_2C-CO_2-NO_2 \text{ (s)} \xrightarrow{\Delta H = -56.5 \text{ kcal mol}^{-1}} 2\ CO_2 \text{ (g)} + 2\ NO_2 \text{ (g)}$$

In order to evaluate the kinetic stability of covalently bound $O_2N-O_2C-CO_2-NO_2$ it was decided to compute a two dimensional potential energy hypersurface

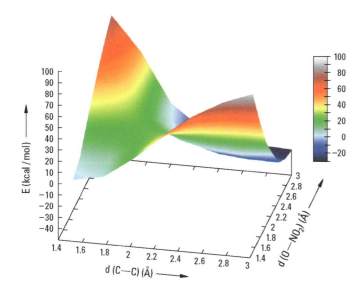

Fig. 4.7: Potential energy hypersurface for the simultaneous dissociation of $O_2N-O_2C-CO_2-NO_2$ into CO_2 and NO_2 at B3LYP/6–31G* level of theory.

Fig. 4.8: Transition state structure for the simultaneous dissociation of $O_2N-O_2C-CO_2-NO_2$ into CO_2 and NO_2 at CBS-4M level of theory (NIMAG = 1, $v_1 = -817$ cm^{-1}, $d\,(C-C) = 2.33$ Å, $d\,(O-NO_2) = 1.91$ Å).

(Fig. 4.7) at B3LYP/6–31G* level of theory. As one can see, the simultaneous dissociation of $O_2N-O_2C-CO_2-NO_2$ into CO_2 and NO_2 has a relatively high activation barrier (Fig. 4.7). The transition state (Fig. 4.8) was located 37.1 kcal mol^{-1} (CBS-4M) above the dinitrate ester. It is interesting that in agreement with Hammond's postulates the transition state lies more towards the higher-energy starting material.

The electrostatic potential (ESP) of covalent $O_2N-O_2C-CO_2-NO_2$ was computed at the optimized structure at the B3LYP/6–31G(d) level of theory. Figure 4.9 shows the electrostatic potential for the 0.001 electron/bohr^3 isosurface of electron density evaluated at the B3LYP level of theory. The colors range from −0.06 to +0.06 hartrees with green denoting extremely electron-deficient regions ($V(r) > 0.06$ hartree) and red denoting electron-rich regions ($V(r) < -0.06$ hartrees). It has recently been proven by Politzer, Murray et al. and extensively used by Rice et al. (see Ch. 8) [39–44]

Fig. 4.9: Electrostatic potential of $O_2N-O_2C-CO_2-NO_2$ and oxalic acid (B3LYP/6–31G(d), 0.001 e bohr^{-3} isosurface, energy values −0.06 H to +0.06 H); color coding: red (very negative), orange (negative), yellow (slightly negative), green (neutral), turquoise (slightly positive), light blue (positive), dark blue (very positive).

that the patterns of the computed electrostatic potential on the surface of molecules can generally be related to the sensitivity of the bulk material. The electrostatic potential at any point r is given by the following equation in which Z_A is the charge on nucleus A, located at R_A.

$$V(r) = \sum \frac{Z_A}{|R_A-r|} - \int \frac{\rho(r')}{|r'-r|} dr'$$

Politzer et al. were able to show that impact sensitivity can be expressed as a function of the extend of this anomalous reversal of the strengths of the positive and negative surface potentials. In most nitro ($-NO_2$) and nitrato ($-O-NO_2$) systems the regions of positive potential are stronger than the negative, contrary to the usual situation. This atypical imbalance between stronger positive regions and weaker negative regions can be related to the impact sensitivities. The calculated electrostatic potential of $O_2N-O_2C-CO_2-NO_2$ (Fig. 4.9) shows strong positive regions over the nitro ($-NO_2$) groups with the positive areas extending into the $O-NO_2$ region (oxygen$-NO_2$ bond). Furthermore, there is also a strong positive region over the relatively weak C−C bond. This is in good accord with the labile $O-NO_2$ and C−C bonds and also accounts for the easy bond cleavage. In comparison, free oxalic acid (Fig. 4.9) does not show any positive regions over the bonds in the molecule.

From this computational study the following conclusions can be drawn:
1. Covalently bound and ionic nitronium and nitrosonium oxalate were researched with respect to their potential use as energetic materials or oxidizers for solid rocket motors. None of these compounds can be expected to be a good high explosive. However, the covalent molecule oxalic acid dinitrate ester, $O_2N-O_2C-CO_2-NO_2$, was identified to be a potentially interesting oxidizer.
2. The computed specific impulse of a $O_2N-O_2C-CO_2-NO_2$/Al formulation (80:20) is comparable to that of a conventional AP/Al (70:30) formulation, however, it is free of toxic perchlorate or any halogen.
3. Oxalic acid dinitrate ester, $O_2N-O_2C-CO_2-NO_2$ is metastable with respect to decomposition into CO_2 and NO_2. The reaction barrier (transition state) for the monomolecular dissociation was calculated to be 37 kcal mol^{-1} at CBS-4M level of theory.
4. The computed electrostatic potential of $O_2N-O_2C-CO_2-NO_2$ shows strong positive areas at the 0.001 s bohr^{-3} isosurface over the O—NO_2 and C—C bonds indicating the relative weakness of the particular bonds.

The results obtained in this study should encourage synthetic work in order to prepare oxalic acid dinitrate ester, $O_2N-O_2C-CO_2-NO_2$ on a laboratory scale and to experimentally evaluate its properties, first and foremost its thermal stability.

Literatur
J. Lavoie, C.-F. Petre, C. Dubois, *Propellants Explos. Pyrotech.* **2018**, *43*, 879–892.

4.2.5 Example: EXPLO5 Calculation of the Gun Propellant Properties of Single, Double and Triple Base Propellants

In order to elucidate the different properties of single, double and triple base gun propellants and to compare these values with the ones obtained for a new high-nitrogen propellant (e.g. NILE, see above), we calculated the relevant combustion parameters using the EXPLO5 code under the assumption of isochoric combustion. The results are summarized in Table 4.15.

The general trend in terms of performance can clearly be seen in the calculated results shown in Table 4.15. While a double-base propellant performs much better than a single-base formulation (much higher force and pressure), the combustion temperature is also considerably higher (over 360 K) which causes increasing erosion problems. The triple-base propellant on the other hand shows performance in between the single- and double-base propellant while having a combustion temperature only slightly higher (137 K) than that of the single base propellant.

The experimental formulation of a new high-nitrogen propellant with 85% TAGzT (triaminoguanidinium azotetrazolate) and 15% NC has the lowest combus-

Tab. 4.15: Computed (EXPLO5) gun propellant parameters for a single, double and triple base propellant in comparison with a new high-nitrogen formulation (isochoric conditions).

propellant	ρ / g cm^{-3}	loading density / g cm^{-3}	f_E / kJ g^{-1}	b_E / cm^3 g^{-1}	$T_{comb.}$ / K	p_{max} / bar	N_2 / CO
NC (12.5)	1.66	0.2	1.05	1.08	3119	2674	0.26
NC : NG (50 : 50)	1.63	0.2	1.21	1.02	3987	3035	0.59
NC : NG : NQ (25 : 25 : 50)	1.70	0.2	1.13	1.06	3256	2875	1.25
TAGzT : NC (85 : 15)	1.63	0.2	1.09	1.22	2525	2875	6.07

tion temperature of all shown formulations (nearly 600 K below $T_{comb.}$ of the single-base propellant). Furthermore, the performance of the new high-nitrogen formulation can be expected to be similar (e.g. same maximum pressure as a triple-base propellant) to a triple-base propellant. Equally important, the N_2/CO ratio of a triple-base propellant increases from 1.25 to 6.07. This, together with the significantly reduced combustion temperature should help to drastically reduce erosion problems. Initial field studies using TAGzT based propellants for large caliber (105 Howitzer) guns have indicated that the life-time of a gun barrel could possibly be increased by a factor of up to 4. This would justify the (still) higher cost for high-nitrogen formulations because of the increased life-time and therefore drastically reduced cost of the gun system.

A large number of empirical, numerical, semi-empirical and quantum-chemical methods for the prediction of various explosive properties such as:
- Detonation performances
- Impact, shock, friction and electrostatic discharge sensitivities
- Thermal stabilities
- Densities
- Melting points
- Reaction enthalpies
- ...

have been published by M. H. Keshavarz et al. from the Malek-ashtar University of Technology, Shahin-shahr, Iran.

See for example (and the cited literature therein):
(a) Keshavarz, Mohammad Hossein, Explosive Materials (2011), 179–201.
(b) Keshavarz, Mohammad Hossein, Explosive Materials (2011), 103–123.
(c) Keshavarz, Mohammad Hossein; Shokrolahi, Arash; Esmailpoor, Karim; Zali, Abbas; Hafizi, Hamid Reza; Azamiamehraban, Jamshid, Hanneng Cailiao (2008), 16(1), 113–120.
(d) Keshavarz, Mohammad Hossein; Esmailpour, Karim; Zamani, Mehdi; Roknabadi, Akbar Gholami, Propellants, Explosives, Pyrotechnics (2015), in press.

(e) Rahmani, Mehdi; Vahedi, Mohamad Kazem; Ahmadi-Rudi, Behzad; Abasi, Saeed; Keshavarz, Mohammad Hossein, International Journal of Energetic Materials and Chemical Propulsion (2014), 13(3), 229–250.
(f) Keshavarz, Mohammad Hossein; Soury, Hossein; Motamedoshariati, Hadi; Dashtizadeh, Ahmad, Structural Chemistry (2015), 26(2), 455–466.
(g) Oftadeh, Mohsen; Keshavarz, Mohammad Hossein; Khodadadi, Razieh, Central European Journal of Energetic Materials (2014), 11(1), 143–156.
(h) Keshavarz, Mohammad Hossein, Propellants, Explosives, Pyrotechnics (2015), 40(1), 150–155.
(i) Keshavarz, Mohammad Hossein; Zamani, Ahmad; Shafiee, Mehdi Propellants, Explosives, Pyrotechnics (2014), 39(5), 749–754.
(j) Keshavarz, Mohammad Hossein; Pouretedal, Hamid Reza; Ghaedsharafi, Ali Reza; Taghizadeh, Seyed Ehsan, Propellants, Explosives, Pyrotechnics (2014), 39(6), 815–818.
(k) Keshavarz, Mohammad Hossein; Pouretedal, Hamid Reza; Ghaedsharafi, Ali Reza; Taghizadeh, Seyed Ehsan, Propellants, Explosives, Pyrotechnics (2014), 39(6), 815–818.
(l) Keshavarz, Mohammad Hossein; Motamedoshariati, Hadi; Moghayadnia, Reza; Ghanbarzadeh, Majid; Azarniamehraban, Jamshid, Propellants, Explosives, Pyrotechnics (2014), 39(1), 95–101.
(m) Keshavarz, Mohammad Hossein; Seif, Farhad; Soury, Hossein, Propellants, Explosives, Pyrotechnics (2014), 39(2), 284–288.
(n) Rahmani, Mehdi; Ahmadi-rudi, Behzad; Mahmoodnejad, Mahmood Reza; Senokesh, Akbar Jafari; Keshavarz, Mohammad Hossein, International Journal of Energetic Materials and Chemical Propulsion (2013), 12(1), 41–60.
(o) Keshavarz, Mohammad Hossein, Propellants, Explosives, Pyrotechnics (2013), 38(6), 754–760.
(p) Keshavarz, Mohammad Hossein; Seif, Farhad, Propellants, Explosives, Pyrotechnics (2013), 38(5), 709–714.
(q) Keshavarz, Mohammad H.; Motamedoshariati, Hadi; Moghayadnia, Reza; Ghanbarzadeh, Majid; Azarniamehraban, Jamshid, Propellants, Explosives, Pyrotechnics (2013), 38(1), 95–102.
(r) Keshavarz, Mohammad Hossein, Propellants, Explosives, Pyrotechnics (2012), 37(4), 489–497.
(s) Rahimi, R.; Keshavarz, M. H.; Akbarzadeh, A. R., *Central Europ. J. of Energ. Mat.*, **2016**, *13*, 73.

4.2.6 Semiempirical Calculations (EMDB)

The ability to reliably predict the performance and thermochemical properties of a new energetic material from its molecular structure is highly desirable. Several thermochemical equilibrium codes, such as CHEETAH, ICT, and EXPLO5 (which utilize suitable equations of state (EOS)), as well as different empirical packages such as EDPHT (which use calibrated empirical methods), have been introduced in order to determine the performance and thermochemical properties of different EMs.

EMDB_1 is a new professional package in the area of EMs, which calculates more than thirty physicochemical and detonation parameters for different pure explosives or energetic formulations (for C-H-N-O-F-Cl-Al-Br-I-S compounds). The unique property of the EMDB code is that, in contrast to other codes, it does not require the density or enthalpy of formation as an input but estimates these values on the basis of the molecular structure.

The predecessor of EMDB is the improved version of the EDPHT [82] program package. In the viewpoint of capabilities, methods, and appearance, EMDB is therefore the improved, updated, and advanced successor of the EDPHT code. Many of correlations have been modified and upgraded based on the latest literature, which has resulted in a significant improvement in the accuracy and precision of the results which are obtained using EMDB in comparison with the values obtained from the EDPHT code.

Here we want to look at the results of performance (VoD) calculations which have each been calculated using the EMDB, EXPLO5, and CHEETAH codes with the experimentally determined values. We also calculated the impact and friction sensitivities using the EMDB code and compared the calculated values with measured values.

All calculations were carried out on a desktop PC using the EXPLO5_6.03 and EMDM_1.0 codes [83]. The results of CHEETAH_8.0 calculations were taken from the literature [84].

Table 4.16 summarizes the calculated and measured detonation velocities of ten well established and new explosives. The molecular structures of the newer explosives are shown in Fig. 4.10.

Inspection of the calculated and measured VoD values reveals that there is generally reasonably good agreement. Both, the thermodynamic programs (CHEETAH and EXPLO5) as well as the empirical code (EMDB) predict VoD values within a few percent of the experimental values.

Table 4.17 shows the calculated (EMDB) and measured sensitivity parameters for impact and friction. These values can only be predicted using the EMDB code, but not using one of the thermodynamic programs. The EMDB predicted values clearly show the correct trend in terms of the sensitivity categories: very sensitive (BTNPM), moderately sensitive (RDX), and insensitive (NTO) impact sensitivity. The correct trend is also predicted for the friction sensitivities using the EMDB code.

Both the thermodynamic programs (EXPLO5 and CHEETAH) and the empirical code (EMDB) predict VoD values which agree well with the experimental data. In addition, the EMDB code can predict many more properties. For example, safety parameters, phase transition thermochemical data, as well as performance and thermodynamic values.

Tab. 4.16: Calculated and measured detonation velocities (in m s^{-1}) [a] [85]

	density / g cm^{-3}	EMDB_1.0	EXPLO5_6.03	CHEETAH_8.0	LASEM	measured @ TMD
TNT	1.65	7230 (+3 %)	6878 (−2 %)	7192 (+2 %)	6990 ± 230	7026 ± 119
HNS	1.74	7620 (+6 %)	7209 (±0 %)	7499 (+4 %)	7200 ± 210	7200 ± 71
NTO	1.93	8080 (−3 %)	8601 (+3 %)	8656 (+4 %)	88 300 ± 250	8335 ± 120
RDX	1.80	8670 (−2 %)	8919 (+1 %)	8803 (±0 %)	8850 ± 190	8833 ± 64
ε-CL-20	2.04	9600 (±0 %)	9882 (+3 %)	9833 (+3 %)	9560 ± 240	9570
TKX-50	1.88	9140 (−3 %)	9995 (+6 %)	9735 (+3 %)	9560 ± 280	9432
BDNAPM	1.80	8090	8220	8171	8630 ± 210	
BTNPM	1.93	9020	9348	9276	9910 ± 310	
TKX-55	1.84	7860	7666	7548	8230 ± 260	
DAAF	1.75	7960 (−2 %)	8163 (+1 %)	8124 (±0 %)	8050 ± 260	8110 ± 30

[a] The percentage values in parentheses refer to the deviation of the calculated from the measured VoD values.

Fig. 4.10: Molecular structures of some more recent explosives.

Tab. 4.17: Measured and calculated (EMDB) impact (IS) and friction (FS) sensitivities [85]

	IS / J (exptl)	IS / J (EMDB_1.0).	FS / N (exptl)	FS / N (EMDB_1.0)
TNT	15	33	> 353	210
HNS	5	20	> 240	227
NTO	> 120	39	> 353	144
RDX	7.5	5	120	156
ε-CL-20	4	3	48	128
TKX-50	20	15	120	82
BDNAPM	11	39	> 360	166
BTNPM	4	3	144	160
TKX-55	5	12.5	> 360	229
DAAF	7	17	> 360	175

Check your knowledge: See Chapter 14 for Study Questions.

5 Initiation

5.1 Introduction

An explosive can be initiated using different stimuli (e.g. heat or shock) and can then either ignite, deflagrate and turn into a detonation, or directly detonate, if it is initiated using a strong shock (Fig. 5.1). The ignition occurs at an ignition temperature which is characteristic for individual substances, if the linear loss of heat of the surroundings is smaller than the heat generated through the exothermic reaction.

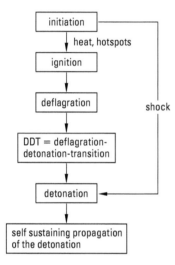

Fig. 5.1: Transition from initiation to detonation.

Generally, it can be said that the **initiation** of explosives is mostly a **thermal process**. However, the ignition (initiation) of an explosive can also occur by impact linking via the low frequency vibrational modes (doorway modes) between 200 and 600 cm^{-1}. In cases where the initiation occurs through a shock-wave, we also observe strong warming-up through adiabatic compression. If mechanical (impact, friction) or electrostatic (ESD) mechanisms are the cause, it can be assumed, that the mechanical or electric energy is also first converted into heat. This occurs, for example, during the formation of so-called **hotspots**. Here, there are small gas bubbles (0.1–10 µm), which are then strongly heated (up to 900 °C) as well through adiabatic compression and therefore initiate the explosive. These can either be gas bubbles in liquids or in solids. The larger the difference in pressure between the original pressure (p_1) and the final pressure on compression p_2, the higher the jump in temperature:

$$T_2 = T_1 \left(\frac{p_2}{p_1}\right)^{\frac{\gamma-1}{\gamma}}$$

Another type of possible hotspots are small, very hard crystals or crystal needles. Prior to breaking the crystal, energy must be applied either through friction or through pressure, in order to bring charges of the same sign closer together. This energy is released again on breaking of the crystals and can result in the formation of a hotspot. However, not every hotspot results in an ignition and finally in detonation. When e.g. the energy is released to the surroundings without resulting in an initiation. Generally it can be said, that the temperature in a hotspot must be at least 430 °C, in order to be sufficient enough to cause the initiation of a secondary explosive. Consequently, explosives with melting points below 430 °C can not be initiated by hotspots. On the other hand, liquids (e.g. NG) can be initiated through hotspots (gas bubbles of dissolved gases) as well through adiabatic compression. On average hotspots exist only for 10^{-3}–10^{-5} s.

When crystals rub together, hotspots can form only through **friction**, as a result of the friction heat and the (in comparison to metals) low heat conductivity.

Usually it is assumed, that on **impact** most primary explosives are initiated by hotspots from inter-crystalline friction. For secondary explosives, as a general rule, initiation by impact results in the formation of hotspots, which originate (adiabatic compression) from gas bubbles between the crystals. Such hotspots only exist for approximately 10^{-6} s.

The thermal energy of hot spots must be efficiently transferred to appropriate molecular vibrational modes, a process called "up-pumping", if bond-breaking and subsequent exothermic decomposition and detonation are to be achieved. The term "trigger linkage" has been applied to the key bond or bonds that are initially ruptured. Any dissipation of hot spot energy, for instance by diffusion, will lessen the likelihood of these processes. Thus Kamlet suggested that free rotation around the trigger linkage (e.g. C-NO_2, N-NO_2, O-NO_2 can have a desensitizing effect, since it uses energy that could otherwise go into bond-breaking vibrational modes. Furthermore, Kamlet recognized the significance of decomposition steps that follow the initial bond rupture, for example the autocatalysis (by NO_2 radicals) for the decomposition of nitramines.

The fact that solids composed of very small particles are less sensitive can now be explained on the grounds that these will have smaller hot spots and thus require higher temperatures for ignition.

It should also be noted that initiation of detonation can occur even in a homogeneous (defect-free) solid (which doesn't really exist). This can occur, for example, if there is efficient anharmonic coupling to channel energy from lattice into the critical molecular vibrations.

Finally, let us consider the situation in TATB which is well known for its remarkable insensitivity. An outstanding feature of TATB is the strong and extensive hydrogen bonding, both inter- and intramolecular, between the NO_2 and NH_2 groups. This would be anticipated from its molecular structure, and has been confirmed by X-ray diffraction. The intermolecular H-bonding gives rise, in the solid, to a two-dimensional network, to which is attributed the relatively high thermal conductivity of TATB. In terms of the hot spot concept, this is a desensitizing factor, since it results is a more rapid dissipation of their energy through thermal diffusion. This is also a reason why TATB requires a high ignition temperature. On the other hand, the strong intermolecular H-bonding should increase the NO_2 rotational barrier, an expectation for which DFT calculations provide support. By Kamlet's reasoning (see above), this should have a sensitizing effect, since it diminishes the loss of energy through rotation and makes more of it available for the key NO_2 vibrational modes. It is believed that $C-NO_2$ homolysis is the first or one of the first steps in TATB initiation.

5.2 Ignition and Initiation of Energetic Materials

As was mentioned above, the ignition of energetic materials and in particular the initiation of explosives can occur in manifold ways and depending on the output of the energetic material, the initiating system is called a detonator (shockwave) or an igniter (flame) (Fig. 5.2). The simplest method is the direct ignition by flame. This was the method which was used in the past to ignite black powder by a burning fuse. Flame as a simple initiating impulse is still used to initiate energetic materials, for example, in fuse-type blasting caps (Figure 5.3, A).

Another method similar to flame as an impulse is to ignite the explosive by heat (Figure 5.3, B + C). A bridgewire which is heated by an electric current is either in direct contact with the primary explosive (hot-wire initiator), or first it is in contact with a pyrotechnical composition which then initiates the primary explosive by flame. This second type, called an electric match, is the most common initiation method in blasting caps worldwide. However, there is also a type of detonator

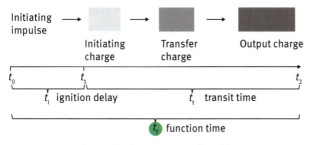

Fig. 5.2: Schematic presentation of an initiation chain.

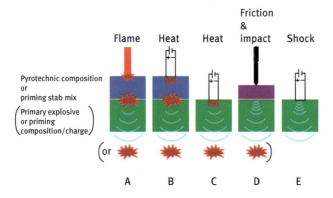

Fig. 5.3: Schematic representation of common initiation and ignition methods of energetic materials. A) Fuse type. B) Electric match type. C) Hot-wire type. D) Percussion / Stab type. E) Exploding-bridgewire type.

known which achieves detonation without a primary explosive. This type of detonator is called NPED (Non Primary Explosives Detonator) and uses the ability of PETN to undergo a DDT after ignition by a pyrotechnic composition. NPEDs also represent a form of electric match type detonators, but provide a higher degree of safety due to the absence of the sensitive primary explosive.

In contrast, the initiation of bombs and grenades, or the ignition of gun propellants is generally achieved by an impact or friction impulse (Figure 5.3, D). Independent of the desired output signal, the primary explosive or priming charge must exhibit a high sensitivity towards friction to ensure reliable initiation. However, it hinders the handling of the explosive and increases the risk of unintended initiation. For detonators like the M55 stab detonator, lead azide based formulations are commonly used for the stab mix, followed by lead azide as the primary explosive. While for percussion caps which only deflagrate, lead styphnate based (SINOXID, NOL-60) and also lead-free (SINTOX) formulations are used. Often tetrazene is added to increase the friction sensitivity, in order to achieve a better ignitability.

Exploding-bridgewire (EBW), and exploding foil initiators (EFI) are relatively safe and very reliable electric type initiators (Figure 5.3, E). During their work for the Manhattan Project, Luis Alvarez and Lawrence Johnston invented EBWs because of the need for a precise and highly reliable detonator system. The initiating impulse is a shockwave which is formed by the explosion of a thin bridgewire. The gold or platinum wire is heated very quickly by the passing high current (such high currents are supplied by low-inductance, high voltage capacitors). While the metal vaporizes, the electrical resistance of the bridgewire increases rapidly and the current decreases. However, a lot of current still passes through the high density metal vapor, and so heating is continued up to the point where inertia is overcome. This results in explosive expansion of the metal vapor. Further effects amplify the shock. This shockwave finally initiates the secondary explosive which – due to practical reasons – is PETN in applications. The advantage of EBWs over classic-type detonators (e.g. hot-wire, stab) is their precise and consistent function time (variations of ≤ 0.1 µs).

However, all of the above-mentioned initiating systems have some disadvantages. Stab detonators or percussion caps contain materials which are extremely friction and impact sensitive. Therefore, there exists a high risk of unintended initiation, and the handling of these materials is problematic. Furthermore, the electric type detonators are susceptible towards electrostatic discharge, electromagnetic interference and corrosion. In the case of the safe EBWs and EFIs, stray currents would not initiate the detonator but may damage it. A further disadvantage of EBWs and EFIs is the bulky size of the power sources which are required.

5.3 Laser Ignition and Initiation

Due to the disadvantages of conventional initiators, a new ignition and initiation method was introduced at the end of the 60s and beginning of the 70s. The first published laser initiation experiments were reported by the Russian research group of Brish et al. and by the Americans Menichelli and Yang. Brish et al. successfully initiated samples of PETN and LA by laser irradiation. As the radiation source a neodymium glass laser (λ = 1060 nm) operating in the Q-switched mode was used. LA was initiated by a laser pulse with an output power of 10 MW, a pulse length τ of 0.1 µs and energy of 0.5 J. The pulse had a power density at the surface of the explosive sample of up to 8 MW cm^{-2}. For the initiation of PETN, the laser beam had to be focused to increase the power density at the surface. Further investigations with a neodymium glass laser and an additional ruby laser (λ = 694 nm) were made by Brish et al., and four possible initiation mechanisms were considered:

(i) Light impact. This could be excluded because the exerted light pressure was several orders of magnitude lower than the pressures required for shock explosion.
(ii) Electrical breakdown could not fully be excluded but did not fit to all observations during the laser initiation experiments. E.g. no differences in the initiation of explosives were observed for dielectrics and metal powders.
(iii) Photochemical initiation could be explained by a multiquantum photoelectric effect, but the probability for initiation was low and a dependency between the laser wavelength (1064 or 694 nm) and the initiation threshold was not observed. Thus, photochemical initiation does not seem to be the appropriate mechanism for laser initiation.
(iv) Photothermal (or simply called thermal) initiation explains all of the experimental results. The laser radiation heats the explosive and leads to an increase of the pressure inside the sample. The rapid heating of the sample results

in an explosive decomposition described as a conversion of light energy into shockwave energy.

While Brish et al. directly irradiated the explosive sample with the laser beam, Menichelli and Yang coated a glass window with a 1000 Å thick aluminum film. Irradiation of this film generated shockwaves by vaporization of the metal, similar to the functionality of EBWs. [15c, 15d]. The generated shockwave is able to directly initiate secondary explosives. In the work of Menichelli and Yang, a Q-switched ruby laser was successfully used (E = 0.8 to 4.0 J; τ = 25 ns) to initiate PETN, RDX and tetryl.

However, the practical use of these solid-state lasers (ruby and neodymium glass) was limited because of their large size. The advent of commercially available laser diodes (electrically pumped semiconductor laser) afforded new possibilities in the initiation and ignition of energetic materials, although solid-state lasers were and are still used for investigations. One of the first reports on diode laser ignition of explosives and pyrotechnics was published by Kunz and Salas in 1988. A GaAlAs diode laser with a power of 1 W, a wavelength of 830 nm, and a pulse duration of 2–10 ms was used to thermally ignite pentaammine(5-cyano-2H-tetrazolate)cobalt(III) perchlorate (CP, Figure 9) and a pyrotechnic mixture of titanium and potassium perchlorate (Ti/KClO$_4$). In contrast to the experiments of Brish et al. and Menichelli and Yang, the laser beam was directly coupled to the explosive's surface by an optical fiber (core diameter of 100 μm) and not focused. CP and Ti/KClO$_4$ were tested upon laser irradiation in pure and in doped (carbon black, graphite and laser dye) form. Additionally, various particle sizes and compressed densities were investigated. In the work of Kunz and Salas, it was shown that the laser ignition threshold is strongly influenced by the absorbance of the energetic materials. Thus, CP exhibited a considerably lower absorbance at 830 nm than the Ti/KClO$_4$ mixture and therefore a higher energy was necessary for ignition (Ti/KClO$_4$: $E_{crit.}$ = 3 mJ; CP: $E_{crit.}$ = 6–7 mJ). Doping CP with 0.8 % carbon black resulted in a similar absorbance ($\alpha \approx 0.8$) to Ti/KClO$_4$, and also the critical energy could be decreased to a value of 3 mJ similar to Ti/KClO$_4$. Additionally, absorbance measurements demonstrated that the particle size and pressed powder density influences the absorbance (α increases with decreasing particle size and an increase in the density) although the effect on the ignition threshold is low. One reason could be that an increasing density also increases the thermal conductivity and – related to this – the ignition threshold. Although the diode laser systems are relatively cheap, small in size and offer significantly improved safety over electric type ones, laser diodes provide a considerably lower output power than the solid-state lasers. Consequentially, in the past, laser diode ignition exhibited long function times and was not suitable as a replacement for fast-functioning electric type initiators like EBWs.

Various further experiments with laser radiation as an ignition and initiation source have been conducted since the seventies. For example, for the ignition of

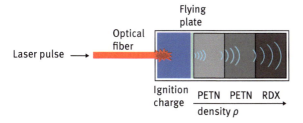

Fig. 5.4: Schematic representation of the laser initiated NPED.

$$\left[\begin{array}{c} H_3N\diagdown\underset{H_3N}{\overset{NH_3}{\diagdown}}\underset{NH_3}{\overset{NH_3}{\diagup}}\underset{N=N}{\overset{N}{\diagdown}}-CN \end{array}\right]^{2+} (ClO_4^-)_2 \qquad \left[\begin{array}{c} H_3N\diagdown\underset{H_3N}{\overset{NH_3}{\diagdown}}\underset{NH_3}{\overset{N=N}{\diagup}}\underset{N=N}{\overset{N}{\diagdown}}-NO_2\\ \underset{N=N}{\overset{N=N}{\diagup}}-NO_2 \end{array}\right]^{+} ClO_4^- \qquad \left[Hg \leftarrow \cdots \left(\begin{array}{c} N=N\\ N-N\\ H \end{array}\right)-\underset{NH}{\overset{NH_2}{\diagup}}\right]^{2+} (ClO_4^-)_2$$

CP BNCP HTMP

Fig. 5.5: Molecular structures of three prominent laser ignitable explosives CP, BNCP and HTMP.

propellants, various pyrotechnic mixtures and metastable intermolecular composites (MICs), the effects of the laser wavelength, gas pressure or dopants on the initiation thresholds, the development of laser initiated detonators, as well as a fundamental understanding of the initiation mechanism were investigated. A remarkably development of a NPED which was successfully initiated by laser irradiation has been achieved. The NPED uses the principle of flying plate detonators and consists of a PETN/graphite/Al mixture as a laser sensitive ignition charge. The ignition charge drives a thin metal plate (called flying plate) which initiates low-density PETN (ρ = 0.7 g cm^{-3}) by shock. This charge is followed by two further charges with higher densities (PETN with a density of 1.0 g cm^{-3} and RDX with a density of 1.6 g cm^{-3}). A schematic representation of the setup is shown in Fig. 5.4. Although no values were given for this detonator, it can be expected that the function times are in the range of milliseconds.

Other recent investigations deal with energetic coordination compounds (ECCs) which exhibit a high sensitivity towards laser irradiation. The investigation of ECCs with respect to their behavior towards laser irradiation started with CP and its promising replacement tetraammine-cis-bis(5-nitro-2H-tetrazolato-N^2)cobalt(III) perchlorate (BNCP). BNCP (Fig. 5.5 and Fig. 5.6) was already reported to exhibit initiating capabilities by Bates in 1986. Since that time, BNCP has been extensively investigated as an initiating explosive which is able to undergo fast DDT after ignition by electric devices or laser irradiation. BNCP exhibits excellent energetic properties, e.g. a detonation velocity of 8.1 km s^{-1} was calculated for BNCP at a density of 1.97 g cm^{-3}. With the advent of more powerful diode lasers, the ignition delay

Fig. 5.6: Synthesis of BNCP according to the literature.

times of BNCP could be decreased considerably and excellent values of 4–5 µs were possible.

At the beginning of the 21st century, Zhilin et al. investigated BNCP and some BNCP analogous tetraamminecobalt(III) perchlorate complexes with 1H-tetrazolate, 5-methyl-1H-tetrazolate, 5-amino-1H-tetrazolate, 5-nitriminotetrazolate, 1,5-diamino-1H-tetrazole, and 5-amino-1-methyl-1H-tetrazole as ligands with respect to laser irradiation ($\tau = 2$ ms, $E = 1.5$ J, $\lambda = 1064$ nm). Only the complexes with 5-nitro-2H-tetrazolate, 5-amino-1H-tetrazolate and 1,5-diamino-1H-tetrazole as ligands detonated after initiation by laser irradiation. The others only combusted or were unable to be ignited.

In order to obtain short ignition delay times, dopants are necessary to increase the absorptance of CP, BNCP or its analogous cobalt(III) complexes in the near infrared region. Therefore, explosives with a higher own absorbance have been investigated. One class of explosives which exhibit a high sensitivity towards laser irradiation re metal complexes with various hydrazinoazoles as ligands. The advantage of this ligand type is the high positive enthalpy of formation for the azole, combined with the low ionization potential of the hydrazine moiety. The following hydrazinoazoles have been investigated as ligands in mercury(II) perchlorate complexes: 3-hydrazino-5-aminopyrazole, 3-hydrazino-5-aminopyrazolone, 3-hydrazino-4-amino-1,2,4-triazole, 3-hydrazino-4-amino-5-methyl-1,2,4-triazole, 3-hydrazino-4-amino-5-mercapto-1,2,4-triazole and 5-hydrazino-1H-tetrazole.[26] The 5-hydrazino-1H-tetrazolemercury(II) perchlorate complex (HTMP, Figure 5.4) exhibited the highest sensitivity towards laser irradiation ($\tau = 30$ ns, $E = 1 \cdot 10^{-5}$ J) of the investigated azoles (e.g. 3-hydrazino-4-aminopyrazole: $\tau = 30$ ns, $E = 2.8 \cdot 10^{-4}$ J). Based on these results, four metal(II) perchlorate complexes (M = Co, Ni, Co, Cd) of the type [M(HATr)$_2$](ClO$_4$)$_2$ with 3-hydrazino-4-amino-1,2,4-triazole (HATr) as the ligand were synthesized and tested towards laser irradiation. All complexes were successfully initiated and the critical initiation energies determined. A correlation between the initiation energy and the ionization potential was proposed to explain the order of the following values:

E: Cu ($1.1 \cdot 10^{-5}$ J) < Cd ($5.03 \cdot 10^{-4}$ J) < Ni ($5.75 \cdot 10^{-4}$ J) < Co ($1.36 \cdot 10^{-3}$ J)

$I_1 + I_2$: Cu (28.02 eV) > Cd (25.90 eV) > Ni (25.78 eV) > Co (24.92 eV)

A mechanism (equation 1 and 2) for the initial stages in the decomposition of the above-mentioned complexes was claimed by Ilyushin et al.:

$$2\,ClO_4^- + M^{2+} \rightarrow ClO_4^{\cdot} + M^+ + ClO_4^- \tag{1}$$

$$ClO_4^- + M^+ \rightarrow ClO_4^{\cdot} + M^0 \tag{2}$$

A highly active perchlorate radical is formed by a laser induced redox reaction between the metal cation and the anion. This radical subsequently oxidizes the organic ligand. The more noble metal cation (higher ionization potential) can be easier reduced. Although the suggested mechanism explains, in agreement with the experimental data, how the ionization potential influences the initiation energy threshold, it doesn't fit with the thermal mechanism proposed by numerous researchers and is therefore doubtful. For example, the mechanism by Ilyushin et al. cannot explain the influence of light absorbing particles like carbon black on the initiation threshold.

To conclude, the results of the investigations which have been carried out so far suggest that laser initiation seems to be a thermal mechanism, although the mechanism itself and the parameters which influence it are still not completely understood. For example, Ilyushin et al. state that there seems to be no dependence between optical properties and the laser initiation threshold for various cobalt(III) complexes which is not plausible for a photothermal process. A thermal initiation mechanism means that laser light is absorbed and converted into heat by internal conversion. At lower laser powers (10^{-1}–10^1 W range), the explosives undergo a DDT after being heated up to the auto-ignition temperature. According to the literature, the explosive is shock initiated at higher power (approximately 10^2 W and more) and DDT no longer takes place. Hafenrichter et al. described four regimes for the laser initiation depending on the power densities:

Region I: Power density is low and the steady state temperature is reached before auto-ignition → no ignition occurs.

Region II: Power density is higher (kW cm^{-2}) and ignition occurs, although a significant amount of energy is conducted away due to thermal conductivity of the materials → slow ignition delay times in the order of milliseconds.

Region III: At even higher power densities (MW cm^{-2}) the auto-ignition temperature is reached quickly before large amounts of heat are conducted away → decreasing ignition delay times in the order of microseconds.

Region IV: At extremely high power densities (GW cm^{-2}), the explosive is shock initiated by laser ablation of the surface → ignition delay times in the order of nanoseconds.

While Region II can be compared with the hot-wire initiation of primary explosives or pyrotechnic compositions, the laser power densities in region IV also make it possible to directly shock initiate secondary explosives by laser irradiation. The

Tab. 5.1: Comparison of electric and laser-type detonators.

Parameter	Hot-Wire	EBW	EFI	Nd:YAG Laser	Diode Laser
Current [A]	5	500	3000	10^1–10^2 [a]	10
Voltage [V]	20	500	1500	10^0–10^2 [a]	2
Power [W]	1	10^5	$3 \cdot 10^6$	$\geq 10^2$	1–10
Function Time [s]	10^{-3}	10^{-6}	10^{-7}	10^{-7}–10^{-6}	10^{-6}–10^{-3}
Initiating Charge	LA, LS, PETN	PETN	PETN, HNS	PETN, RDX, HMX, Tetryl	BNCP, CP, Pyrotechnic Mixtures

[a] Current and voltage depends on the used laser pumping source (e.g. flash lamp, arc lamp, laser diode)

laser power densities of Region IV are achieved by solid-state lasers with laser powers of at least 100 W. In contrast, laser diodes (~ 1–10 W) only provide power densities which fall into the regions II and III. However, more powerful laser diodes have been gradually developed and therefore, laser diode initiators (LDI) have become more and more attractive for applications due to their small size and the low cost of the diodes.

All in all, laser initiation provides several advantages compared to the conventional initiation by electric devices. However, it has to be differentiated between hot-wire, exploding-bridgewire and exploding foil initiators. A comparison of the electric type and typical laser type (e.g. Nd:YAG [$Y_{2.97}Nd_{0.03}Al_5O_{12}$] solid-state laser and GaAlAs or InGaAs diode laser) initiators is given in Table 5.1. The advantages and disadvantages are discussed based on the literature.

Hot-wire initiators are cheap to produce and one of the most manufactured initiators. Hot-wires are operated with low current (~ 5 A) and voltage (~ 20 V) and therefore only a small current source is necessary. On the other hand, hot-wires are susceptible towards unintended initiation by electric impulses, e.g. electrostatic discharge, because of their low threshold. In addition, primary explosives (sensitive toward thermal and mechanical stimuli) are required for the generation of a shockwave. Depending on the application, the slow and less consistent function times in the range of milliseconds can be neglected. EBWs and EFIs exhibit considerably faster function times with very minor standard deviations (down to 25 ns). Furthermore, these two types of detonators do not consist of a primary explosive, which greatly increases the safety. However, a main disadvantage of the EBWs and EFIs is the bulky size of their power sources and the heavy cables which are required, which makes them inapplicable for many applications. Therefore, laser ignition and initiation provides some advantages, such as the high isolation of the energetic material which cannot be achieved with electric type detonators. Unintended initiation by electric impulses is nearly impossible because the induced currents are not able to produce a laser pulse exceeding the initiation energy threshold. In addition, a bridgewire which is susceptible to corrosion is missing, too, and therefore considerably reduces the risk of failures in initiation. Furthermore, the currents which are required for laser initiation are in the range of hot-wire detonators, but the function times can achieve the values of EBWs. In addition, the direct initiation of secondary explosives (e.g. RDX) by solid-state lasers offers new possibilities and considerably increases the safety.

Although solid state-lasers provide a considerably higher output power and can therefore be used similar to the functionality of EBWs and EFIs, the configuration is relatively large compared to diode lasers. Standard diode lasers are only several millimeters in size and can be produced with relatively low cost making them suitable for applications.

Concluding, based on these aspects, the laser initiation of explosives by laser diode radiation seems to be a very attractive and initiation method of the future. However, due to the limited power of laser diodes, the explosives might be thermally ignited and undergo a DDT, in contrast to the shock initiation which results by high-power solid-state laser radiation. The operating mode of laser diode initiators can be compared to that of electric hot-wires but with faster function times as has been demonstrated in the literature.

5.4 Electric Detonators

In general, detonators (or blasting caps) are used to initiate high explosives. The output of a detonator is a strong shock. There are four basic types of detonators:
– electric detonators: EBW, EFI (also known as "slapper detonator"), hot-wire detonators
– non-electric detonators: percussion devices, shock tube (90 % HMX, 10 % Al)
– instantaneous
– delay: pyrotechnic, electronic

In contrast to a detonator, an igniter has a flame (= hot flash) as the output. Igniters are used for the initiation of low explosives.

Electric Bridge Wire Detonators (EBW) and Exploding Foil Initiators (EFI) were originally developed for military applications involving nuclear weapons. They have now also found use in numerous non-military applications (e.g. explosive welding, oil exploration, mining, etc.), although they still are not as common as the more popular hot wire initiators. EBWs were invented in the 1940s for the Manhattan project in order to be able to obtain the desired simultaneity for a nuclear device. Later, EFIs were invented by LLNL in 1965 (see below).

EBWs and EFIs are both secondary explosive detonators that require a high amplitude, but short duration electrical impulse to function. The difference is that the secondary explosive in an EBW is directly in contact with the bridgewire, and it is shock-initiated by the exploding wire. In an EFI, the exploding foil accelerates a disc across a gap, and the explosive (not directly in contact with the exploding foil) is initiated by the kinetic energy of the flying disc.

Although electric detonators appear to be much more complicated than simple hot wire devices, the advantages of electric detonators are obvious:
- safety, insensitivity
- reliability
- precision
- repeatability
- simultaneity
- shot to shot reliability (under 5 microseconds).

In Fig. 5.7, a comparison between a typical EBW detonator and a common hot-wire detonator is shown. The major differences are in the brigewires. EBWs generally use gold or platinum wires (inertness), while hot-wire devices use high resistance materials such as Cr-Ni-alloys. The secondary explosive against the bridgewire in an EBW is usually PETN, or less commonly, RDX or a thermite. Hot-wire detonators usually contain LA or LS in contact with the bridge wire. In EBWs, the explosion of the wire starts a detonation without any deflagration (no DDT), as is also the case for a hot-wire detonator

	hot-wire	EBW
1 head	plastic	plastic
2 bridgewire	hi-resist	lo-resist
3 initial pressing	lead azide	PETN
4 output pellet	PETN / RDX	PETN / RDX

Fig. 5.7: Comparison between a common hot-wire and an EBW detonator [V. Varosh, *Propellants Explosives Pyrotech* **1996**, *21*, 150–154].

Exploding Foil Initiators (EFI) have been used since the early 1980s as a safe and reliable method of initiating insensitive secondary explosives in the first stage of an explosive chain. Fig. 5.8 illustrates the major components of an EFI. Tampers can be any rigid dielectric material such as plastic, sapphire etc., and the bridge foil can be made from any conductor (e.g. Cu or Al). The dielectric flyer is usually

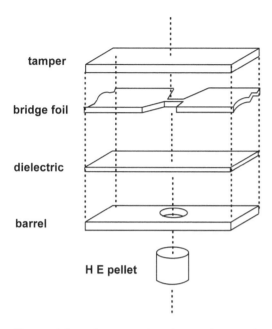

Fig. 5.8: Schematic presentation of an EFI detonator [V. Varosh, *Propellants Explosives Pyrotech* **1996**, *21*, 150–154].

polyimide. The barrel is usually made of a dielectric material, but a conductor can also be used instead. The diameter of the barrel hole equals the length of the bridge foil, approx. 0.2 mm. The above mentioned four components are laminated into one subassembly and clamped against a high density explosive pellet, usually HNS. In operation, a high current explodes the narrow section of the bridge foil, which shears out a disc of a dielectric material which accelerates down the barrel and initiates the high explosive pellet (HNS) by means of kinetic energy.

In summary, both the EBW and EFI detonators, explode because an electric current is heating the conductor which tries to expand, but the conductor is heated

Tab. 5.2: Comparison of the electric characteristics of electric detonators.

	Hot-Wire	EBW	EFI
Operating Current / A	5	500	3000
Voltage / V	20	500	1500
Energy / J	0.2	0.2	0.2
Power / W	1	100,000	3,000,000
Function time / µs	1000	1	0.1

[a] Current and voltage depends on the used laser pumping source (e.g. flash lamp, arc lamp, laser diode)

faster than it can expand. The clear advantage of EFIs is that they are able to initiate HNS, whereas EBWs usually contain thermally less stable PETN or RDX.

Tab. 5.2 summarizes the most important electrical characteristics of EBW and EFI detonators in comparison with conventional hot-wire detonators.

Check your knowledge: See Chapter 14 for Study Questions.

6 Experimental Characterization of Explosives

6.1 Sensitivities

As we have seen in the previous chapter, energetic materials are often initiated using thermal processes. However, the explosion stimuli could also come from mechanical or electrostatic sources. Therefore, it is important to know the exact sensitivities for explosive compounds. The important values that have to be determined are:
1. the impact sensitivity
2. the friction sensitivity
3. the electrostatic sensitivity (ESD) and
4. the thermal sensitivity

Testing the response of solid, liquid or pasty substances to impact, friction and thermal stimuli is required in various standards such as EEC, Official Journal of the European Communities as well as UN Recommendations on the Transport of Dangerous Goods, 13.4.2 Test 3(a)(ii) BAM drop hammer.

The **impact sensitivity** of solid, liquid or gelatinized explosives is determined by using the drophammer method (Tab. 6.1a, Tab. 6.1b). The drophammer essentially consists of a cast steel block with a cast base, a round anvil, a column fixed at the steel block, hardened, smoothed guide bars and the drop weight with retaining and releasing device. The heavy iron block is essential in order to adsorb the shock waves caused by the falling weight. Both guide bars are attached to the column with three brackets. An adjustable metre rule allows an exact measurement of the drop height.

In this test the sample (approx. 40 mg) to be investigated is placed in the plunger assembly, consisting of two steel rollers, a hollow steel collar and a centering ring for fixation (Fig. 6.1). The assembly is placed onto a small anvil. The impact energy (energy = work × distance = mass × acceleration × distance) can be varied by changing the drop-height (approx. 0.1–1 m) and the mass (approx. 0.1–10 kg). The minimum impact energy is determined by looking at which one had at least one out of five samples explode. It is important that while determining the impact sensitivity of crystals the size of the crystals must be given; smaller crystals are generally less impact sensitive than larger crystals of the same substance.

Tab. 6.1a: Impact Sensitivity Values (if more than one value is reported, the literature disagrees on the friction value of this particular compound).

compound	IS (J)	compound	IS (J)
5-ADP, 5-amino-3,4-dinitropyrazole	23	3,6-bis(Tetrazol-5-yl)-1,2,4,5-tetrazine	6
ADNQ, Ammonium dinitroguanidine	10	5,5'-Bis(1H-tetrazolyl)amine	> 30
Silver nitrotetrazolate	1	BNFF-1, 3,4-Bis(4-nitro-1,2,5-oxadiazol-3-yl)-1,2,5-oxadiazole	11.5

Name	N content	Structure
1-Amino-3,6-dinitropyrazolo[4,3-c]pyrazole	7	
BTATz, 3,6-Bis(1H-1,2,3,4-tetrazol-5-yl-amino)-s-tetrazine	14	
3-Amino-4-nitrofurazan	7	
1,5-Diamino-1H-tetrazolium perchlorate	10	
Ammonium nitroformate	2	
1,5-Diaminotetrazolium dinitramide	3	
1,5-Diamino-1H-tetrazolium nitrate	9	
ANTTO, 1-Azido-8-nitrato-2,4,6-trinitro-2,4,6-triazaoctane	4	

Tab. 6.1a (continued)

compound	IS (J)	compound	IS (J)
APX, 1,7-Diamino-1,7-dinitrimino-2,4,6-trinitro-2,4,6-triazaheptane	≥3	Dinitrourea	5
5-Amino-1H-tetrazolium perchlorate	1.5	3(5),4-Dinitropyrazole	20
5-Aminotetrazolium dinitramide	2	Potassium 5-Nitrotetrazolate	10
5-Aminotetrazolium nitrate	10	Potassium tetrazolate	> 100

Diaminoazobistetrazine	1.5	Keto-RDX	5		
5-Nitriminotetrazole	5	BTOx, 2,2,2-Bis(trinitroethyl) Oxalate	10		
Pentanitrobenzene	1.5	TKX-55, 5,5'-Bis(2,4,6-trinitrophenyl)-2,2'-bi(1,3,4-oxadiazole)	5		

Tab. 6.1a (continued)

compound	IS (J)	compound	IS (J)
Sodium 5-Nitrotetrazolate dihydrate	> 30	Bis(3,4,5-trinitropyrazolyl)methane	4
Triaminoguanidinium nitroformate	2	Butanetriol Trinitrate	1
Ammonium Dinitramide	3–5	BuNENA, N-Butyl-N-(2-nitroxyethyl)nitramine	6

Ammonium nitrate	> 40	ε-CL-20	3
Ammonium perchlorate	15	DBX-1, Copper(I) 5-Nitrotetrazolate	0.036
Ammnoum picrate	33	Cyanuric azide	0.15
Azotriazolone	15	DADP, Diacetone diperoxide	5

Tab. 6.1a (continued)

compound	IS (J)	compound	IS (J)
Benzoyl peroxide	5	DAAF, 3,3′-Diamino-4,4′-azoxyfurazan	7
Bis(3,5- dinitro-4-aminopyrazolyl)methane	11	DDNP, 2-Diazonium-4,6-dinitrophenolate	1
Bis(nitramino)triazinone	> 50.5	DEGN, Diethyleneglycol Dinitrate	0.1

6.1 Sensitivities — 195

Name	Value	Name	Value
MAD-X1, Dihydroxylammonium-3,3'-dinitro-5,5'-bis(1,2,4-triazole)-1,1'-diolate	> 40	Diglycerol tetranitrate	1.5
DMDNB, 2,3-Dimethyl-2,3-dinitrobutane	40	TKX-50, Dihydroxylammonium 5,5'-bitetrazole-1,1'-dioxide	20
DNAN, 2,4-Dinitroanisole	> 50	LA	2.5–4

Tab. 6.1a (continued)

compound	IS (J)	compound	IS (J)
DNDA-5, 2,4-Dinitro-2,4-diazapentane	> 29.43	D-Mannitol hexanitrate	1
DNDMOA, Dinitrodimethyloxamide	6	Nitroaminoguanidine	20
DINGU, 1,4-Dinitroglycolurile	5.5	NG, Nitroglycerine	0.2

| Dinitroorthocresol | > 50 | NQ, Nitroguanidine | > 50 |

[structure: Dinitroorthocresol — methylphenol with OH, two NO$_2$ groups]

[structure: NQ — O$_2$N–NH–C(=NH)–NH$_2$]

| TEX, 4,10-Dinitro-2,6,8,12-tetraoxa-4,10-diazaisowurtzitane | 32 | Nitromethane | 40 |

[structure: TEX cage with two NO$_2$ groups on N atoms]

| 2,6-DNT, 2,6-Dinitrotoluene | > 40 | 2-Nitrotoluene | > 40 |

[structure: toluene with two NO$_2$ at 2,6 positions]

[structure: toluene with one NO$_2$ at ortho position]

| DINA, Dioxyethylnitramine Dinitrate | 6 | NTO, 3-Nitro-1,2,4-triazole-5-one | — |

[structure: O$_2$N–O–CH$_2$CH$_2$–N(NO$_2$)–CH$_2$CH$_2$–O–NO$_2$]

[structure: NTO — triazolone with NO$_2$ substituent]

Tab. 6.1a (continued)

compound	IS (J)	compound	IS (J)
DIPEHN, Dipentaerythritol Hexanitrate	4	HMX, Octogen	6.35
ETN, Erythritol Tetranitrate	3	PETN, Pentaerythritol Tetranitrate	2.90
EDD, Ethylenediamine Dinitrate	10	Picramic acid	34

EDNA, Ethylene dinitramine 8 Picric acid

K2DNABT, Potassium 1,1′-Dinitramino-5,5′-bistetrazolate 1

Ethylene glycol dinitrate 0.2 KDNP, Potassium 5,7-dinitro-[2,1,3]-benzoxadiazol-4-olate 3-oxide 10

Ethyltetryl 5

Tab. 6.1a (continued)

compound	IS (J)	compound	IS (J)
Fox-7, 1,1-diamino-2,2-dinitroethene	15–40	PYX, 3,5-dinitro2,6-bispicrylmino pyridine	8
Fox-12, Guanylurea Dinitramide	31	Silver azide	1.18
Glycerol 1,3-Dinitrate	1.5	Silver fulminate	~0.8
Hexamethylenetetramine Dinitrate	15	Styphnic acid	7.4

Name	Value	Structure
Hexanitroazobenzene	8.57	(2,2',4,4',6,6'-hexanitroazobenzene)
Tacot	69	
2,4,6,2',4',6'-Hexanitrobiphenyl	2–20	
TATP, Triacetonetriperoxide	0.3	
HDNP, 2,4,6,2',4',6'-Hexanitrodiphenylamine	7.5	
2,3,4,6-Tetranitroaniline	6	

Tab. 6.1a (continued)

compound	IS (J)	compound	IS (J)
2,4,6,2',4',6'-Hexanitrodiphenylsulfide	6	Sorguyl	2.04
2,4,6,2',4',6'-Hexanitrodiphenylsulfone	3.86	Tetrazene	1

HNO, Hexanitrooxanilide	7.50–14.2	Tetryl	3
HNS, Hexanitrostilbene	5	TAGN, triaminoguanidinium nitrate	4
RDX, Hexogen	7.5	TATB, 1,3,5-Triamino-2,4,6-trinitrobenzene	50
HMTD	2	TATNB, 1,3,5-Triazido-2,4,6-trinitrobenzene	< 4.9

Tab. 6.1a (continued)

compound	IS (J)	compound	IS (J)
Hydrazinium Nitrate $N_2H_5^+\ NO_3^-$	7.4	Trinitroaniline	15
Hydrazinium Nitroformate $N_2H_5^+\ C(NO_2)_3^-$	4	Trinitroanisole	20
Hydrazinium Perchlorate $N_2H_5^+\ ClO_4^-$	2	TNAZ, Trinitroazetidine	6

Isosorbitol Dinitrate	15	Trinitrobenzoic acid	10
		2,4,6-Trinitrocresol	12
Pentryl	4	TNX, trinitroxylene	10

Tab. 6.1a (continued)

compound	IS (J)	compound	IS (J)
Trinitropyridine	4.5–6.5	Uronium Nitrate	> 40
TNPyOx, Trinitropyridine-N-oxide	1.5–3.0		
TNT	15		

(Values taken from: T. M. Klapötke, Energetic Materials Encyclopedia, de Gruyter, Berlin, 2nd edn., 2019).

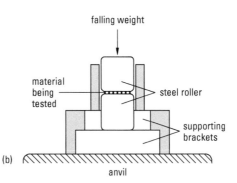

Fig. 6.1: Drophammer in accordance with the BAM (a) and detailed sketch of the drop-hammer test with the two steel rollers, a hollow steel collar and a centring ring for fixation (b).

Fig. 6.2: Friction tester according to BAM regulations.

In order to determine the **friction sensitivity** according to the BAM regulations, the sample is placed onto a rough porcelain plate (25 × 25 × 5 mm). This plate is clamped onto the moving platform of the friction apparatus (Fig. 6.2). The friction force between the moving porcelain plate and a static porcelain peg (10 × 15 mm) (curvature radius 10 mm) causing sample initiation is determined. A set of 13 weights supplied with the instrument allow to achieve friction forces in the range of 0.5–360 N (*N.B.* The force applied (at end points of the lever) is proportional to the ratio of the length of the lever arm measured between the fulcrum and application point of the force applied at each end of the lever. This test method is applicable both for sensitive primary explosives and less sensitive high explosives. The porcelain plate moves back and forth under the porcelain peg. The minimum force of friction is determined. During this process at least one sample out of six ignites, turns black, makes a cracking sound or explodes. As discussed above for the impact sensitivity, before determining the friction sensitivity, the sample should be sieved in order to perform the test on a sample with a uniform and defined crystal size.

Tab. 6.1b: Friction Sensitivity Values (if more than one value is reported, the literature disagrees on the friction value of this particular compound).

compound	FS (N)	compound	FS (N)
ADN	64 → > 350	EDD	> 353
AN	> 360	EDNA	47.4
AP	> 100 → > 363	EGDN	> 360
Ammonium picrate	unaffected in friction pendulum test	Ethyltetryl	> 353
Benzoyl peroxide	120, 240	Fox-7	216, > 350
DNAM	216, > 360	Fox-12	> 352
BTOx	> 360	GAP	> 352
TKX-55	> 360	Guanidinium nitrate	> 353
BTNPM	144	Hexamethylenetetramine dinitrate	240
CL-20	28–96	HNDP	> 353
DBX-1	0.1	hexanitroethane	240
Cyanuric azide	0.1, < 0.5	HNS	192, 240, > 360
DADP	5, 1.75, 2.99, 49 (crude)	RDX	120, 13.73, 250, > 360
DAAF	> 360	HMTD	< 5, 0.63, 0.1, 0.01
LLM-105/ANPZ-O	360	Hydrazinium nitroformate	16 → 36, 12 – → 14, 25
DDNP5 – 22		Hydrazinium perchlorate	> 10
TKX-50	120, 50, 40	Isosorbitol dinitrate	> 160
MAD-X1	> 360	LA	0.1 → 1.0, 10
DNAM	170, > 360	LS	< 1, 0.1, 1.45
DNBF	ignites @ 4.5 J but not @ 0.45 J	Basic LS	< 1
DINGU	20 → 300	Mannitol hexanitrate	30
TEX	161.3, > 360	MF	5.3, 6.48
DNT	> 360	Nitroaminoguanidine	144, 240
DIPEHN	less sensitive than PETN	NC	353
ETN	~50	NG	360, > 353, 112

Tab. 6.1b (continued)

compound	FS (N)	compound	FS (N)
NQ	> 360	Picryl chloride	353
NM	> 360	Trinitrocresol	353
NTO	> 353, > 360	Trinitropyridine	> 353
Octogen/HMX	120 → 240	TNT	353, > 360
PETN	80, 73, 60	Urea nitrate	353, > 360
Picramic acid	> 353	$Ni(N_2H_4)_3(NO_3)_2$	same as silver azide
PA	> 363, explodes if exposed to friction between plates of iron, steel or stone	Ammonium nitroformate	96
AMMO/polyAMMO	40 (monomer)	Aminotetrazolium perchlorate	8
PVN	196	Aminotetrazolium dinitramide	20
KDN	> 360	Aminotetrazolium nitrate	> 324 → > 360
KDNBF	same as LS, 38	H_2BTA	360
KDNP	~10	Diaminotetrazolium perchlorate	60
PYX	280, 360	Diaminotetrazolium dinitramide	5
Silver azide	?	Diaminotetrazolium nitrate	192
Styphnic acid	> 353	dinitrourea	76
TATP	< 5, 1.6, 0.2 (crude)		
Sorguyl	54		
Tetrazene	8.57		
Tetryl	360, 353		
$TAG^+NO_3^-$	120, > 120, 216		
TATB	353, 360		
TNEF	96		
Tirnitroaniline	353		
Trinitroanisole	353		
TNAZ	110, 160, 324		
Trinitrobenzene	353		
Trinitrobenzoic acid	353		

(Values taken from: T. M. Klapötke, Energetic Materials Encyclopedia, de Gruyter, Berlin, 2nd edn., **2019**.

Fig. 6.3: Electrostatic test equipment.

Electrostatic discharge is one of the most frequent and the least characterized cause of accidental explosions of energetic materials. It is therefore imperative in R&D, manufacture, processing, loading or demilitarization to have reliable data on electrostatic spark sensitivities of energetic materials. The **electrostatic sensitivity** (electrostatic discharge sensitivity, ESD) is determined by using an ESD test apparatus. Different spark energies (usually between 0.001 and 20 J (Fig. 6.3) can be applied, by using variable capacitive resistances C (in Farad, F) and loading voltages V (in Volt, V):

$$E = \frac{1}{2}CU^2$$

This is a particularly important test for the safe handling of explosives since the human body can be electrically charged (depending on the type of clothing worn and humidity etc.) which on discharge can cause spark formation. Typical values for the human body are (Fig. 6.3a):

C = 0.0001–0.0004 µF
U = 10000 V
E = 0.005–0.02 J.

For capacitive discharge calculations with C[F] and Q[C], V[V] and E[J], the following can be written:

$$C = \frac{Q}{V} \quad \text{and} \quad E = \frac{1}{2}CV^2 = \frac{1}{2}QV = \frac{Q^2}{2C}$$

Fig. 6.3a: ESD test data reference (according to OZM).

Fig. 6.4: Dependence of the ESD sensitivity on the particle size for hexogen (RDX) and octogen (HMX).

As is shown in Figure 6.4, the ESD values are greatly influenced by the particle size; in these tests the sample must be carefully sieved prior to measuring. The finer the powder of a particular sample is, the higher the ESD sensitivity values are. Usually, compounds with ESD 0.1 J are classified as sensitive, those with ESD 0.1 J as insensitive.

Table 6.1c shows a summary of typical values for the impact, friction and electrostatic sensitivities of some primary and secondary explosives.

In accordance with the UN guidelines for the transport of dangerous/hazardous goods, the substances in Table 6.2 have been classified based on their impact and friction sensitivities.

For the thorough classification of a substance (which is necessary for safe handling), the **thermal stability** must also be determined. The first indication of the thermal stability of energetic compounds or formulations can be obtained from

Tab. 6.1c: Values for the impact, friction and electrostatic sensitivities of some primary and secondary explosives.

explosive	impact sensitivity / J	friction sensitivity / N	ESD sensitivity / J
primary explosive			
$Pb(N_3)_2$	2.5–4	0.1–1	0.005
secondary explosive			
TNT	15	> 353	0.46 – 0.57
RDX	7.5	120	0.15 – 0.20
β-HMX	7.4	120	0.21 – 0.23
PETN	3	60	0.19
NQ	> 49	> 353	0.60
TATB	50	> 353	2.5 – 4.24

Tab. 6.2: UN classification for the transport of dangerous goods.

	impact sensitivity / J	friction sensitivity / N
insensitive	> 40	> 360
less sensitive	35–40	ca. 360
sensitive	4–35	80–360
very sensitive	< 4	10– 80
extremely sensitive		< 10

Fig. 6.5: Schematic construction of a steel sleeve test.

DSC data (see Fig. 2.18b, Ch. 2.5.2). The so-called steel-sleeve test (or Koenen test) is used particularly for assessing the transport safety. In this test, the substance is filled into a steel sleeve (internal diameter: 24 mm, length: 75 mm, wall thickness: 0.5 mm, V = 25 mL) up to a height of 15 mm beneath the top edge and then the sleeve is closed with a nozzle plate. Nozzle plates are available with an orifice of

8 mm

10 mm

Fig. 6.6: Results of steel sleeve tests for a new explosive with 8 and 10 mm holes in the nozzle plate.

Tab. 6.3: UN hazard classification.

UN classification	hazard potential
1.1	mass detonation with possible formation of fragments
1.2 (18)	no mass detonation, whereby most fragments land within the stated distance (in m)
1.2 (12)	
1.2 (08)	
1.2 (04)	
1.3	mass fire
1.4	moderate fire

1.0–20 mm. The steel sleeve is attached to the nozzle plate with a two-part threaded socket. It is then heated simultaneously by four burners. By varying the orifice diameters, the critical diameter can be determined. It is the diameter at which the pressure increase on burning and the subsequent explosion destroys the sleeve into at least four small splinters. Figure 6.5 schematically shows the construction of a steel sleeve test, and Figure 6.6 shows the results of some tests.

According to UN guidelines, all energetic materials can be divided into four main categories based on their potential danger/hazard. These UN classifications are summarized in Table 6.3.

6.2 Long-Term Stabilities

Simple **DSC measurements** (Differential Scanning Calorimetry) are an ideal method to quickly investigate the thermal stability of a substance while only using very small quantities of substance, which is particularly important when dealing with dangerous samples. During the DSC measurements, heating rates of 5 °C min^{-1} are commonly used to give an initial indication of the thermal stability and decomposition temperature (see Ch. 2.5.2). However, it is necessary to research the long-term stability by using additional measurements. In this context, the so-called isotherm safety calorimeter can be used for example. They can investigate the thermal stability of a substance at a certain temperature (e.g. 30 °C below the decomposition temperature) for 48 hours or longer. In isoperibolic long-term evaluations (approximately isothermal; heat exchange), the stability of substances or mixtures of substances in a sample cell (RADEX cell) can be investigated, with or without an additional pressure sensor.

Figure 6.7 shows an example of DSC plots of a highly energetic material (ammonium 1-methyl-5-nitriminotetrazolate) using different heat rates, while Figure 6.8 shows the long-term stability of various other compounds in a **FlexyTSC** safety calorimeter.

For the unambiguous determination of the thermal stability of a substance or mixture of substances, it is generally necessary to determine the heat flow rate P (in µW g^{-1}). The substance can be considered to be thermally stable at a certain temperature, if the heat flow corresponds to a value not higher than 300 µW g^{-1} over a time period of 7 days. Figure 6.9 shows an example of such an investigation for diaminotetrazolium nitrate (HDAT nitrate, Fig. 6.10) at a temperature of 89 °C. Here it can be observed that the substance is thermally stable at this temperature.

The so-called **Kneisl-Test** is a simple method which can be used for the estimation of the long-term thermal stability, particularly for primary explosives. A de-

Fig. 6.7: DSC plot for ammonium 1-methyl-5-nitriminotetrazolate using different heat rates (up = exo).

Fig. 6.8: Isothermal long-term measurement (48 h) for TTD, BTAT and BTHC without decomposition.

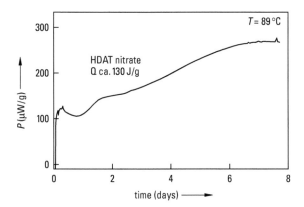

Fig. 6.9: Heat flow for HDAT nitrate at 89 °C.

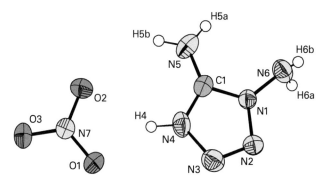

Fig. 6.10: Molecular structure of HDAT nitrate.

fined quantity (e.g. 100 mg) of the substance under investigation is air sealed in a glass ampoule. This ampoule is then placed in an oven for 100 h at the temperature for which thermal stability should be determined. The ampoule is then opened, the remaining mass of substance weighed and any gaseous decomposed products which may have been formed are analyzed (IR, MS, GC-MS). In order to "pass" the test, the mass lost due to decomposition should not exceed 2% of the original sample mass.

6.3 Insensitive Munitions

The term insensitive munition (IM, see STANAG 4439) is used to describe munition which is particularly safe to handle and which is difficult to initiate accidentally, but at the same time has the power and reliability to fulfil the requirements necessary to complete the mission. The insensitivity of munition can be tested and classified into six categories (Tab. 6.4).

Six tests are performed on each substance in order to classify them into one of these categories. The results of these tests can range from "no reaction" to "complete detonation". Those tests are briefly summarized in Table 6.5.

Tab. 6.4: Categories for insensitive munition.

	German	English
NR	Keine Reaktion	No reaction
V	Abbrand	Burning
IV	Deflagration	Deflagration
III	Explosion	Explosion
II	Teilweise Detonation	Partial detonation
I	Vollständige Detonation	Detonation

Fig. 6.11: Simple construction of an FCO test (fast cook-off test).

Tab. 6.5: IM tests.

scenario	IM test	test procedure	allowed response for IM (see Tab. 6.4)
small arms attacks	BI	attack (bullet impact) with three 12.7 mm bullets with 850 m s^{-1}	V – burning
fragmenting munitions attacks	FI	attack with five 16–250 g-fragments with 2530 m s^{-1}	V – burning
magazine, store, plane, ship or truck in fuel fire	FCO	fast cook-off-test, simulation of a fuel fire (see Fig. 6.11)	V – burning
fire in adjacent magazine, store or vehicle	SCO	slow cook-off-test (SCO) with a heat rate of 3.3 °C / h	V – burning
shaped charge weapon attack	SCI	impact of a shaped charge between 50 und 62 mm	III – no detonation
detonation in magazine, plane, truck etc.	SR	reaction to detonation of a neighboring charge (sympathetic reaction)	III – no detonation

Since this area is so important, NATO has founded a Munitions Safety Information Analysis Center (MSIAC).

6.4 Gap Test

In the so-called gap test the sensitivity towards shockwaves is determined [36]. The apparatus for the gap test (Fig. 6.12) is used to measure the ability of a material to propagate a detonation by subjecting it to a detonating booster charge under confinement in a polycarbonate tube. The gap test enables the determination of the minimum shock wave pressure that can cause complete detonation of the tested explosive. The explosive under investigation is subjected to the action of the shock wave of a known pressure. Such a wave is generated by a booster and a shock wave pressure attenuator. Whether or not the shock wave causes the complete detonation of the explosive can be concluded on the basis of the mechanical effects produced after the detonation of the explosive: hole cutting in a steel plate, dent depth in a witness steel block or compression of a copper cylinder. In this test, the gap medium (usually water) stops flying particles and direct heat transmission, serving as a heat filter. Consequently, the shock wave is the only energy transmitted to the explosive. If other factors are constant, the possibility of the transmission of the detonation is determined mostly by both the sensitivity of the acceptor

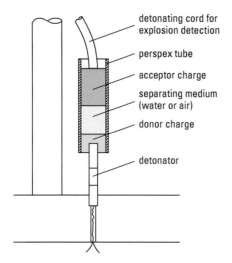

Fig. 6.12: Schematic set-up of a gap test.

charge and by the initiating strength of the donor charge. The gap test method is based on the determination of the distance between the donor and the acceptor charge of given masses at which transmission or failure of detonation occurs.

Figure 6.12 schematically shows the set-up of the gap test. Typical values for the donor charge are 10 g RDX (with 0.6 g PETN detonator). The medium is water (or alternatively air), the confinement is a Plexiglass tube of approx. 20 mm diameter. With such an experimental set-up, values for the initiation limits are approx. 25–27 mm for PA and 21–23 mm for Tetryl. In contrast, the value for TNT (less impact sensitive) is only 7–8 mm. The extremely insensitive compound FOX-12 only shows a response at 2 mm under similar conditions.

6.5 Classification

If new energetic materials are synthesized in a research laboratory and are sent to other institutes for further analysis, various stability tests have to be carried out, so that the materials can initially be classified (IHC, **interim hazard classification**) as being at least hazard class "1.1 D" (see Ch. 6.1). The requirements to "pass" the necessary tests (UN 3a–UN 3d) are summarized in Table 6.6. A positive test results (+) means that the substance did *not* pass the test.

If a sample does not pass test 3c, but is shown to be thermally stable using a DSC, test 4a must additionally be performed. If the substance is thermally unstable, it can not be transported. In the case of a sample not passing tests 3a, 3b or 3c, test 4b(ii) must be performed (Tab. 6.7). A preliminary transportation certificate can then only be issued if the sample passes tests 3a–3d or 4a and 4b(ii).

Tab. 6.6: Required UN tests 3a–3d to obtain an IHC (interim hazard classification).

UN test	test	conditions	+
UN 3a	impact sensitivity	5 tests; positive, if 50 % of the tests give positive results	≤ 3.5 J
UN 3b	friction sensitivity	5 tests; positive, if one out of five test is positive	≤ 184 N, visible sparks, visible flames, explosion heard, clear crackling noise
UN 3c	thermal stability	75 °C, 48 h	visible examination: – color change, – explosion, – ignition (burning)
			weight loss (except adsorbed moisture)
			RADEX: self-heating max. 3 °C (RADEX, see Ch. 6.2)
UN 3d	small-scale-burning test (see Fig. 6.13)	sawdust soaked in kerosene, not confined	explosion or detonation

Fig. 6.13: Experimental set-up for a small-scale-burning test (UN 3d).

Tab. 6.7: UN tests 4a and 4b(ii) to establish an IHC (interim hazard classification).

UN test	test	conditions	+
UN 4a	thermal stability of packed and unpacked energetic materials	75 °C, 48 h	visible examination: – color change, – explosion, – ignition (burning)
			weight loss (excluding adsorbed moisture)
			RADEX: Self-heating max. 3 °C (RADEX, see Ch. 6.2)
UN 4b(ii)	free fall from 12 m height	12 m free fall (see Fig. 6.14)	fire, explosion or detonation (damage of the packaging does not count as a "fail")

Fig. 6.14: Set-up for the free fall test (UN 4b(ii)).

6.6 Trauzl Test

The lead block test (or Trauzl test) is a method which can be used to determine the strength of an explosive. To perform this test, 10 g of the explosive under investigation is wrapped in tin foil and placed into the central borehole (125 mm deep, 25 mm in diameter) of a massive lead cylinder (20 cm in diameter, 20 cm long). A blasting cap with an electrical primer is introduced into the center of the explosive charge, and the remaining free space is filled with quartz sand (Fig. 6.15 and 6.16).

Fig. 6.15: Trauzl (lead block) test: 1 sand stemming, 2 test sample, 3 lead block, V original volume of the cavity (61 cm³).

Fig. 6.16: Lead block for Trauzl test before (left) and after (right) the test.

After the explosion, the volume of the resulting cavity is determined by filling with water. A volume of 61 cm³ (which is the original volume of the cavity) is deducted from the result which is obtained. Typical lead block excavation values are (per 10 g of explosive, given in cm³):

PETN   520
RDX   480
TNT   300
AP   190
AN   180

Tab. 6.8: Values of V+ and V− which are used to account for the presence of certain molecular moieties.

Molecular moiety	V+ (cm^{-3})	V− (cm^{-3})
R-(ONO$_2$)$_x$ x = 1, 2	1.0	
R-(ONO$_2$)$_x$ x ≥ 3	0.5	
R-(NNO$_2$)$_x$ x = 1, 2	0.5	
Ph-(NO$_2$)$_x$ x ≤ 2	0.8	
H$_2$N-C(=O)-NH-R		1.0
phenyl-(OH)$_x$ or phenyl-(ONH$_4$)$_x$		0.5 x
phenyl-(NH$_2$)$_x$ or phenyl-(NHR)$_x$		0.4 x
phenyl-(OR)$_x$		0.2 x
phenyl-(COOH)$_x$		0.9 x

Quite recently, Keshavarz et al. introduced a new model for the prediction of the strength – and therefore power index – of pure explosives and mixtures containing C$_a$H$_b$N$_c$O$_d$ high energy materials, on the basis of volume expansions in the Trauzl lead block [M. Kamalvand, M. H. Keshavarz, M. Jafari, *Propellants, Explos. Pyrotech.* **2015**, *40*, 551–557; M. Jafari, M. Kamalvand, M. H. Keshavarz, S. Farrashi, *Z. Anorg. Allg. Chem.*, in press].

The optimized correlation has the following form:

$$\Delta V_{Trauzl} = 578.8 - 175.4(a/d) - 88.88(b/d) + 140.3(V+) - 122.4(V-) \quad (1)$$

The V+ and V− are raising and lowering correction functions, respectively.

Nitrate and nitramine groups are two important substituents in energetic materials. The energy released during the detonation of common explosives is mainly derived from the oxidation of fuel atoms, i.e. carbon and hydrogen, by oxygen atoms which can be provided by −NO$_2$ groups during the initial stage of a detonation reaction. In contrast, the oxygen of carbonyl and hydroxyl groups cannot lead to a net increase in the energy, because the energies of their cleavage and recombination are equivalent. In other words, nitrate and nitramine groups can increase the strength of a high energy molecule by increasing the value of Q_{det}. The predicted ΔV_{Trauzl} on the basis of a/d and b/d was also underestimated in the case of aliphatic energetic compounds, therefore the raising parameter, i.e. V+, for correcting the predicted values of ΔV_{Trauzl} must be introduced. Molecular moieties as well as their corresponding values for V+ are given in Table 6.8.

Strong intermolecular attractions – especially hydrogen bonding – can decrease the energy content of energetic compounds and increase their thermodynamic stability. Some polar groups, such as −COOH, −OH, and NH$_2$, in aromatic energetic molecules decrease the value of Q_{det}. Similarly, the strength of these energetic materials is overestimated on the basis of a/d and b/d. Thus, a lowering

Tab. 6.9: Experimentally obtained and calculated excavation values for some pure energetic compounds.

explosive	exptl. ΔV_{Trauzl} (cm³ / 10g)	calcd. ΔV_{Trauzl} (cm³ / 10g)
NQ	302	313
NG	520	541
HMX	428	472
PETN	520	517
TATB	175	168
TNT	300	300
HNS	301	330
RDX	480	472

parameter, $V-$, should be included in Eq. (1) for correcting the predicted ΔV_{Trauzl}. Another molecular fragment, which can lead to the formation of significant intermolecular hydrogen bonding, is the (NH_2)-CO-(NH)-fragment. The lowering parameter, $V-$, is assigned to account for the presence of this fragment in energetic molecules. Table 6.8 contains the functional groups and molecular moieties, which can be used to determine the values of $V-$.

Tab. 6.9 gives the experimentally obtained and calculated excavation values for some pure energetic compounds.

6.7 Aging of Energetic Materials

The aging phenomenon of energetic materials is a very important aspect, since it aging can result in the energetic material exhibiting different performance properties such as the heat of reaction and reaction kinetic parameters. Obviously, to be of practical importance and usage, energetic materials must show an extremely slow decomposition rate under ambient conditions. However, this stability has the consequence of making it essentially impossible to investigate the effects of aging on an energetic material using experimental methods over a timespan of decades. In order to overcome this problem, theoretical prediction methods such as the Arrhenius approach can be used. The Arrhenius approach predicts the shelf-life of a chosen energetic material based on the ratio of reaction rates between the storage temperature (T_1) and the elevated temperature (T_2) which is selected as the aging temperature and at which the accelerated aging experiment is performed. The Kissinger evaluation uses the Arrhenius approach as shown below:

$$k(T) = Z \cdot \exp\left(-\frac{Ea}{R \cdot T}\right)$$ k is the reaction rate constant which describes the Arrhenius T dependence

$$\frac{d\alpha(T)}{dt} = k(T) \cdot f(\alpha) \quad \text{which corresponds to the general rate expression}$$

in which the nth order reaction $f(\alpha)$ is given as follows:

$$f(\alpha) = (1-\alpha)^n$$

The degree of reaction (conversion α) is a constant value at the peak maximum in a set of DSC curves and it is independent of the linear heating rate h (see Fig. 6.17). Therefore, if the value of conversion α is obtained and the second derivative thereof calculated, then an equation can be generated which relates the heating rate h, peak maximum temperature T_P, conversion at peak maximum α_P and the Arrhenius activation parameters Z and E_a.

$$\frac{d\left(\frac{d\alpha(T)}{dt}\right)}{dt} = \frac{dk(T)}{dt} \cdot f(\alpha) + k(T) \cdot \frac{df(\alpha)}{dt} \cdot \frac{d\alpha(T)}{dt}$$

$$\frac{d\left(\frac{d\alpha(T_P)}{dt}\right)}{dt} = \frac{dk(T_P)}{dt} \cdot f(\alpha(T_P)) + k(T_P) \cdot \frac{df(\alpha(T_P))}{dt} \cdot \frac{d\alpha(T_P)}{dt} = 0$$

$$\frac{dk(T_P)}{dt} \cdot f(\alpha(T_P)) = -k(T_P) \cdot \frac{df(\alpha(T_P))}{dt} \cdot \frac{d\alpha(T_P)}{dt}$$

$$Z \cdot \exp\left(-\frac{E_a}{R \cdot T_P}\right) \cdot \frac{E_a}{R \cdot T_P^2} \cdot \frac{dT_P}{dt} \cdot f(\alpha(T_P)) = -Z \cdot \exp\left(-\frac{E_a}{R \cdot T_P}\right) \cdot \frac{df(\alpha(T_P))}{dt} \cdot \frac{d\alpha(T_P)}{dt}$$

$$k(T_P) \cdot \frac{E_a}{R \cdot T_P^2} \cdot \frac{dT_P}{dt} \cdot f(\alpha(T_P)) = -Z \cdot \exp\left(-\frac{E_a}{R \cdot T_P}\right) \cdot \frac{df(\alpha(T_P))}{dt} \cdot k(T_P) \cdot f(\alpha(T_P))$$

$$\frac{E_a}{R \cdot T_P^2} \cdot h = -Z \cdot \exp\left(-\frac{E_a}{R \cdot T_P}\right) \cdot \frac{df(\alpha(T_P))}{dt}$$

$$\frac{h}{T_P^2} = -Z \cdot \frac{E_a}{R} \cdot \frac{df(\alpha(T_P))}{dt} \cdot \exp\left(-\frac{E_a}{R \cdot T_P}\right)$$

$$\ln\left(\frac{h}{T_P^2}\right) = \ln\left(-Z \cdot \frac{E_a}{R} \cdot \frac{df(\alpha(T_P))}{dt}\right) - \frac{E_a}{R \cdot T_P}$$

$$\ln\left(\frac{h}{T_P^2}\right) = \ln(Ak) - \frac{E_a}{R \cdot T_P} \quad \text{general Kissinger equation}$$

$$Ak = -Z \cdot \frac{E_a}{R} \cdot \frac{df(\alpha(T_P))}{dt}$$

Fig. 6.17: Non-isothermal DSC thermogram for (a) 97.5 % RDX and (b) 95 % HMX.

For a reaction of nth order:

$$\ln\left(\frac{h}{T_P^2}\right) = \ln\left(n \cdot (1-\alpha_P)^{(n-1)} \cdot \frac{R}{Ea} \cdot Z\right) - \frac{Ea}{R \cdot T_P}$$

$$Ak = n \cdot (1-\alpha_P)^{(n-1)} \cdot \frac{R}{Ea} \cdot Z$$

$$Zk = n \cdot (1-\alpha_P)^{(n-1)} \cdot Z$$

For a reaction of first order, n = 1:

$$\ln\left(\frac{h}{T_P^2}\right) = \ln\left(\frac{R}{Ea} \cdot Z\right) - \frac{Ea}{R \cdot T_P}$$

$$Ak = \frac{R}{Ea} \cdot Z$$

$$Zk = Z$$

Procedure for prediction with the case first order decomposition of the substance:

If the values of $\ln\left(\frac{h}{T_P^2}\right)$ are plotted against $1/T_P$ in 1/Kelvin, the activation energy E_a can be obtained from the slope of the plot and the ordinate section (where the slope of the graph intersects the y axis) gives $\ln\left(\frac{R}{Ea} \cdot Z\right)$, from which Z can be calculated.

Via Arrhenius the reaction rate constants at wished temperatures are available.

Via the ratios of rate constants at different temperatures, the times can be calculated, reaction rate constant and time are reciprocal connected.

From the principle of determination this is valid for 50 % conversion of the substance.

$$\frac{k(T_1)}{k(T_2)} = \frac{Z \cdot \exp\left(-\frac{Ea}{R \cdot T_1}\right)}{Z \cdot \exp\left(-\frac{Ea}{R \cdot T_2}\right)} = \frac{t(T_2)}{t(T_1)}$$

$$\boxed{t(T_2) = t(T_1) \cdot \exp\left(-\frac{Ea}{R} \cdot \left(\frac{1}{T_1} - \frac{1}{T_2}\right)\right)}$$

Whereby t_1 and t_2 are the shelf-lives at temperatures T_1 and T_2, respectively, and k_1 and k_2 are the reaction rates at temperatures T_1 and T_2, respectively.

A word of caution has to be given though, since it has been reported in the literature that this Arrhenius approach generally over-predicts the shelf-life of an energetic material.

A general strategy which can be used in order to predict the shelf-life of an energetic material (or a formulation of energetic materials) involves the following steps:

- Artificial aging of the energetic material at elevated temperatures for defined periods of time
- Determination of various energetic material properties after the artificial ageing process
- Extrapolation of these results to estimate the effect of exposure for longer periods of time at lower temperatures based on kinetic models, e.g.

$$k = A \cdot e^{-\frac{E_A}{kT}}$$

$$F = e^{\left(\frac{E_A}{R} \cdot \frac{\Delta T_F}{T_T^2}\right)}$$

$$t_E = t_T \cdot F^{\frac{T_T - T_E}{dT_F}} \cdot \frac{1}{365.25}$$

where: T_E = service temperature
T_T = test temperature
F = factor by which the rate changes for every ΔT_F change in temperature
t_E = estimated lifetime (in years)
t_T = duration of aging test (in days)

The most important test methods used for assessing the chemical stability of high explosives (HE) and double-base propellants, as well as for determining compatibility are summarized below:

Chemical stability
Main test methods for HEs and primary explosives:
- Vacuum stability test (VST)
- Mass Loss
- Heat Flow Calorimetry (HFC)

Main test methods for double base propellants:
- Stabilizer depletion
- Bergman-Junk test
- Abel test

Compatibility
- Vacuum Stability Test (VST)
- Heat Flow Calorimetry (HFC)
- DSC and TGA
- Stabilizer depletion

In the **VST**, usually 2.5 g of sample are heated to 100 °C in vacuum and kept at this temperature for 40 hrs. The volume of gas evolved by 1 g of the sample is calculated at the end of the test. The thresholds are usually defined as ≤ 2 mL/g for HE, propellants and pyrotechnics. Typical values for RDX are 0.2 mL/g and for Comp-B 0.04 mL/g.

In the **HFC**, the heat flow from a decomposing explosive at a constant temperature is determined over a given period of time. The total heat generation is determined and must be below a specific threshold value in order for storage safety to be ensured, e.g. for 10 years at 25 °C:

temperature / °C	test time / days	threshold / µW g^{-1}
60	123	9.8
70	34.8	34.5
80	10.6	114
90	3.43	350

In the **Mass Loss** test, 2–5 g of a sample are heated for a specific time. The sample mass is measured in intervals until autocatalytic reaction or 3% mass loss is reached. The typical criterion which is used to predict that an energetic material will be stable during storage for 15 years at 25 °C is a mass loss of < 3% over a period of 16.2 days at 80 °C.

Fig. 6.18 demonstrates the behavior of ADN in the mass loss test.

Fig. 6.18: Mass loss of ADN (isothermal TG curve, vacuum, 52 °C), y-axis weight in %, x-axis: time in s.

In the **stabilizer depletion** test, the stabilizer gets extracted and analysed by GC or HPLC.

In the **Bergman-Junk** (BeJu™ by OZM) test, the test apparatus is used to determine the chemical stability of firearm and rocket propellants, and does so by the qualitative analysis of gaseous products evolved during the thermal decomposition of a sample. The sample is heated in a heavy, aluminum heating block at 120 or 132 °C. Gaseous products evolved during the thermal decomposition are absorbed into a glass adaptor filled with water. The amount of gaseous products is then determined by quantitative volumetric analysis of the resulting aqueous solution. The typical sample mass is 5 g, and the temperature ranges of the test are between 120–132 °C with a testing period of up to 6 hours.

The **Abel test** (ABT™ by OZM) apparatus is traditionally used to determine the thermal stability of firearm and rocket propellants by measuring the time interval needed to evolve nitrogen oxide gasses from the heated sample. Iodide Starch Indicator Papers are used to detect the presence of decomposition gases. The test temperature is typically set between 60–82 °C. Nitrate esters decompose by evolving nitrogen oxide gases which are detected by the standardized color change of the Iodide Starch Indicator Paper. The period of time between inserting the sample into the heating block and the color change of the indicator paper occurring is the test time. The typical sample mass is 1–3 g and the temperature used ranges between 60–82 °C with a testing period of up to 1 hour.

The **chemical compatibility** of an energetic material can be tested using the VST, HFC or DSC and/or TGA methods. These measurements are performed on 1:1 mixtures of the energetic material and material against which the compatibility is tested, and a heating rate of 2 K/min is used. The following table summarizes the compatibility criteria:

	sample mass	HR, T_{max}	change in	compatible	doubtful	incompatible
DSC	1–10 mg	2 K/min, 700 °C	peak max.	$\Delta < 4\,°C$	$4\,°C < \Delta < 20\,°C$	$\Delta > 20\,°C$
TGA	1–10 mg	2 K, 1100 °C	weight loss	$\Delta < 4\,°C$	$4\,°C < \Delta < 20\,°C$	$\Delta > 20\,°C$

Organic Explosives

When considering only pure organic explosives, it can be generalized that the chemical aging frequently begins with the hemolysis of the weakest bond within the molecule (unimolecular decay), which is then subsequently followed by and accompanied by self-accelerating parallel reactions which involve the decomposition products which are formed. Therefore, the activation energy corresponds approximately to the bond energy of the weakest bond in the molecule and is consequently a good indicator of the thermal stability of a specific explosive. It has been shown that explosives of this type can be very approximately categorized into two groups: those with an activation energy to decomposition of above 170 kJ/mol and those with a value below 155 kJ/mol. The former is found to be stable for thousands of years under ambient conditions, whereas the latter show limited chemical stability, and the stability of these compounds must be more closely investigated. Table 6.10 gives an overview of many commonly used explosives including aromatic and aliphatic nitro compounds, secondary nitramines, and organic azides, which can be categorized as being very stable, whereas aliphatic nitrate esters show a much lower chemical stability, as do organic peroxides, perchlorates, and nitrogen-rich heterocycles.

Tab. 6.10: Activation energies for decomposition and the corresponding bond which undergoes hemolytic cleavage for various classes of organic explosives.

substance class	typical representative	functional group	homolytic bond cleavage	E_A/kJ mol^{-1}	chemical stability
aromatic nitro comp.	TNT, PA, TATB, HNS	AR–NO$_2$	C–NO$_2$	190–290	excellent
aliphatic nitro	nitromethane	R–NO$_2$	C–NO$_2$	230–260	excellent
aliphatic nitramines	RDX, HMX, NQ	R$_2$CN–NO$_2$	N–NO$_2$	170–200	excellent
aliphatic azides	GAP, Et–N$_3$	R$_3$C–N$_3$	N–N$_2$	165–170	mostly good
aliphatic nitrate esters	NC, NG, PETN	R–O–NO$_2$	O–NO$_2$	155–190	poor: NC, NG good: PETN

Inorganic Explosives

There are also many explosive compounds which are inorganic compounds (*e.g.* azides, acetylides, and fulminates of lead, silver, and mercury) and the chemical stabilities of these classes of compounds range from being chemically very stable (lead azide) to very unstable (mercury fulminate reacts with metals in a moist atmosphere). TNT is one of the rare examples of explosives which is used as a pure compound and also in composite explosives (blends). These blends may consist of the explosive compound as well as other explosive compound, or with inert materials such as binders. Even in such composite explosives, the chemical aging behavior of the final product is still largely dominated by the aging behavior of the main explosive compound.

High explosives retain the excellent chemically stability of their major components (aromatic nitro compounds and/or secondary nitramines).

Modern and currently used polymer-bonded rocket propellants only contain stable explosive components (*e.g.* nitramines and/or ammonium nitrate and aluminum), however, most gun propellants and also many rocket propellants, show a considerably lower chemical stability since they are based on aliphatic nitric acid esters (nitrate esters) such as nitrocellulose (cellulose nitrates with a nitrogen content between 12.2 and 13.4 mass %). In contrast, pyrotechnics – which are mixtures of oxidizers and fuels – show significantly different chemical aging behavior, in which the chemical degradation is mainly the result of oxidation and hydrolysis of the fuel (*e.g.* hydrolysis/corrosion of Mg by moisture, or oxidation of Al, Mg and Ti by oxygen).

Propellants

In this section, the problem of chemical aging with respect to propellants will be considered in more detail, since it is or particular importance for this class of compounds. As was mentioned above, chemical aging of propellants begins with homolytic breaking of the weak $O-NO_2$ bond of the aliphatic nitrate esters (*e.g.* nitrocellulose NC and nitroglycerine NG), which results in the formation of nitrogen dioxide and the corresponding alkoxyl radical:

$$R-O-NO \rightarrow R-O\cdot + \cdot NO_2 \qquad (1)$$

The free radicals which are generated in this process are highly reactive and immediately undergo consecutive reactions with nearby nitrate ester molecules. In case of NC, the backbone alkoxyl radical $R.O\cdot$ also performs internal destabilization reaction by splitting off small stable molecules:

$$\begin{array}{l} R-O-NO_2 + R-O\cdot \\ \\ R-O-NO_2 + \cdot NO_2 \end{array} \searrow \begin{array}{l} N_2, N_2O, \cdot NO, \cdot NO_2 \\ H_2O, H_2, CO_2, CO \\ C_2H_2O_4, \text{other fragments} \end{array} \qquad (2)$$

A further main decomposition pathway is the neutral to acidic hydrolysis of the nitrate esters. This reaction is catalyzed by moisture and residual acids, H_3O^+. Residual acids may be present if they were not completely removed after the nitrate ester synthesis, or as a result of the formation of water and acids formed during decomposition:

$$R-O-NO_2 + 2\,H_2O\,(H_3O^+) \rightarrow R-OH + HNO_3 \qquad (3)$$

Furthermore, another possible decomposition reaction which must be mentioned is the enhanced hydrolysis caused by the interaction between nitrate groups and N_2O_4. This proceeds by the initial conversion of the nitrate group to the nitrite group R—O—NO, followed by the subsequent hydrolysis of the O—NO bond. This reaction was found to occur with nitrogylcerine and a significantly lower activation energy of 71 kJ/mol was found for hydrolysis of the O—NO bond in the nitrite group compared to 100 kJ/mol for the hydrolysis of the O—NO_2 bond of the nitrate group. Therefore, this means that at lower temperature, this pathway can be a dominant decomposition route.

Some reaction products of equations (1) and (2) undergo further reaction in presence of moisture and oxygen as is shown by the following equations:

$$2 \cdot NO + O_2 \rightarrow 2 \cdot NO_2 \rightarrow N_2O_4 \tag{4}$$

$$\cdot NO + \cdot NO_2 + H_2O \rightarrow 2\, HNO_2 \tag{5}$$

$$3 \cdot NO_2 + H_2O \rightarrow 2\, HNO_3 + \cdot NO \tag{6}$$

The radicals and acids formed by reactions (1)–(6) strongly self-accelerate both radical (2) and hydrolytic (3) decomposition of the nitrate esters. This accelerating action is called autocatalysis. Although the primary homolytic reaction (1) cannot be suppressed, the consecutive reactions (2)–(3) can be slowed down nearly to zero by the removal or binding of any acids, nitric oxides, and water from the system. This strategy is used for the stabilization of nitric esters-based explosives. Most commonly, aromatic amines (e.g. diphenylamine, 2-nitro-diphenylamine, *p*-nitro- N-ethylaniline, *p*-nitro-N-methylaniline) and urea derivatives (e.g. Akardite-II (1-methyl-3,3-diphenylurea) and ethyl centralite (1,3-diethyl-1,3-diphenylurea)) are used as stabilizers, since both are capable of binding nitric acid and nitric oxides. A simplified and more phenomenological picture of the chemical aging of propellants is given in Fig. 6.19.

---

i Check your knowledge: See Chapter 14 for Study Questions.

Literatur

Y. Kim, A. Ambekar, J. J. Yoh, *J Therm Anal Calorim.*, **2018**, *133*, 737–744.

B. Vogelsanger, *Chimia*, **2004**, *58*, 401–408.

M. A. Bohn, *International Annual Conference of ICT*, **1999**, *30*, 16-1–16-45.

M. A. Bohn, *Proceedings of the Workshop on the Microcalorimetry of Energetic Materials*, Leeds, UK, April 7th–9th **1997**, G1–G31.

M. A. Bohn, *International Annual Conference of ICT*, **1997**, *28*, 109-1–109-46.

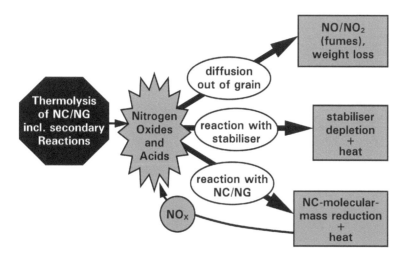

Fig. 6.19: Simplified scheme to illustrate the chemical aging of propellants. In the first step of aging, homolytic breaking of the nitrate ester bonds of nitrocellulose (NC) and nitroglycerine (NG) occurs. This produces highly reactive nitrogen oxides and acids which should be trapped by the stabilizer during consecutive reactions with moisture and oxygen, resulting in stabilizer consumption and evolution of heat. Any nitrogen oxides and acids which are not captured quickly enough can undergo the following fates: either they react further with the nitrate ester component(s) of the propellant (which reduces the molecular mass of the NC and produces more heat and nitrogen oxides, which then accelerates aging of the propellant further), or diffuse out of the propellant (which results in the appearance of 'red fumes'; the leaving nitrogen oxides, together with the other escaping backbone decomposition products such as H_2O, CO_2, CO, and N_2, also cause weight loss).
(Reproduced with kind permission of CHIMIA from *Chimia*, **2004**, *58*, 401–408.)

7 Special Aspects of Explosives

7.1 Shaped Charges

A **shaped charge** is an explosive charge which is shaped to focus the effect of the explosive's energy [37]. In contrast to an explosion of a solid cylindrical explosive charge which creates a relatively wide but only moderately deep crater (Fig. 7.1), a shaped charge consisting of a solid cylinder of explosive (with rotational symmetry) with a conical hollow in one end and a central detonator causes a much deeper though narrower crater (Fig. 7.1). A shaped charge with a metal-lined conical cavity (e.g. Cu or W liner) creates an even deeper and narrower crater (Fig. 7.1). The extremely high pressure generated by the detonation of the explosive drives the liner contained within the hollow cavity inward whereby it collapses upon its central axis. The resulting collision forms and projects a high-velocity jet of metal forward along the axis (see Fig. 7.2). Most of the jet material originates from the innermost

Fig. 7.1: Conventional cylindrical explosive charge, shaped charge, shaped charge with metal liner.
(This diagram is reproduced with slight modification from the original of Prof. Dr. Manfred Held, who is herewith thanked for his permission to reproduce this.)

Fig. 7.2: Jet formation from a metal-lined shaped charge.
(This diagram is reproduced with slight modification from the original of Prof. Dr. Manfred Held, who is herewith thanked for his permission to reproduce this.)

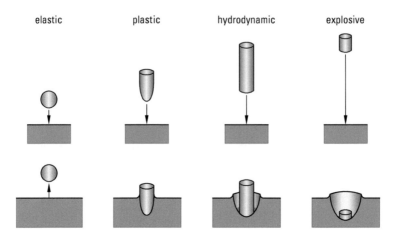

Fig. 7.3: Penetration behavior: elastic, plastic, hydrodynamic, explosive.
(This diagram is reproduced with slight modification from the original of Prof. Dr. Manfred Held, who is herewith thanked for his permission to reproduce this.)

layer of the liner and remains below the melting point solid in the jet. A typical modern lined shaped charge can penetrate armor steel to a depth of seven or more times the diameter of the charge's cone.

Ideally, the liner consists of a malleable metal of high density. For this reason copper is most commonly used. Tungsten and tantalum are also in use because of their extremely high densities, while depleted uranium, though in use for kinetic energy munition (see below) has not found widespread application as liner material for shaped charges. The pyrophoric effects of tantalum (and DU) further enhance the damage caused by the lined shaped charge after penetrating the armor steel.

The velocity of the jet depends on the brisance of the high explosive used and the internal apex angle of the metal liner. The smaller the angle is, the higher is the velocity of the jet. But very small angles can result in jet bifurcation or even failure of the jet to form at all. For this reason a "good" compromise needs to be found (usually 40–90°). In any case, most of the jet formed moves at hypersonic speed with its tip reaching speeds of up to 7–12 km s^{-1}, the tail at a lower velocity of 1–3 km s^{-1} and the slug at still lower velocity (< 1 km s^{-1}). The high velocity of the jet combined with the high density of the material of the metal liner gives the jet a very high kinetic energy. When the jet hits the target, a very high pressure is formed as a result. A typical jet velocity of 10 km s^{-1} results in a pressure of about 200 GPa. The penetration process therefore generates enormous pressures, and therefore it may be considered to be hydrodynamic (Fig. 7.3). At such a pressure the jet and armor may be treated like an incompressible fluid (if their material strengths are ignored) in order to get a good approximation so that the jet penetrates the target (armor steel plate) according to the laws of hydrodynamics like a

Tab. 7.1: Penetration behavior of a projectile as a function of its velocity.

velocity of projectile (km s^{-1})	effect	launch method
< 50	elastic, plastic	mechanical, air rifle
50– 500	plastic	mechanical, air rifle
500– 1 000	plastic or hydrodynamic, target material appears to be very viscos	powder gun
3 000–12 000	hydrodynamic, target material appears to be liquid	explosive acceleration
> 12 000	explosive, evaporation of the colliding solids	explosive acceleration

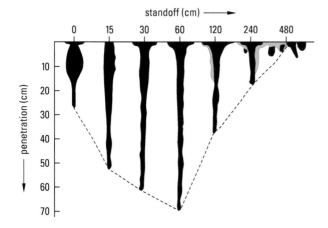

Fig. 7.4: Influence of the standoff distance on the penetration depth of a typical shaped charge with metal liner (diameter: 10 cm, length: 18 cm).
(This diagram is reproduced with slight modification from the original of Prof. Dr. Manfred Held, who is herewith thanked for his permission to reproduce this.)

liquid (Fig. 7.3). The general penetration behavior of a projectile as a function of its velocity is shown in Table 7.1.

Since the jet needs some time and space to fully form, most shaped charges have a long ballistic cap to guarantee the correct standoff distance. When the ballistic cap hits the target the main charge gets detonated while still being at sufficient (for the jet formation) distance from the target surface. If, however, the standoff distance is too large, the jet stretches (due to different tip and tail velocities) and eventually breaks up into particles which dramatically lowers the penetration depth. Figure 7.4 shows the influence of the standoff distance on the penetration depth for a typical metal-lined shaped charge (diameter: 10 cm, length: 18 cm).

Fig. 7.5: Top: Schematic design of a shaped charge with metal liner and wave shaper. Bottom: Design of a warhead for a shoulder-launched anti-tank missile (shaped-charge). (Both diagrams are reproduced with slight modification from the original of Prof. Dr. Manfred Held and Dr. Marcel Holler, who are herewith thanked for their permission to reproduce the figures.)

Because of the extremely high jet velocity the (low) speed with which the shaped charge travels towards its target is relatively unimportant. For this reason, relatively slow and repercussion-free projectiles are often used to carry the shaped charge. The advantage of such systems is that they do not require a big and heavy launch platform which makes the entire system more mobile. Shaped charges are frequently used as warheads in anti-tank missiles and also as gun-fired projectiles, rifle grenades, mines, bomblets, torpedos and as various types of air-, land- or sea-launched guided missiles. Figure 7.5 shows the schematic design of a shaped charge with a metal liner.

The most common shape which is found for liners is conical, with an internal apex angle of 40° to 90°. Different apex angles yield different distributions of jet mass and velocity. Small apex angles can result in jet bifurcation, or even result in the failure of the jet to form at all. This is attributed to the collapse velocity being above a certain threshold – normally slightly higher than the liner material's bulk sound speed.

Liners have been made from many materials, including glass and various metals. The deepest penetrations are achieved with liners constructed from a dense, ductile metal, and a very common choice is copper. For some modern anti-armor weapons, molybdenum has been used.

In early anti-tank weapons, copper was used as a liner material. Later, in the 1970s, tantalum was found to be superior to copper, since it has a much higher density and very high ductility at high strain rates. Other high-density metals and alloys often have drawbacks in terms of price, toxicity, radioactivity (DU), or lack of ductility.

To achieve maximum penetration, pure metals yield the best results, because they show greatest ductility, which delays the breakup of the jet into particles as it stretches.

Rhenium has some interesting properties in this context. For example, it is one of the most dense elements known, and it also has one of the highest boiling points of all the elements.

Rhenium is a ductile, malleable, silvery metal. Ductile means that the metal is capable of being drawn into thin wires, whereas malleable means the metal can be hammered into thin sheets. Rhenium has a density of 21.02 grams per cubic centimeter, a melting point of 3,180 °C, and a boiling point of 5,630 °C. These values are among the highest to be found for any of the chemical elements.

Rhenium is quite dense, which is unusual for a metal. When most metals are heated, they reach a point where they change from being ductile to being brittle. This means that the metals can be worked with at temperatures below that point, but not above it, since above this transition temperature they become brittle, and on attempts to bend or shape them, they break apart. The unusual behavior of rhenium (high ductility) means that it can be heated and recycled many times without breaking apart.

Quite recently, experiments involving reactive liner materials (RMs) have also been performed (*Propellants Explos. Pyrotech.* **2018**, *43*, 955–961). RMs liners were prepared using a mixture of AL and PTFE powder with a mass ratio of approximately 26 % Al and 74 % PTFE. The liner samples were prepared by the following process: mixing, cold pressing, and sintering. The results are promising, but further work is necessary to reach a conclusion with respect to the suitability of RMs liners.

Wide angle cones and other liner shapes such as plates or dishes with apex angles greater than 100° do not jet, but instead give an **explosively formed projectile or explosively formed penetrator (EFP)** (Fig. 7.6). The projectile forms through dynamic plastic flow and has a velocity of 1–3 kms^{-1}. Target penetration is much smaller than that of a jet, but the hole diameter is larger with more armour backspall.

An EFP uses the action of the explosive's detonation wave (and to a lesser extent the propulsive effect of its detonation products) to project and deform a plate or dish of ductile metal (such as copper or tantalum) into a compact, high-velocity projectile, commonly called the slug (Fig. 7.7). This slug (one projectile with homogeneous velocity) is projected towards the target at about two kilometers per second. The main advantage of the EFP over a conventional shaped charge is its effectiveness at very large standoff distances, which is equal to hundreds of times the charge's diameter (perhaps a hundred meters for a practical device) (Fig. 7.8).

Fig. 7.6: Comparison between a shaped charge (a) and an explosively formed projectile (b). (This diagram is reproduced with slight modification from the original of Prof. Dr. Manfred Held, who is herewith thanked for his permission to reproduce this.)

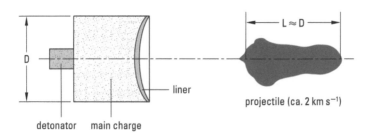

Fig. 7.7: Schematic design of a charge to generate an explosively formed projectile (EFP). (This diagram is reproduced with slight modification from the original of Prof. Dr. Manfred Held, who is herewith thanked for his permission to reproduce this.)

Fig. 7.8: Influence of the standoff distance on the perforation depth for shaped charges (SC) and explosively formed projectiles (EFP).
(This diagram is reproduced with slight modification from the original of Prof. Dr. Manfred Held, who is herewith thanked for his permission to reproduce this.)

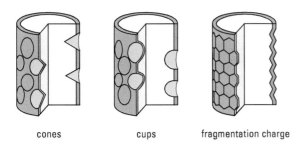

cones   cups   fragmentation charge

Fig. 7.8a: Schematic presentation of a multi-projectile warhead.
(This diagram is reproduced with slight modification from the original of Prof. Dr. Manfred Held, who is herewith thanked for his permission to reproduce this.)

A **fragmenting warhead** is a special type of shaped charge or device that explosively forms projectiles. In such a device many EFPs of the shape of cones or cups are combined into one multi-projectile warhead (Fig. 7.8a). Such devices which send out the projectiles in a cylindrical geometry are particularly suitable for targets in the air.

An **explosive lens** is a very specialized type of shaped charge, which affects the shape of the detonation wave which passes through it in a similar way as light is affected when it is passed through an optical lens. The general design of an explosive lens necessitates the use of several explosive charges. These explosive charges are shaped in a manner which results in a change in the shape of the detonation wave which passes through the shaped charge. In order for this to be achieved, the different explosive charges which the explosive lens is constructed of must have different detonation velocities.

For example, in order to convert a spherically expanding wavefront (Fig. 7.9, top) into a spherically converging wavefront (Fig. 7.9, bottom), in which there is only a single boundary between the explosive with fast detonation velocity and explosive with slower detonation velocity, the boundary shape must be a hyperboloid. Instead of containing a wave shaper, many – particularly older – explosive lenses use two explosives instead which have significantly different detonation velocities (VoD) – but which are in the range from 5000–9000 m s^{-1}. Again, the combination of a correctly chosen pair of explosives – in which one has a high detonation velocity and the other has a low detonation velocity – results in a spherical converging detonation wave.

An explosion usually expands outward (spherically expanding) through the explosives in every direction from the point of its detonation. Therefore, for a spherically converging wavefront, the shape of the shockwave has to be manipulated so that it strikes every point on the surface of the pit (inner blue area, surrounded by a tamper, which is a container that surrounds the pit) at the same time. This can be accomplished using an explosive lens.

The detonator is located on the outside of the lens, and is attached to the explosive which possesses a high detonation velocity. A second explosive – with a

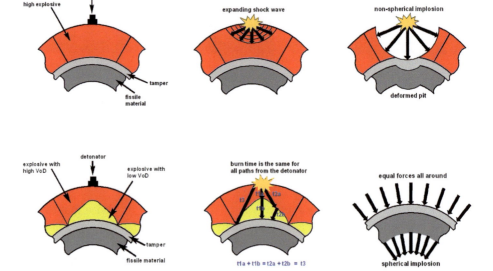

Fig. 7.9: Schematic presentation of a spherically expanding wavefront (top) and of a spherically converging wavefront (bottom), using an explosive lens consisting of two high explosives. One of the explosives has a high detonation velocity, and the other explosive a lower detonation velocity. (Reproduced with kind permission of Phillip R. Hays PhD LT USNR-R, Nuclear/Special Weapons Officer, USS *Oklahoma City* CLG-5 1970-1972 from:
http://www.okieboat.com/How%20nuclear%20weapons%20work.html)

lower detonation velocity – is contained inside the explosive which has a high detonation velocity. The shape of these two explosives is important since they must be shaped to refract the shock wave similar. The inner surface of the low velocity explosive is spherical and rests against the spherical pit. The surface between the two explosives is essentially conical, but with a rounded point. Using the values of the speeds of detonation of the two explosives, the shape which the surface should adopt can be calculated. Ultimately, the explosive lens is used in order to focus the energy of the chemical explosion onto the surface of the pit.

The shock wave of an explosive quickly passes through the explosive with a high detonation velocity until it comes into contact with the explosive with low detonation velocity. At this point, the shock wave then detonates the explosive with the lower detonation velocity. The detonation wave then passes more slowly through the explosive with low detonation velocity. The shock wave reaches different points on the boundary between the two explosives at different times. It arrives first at the point which is closest to the detonator, whereas the last point it reaches is that which is furthest away from the detonator. Since there is more of the explosive with low detonation velocity along the shorter path which the shock wave takes from the detonator to the surface of the pit, this means that the shock wave requires a longer period of time to expand inward. Conversely, there is a larger quantity of the explosive with higher detonation velocity along the longer paths from the detonator to the pit, and this means that the shock wave passes through the longer distance faster. The amounts of the two explosives are calculated so that the shape of the shock wave means that it will travel along all of the paths originating from the detonator and finally reach the surface of the pit in the same amount of time (about 10 to 15 microseconds). This causes the shape of the shock wave to be that of a spherical implosion, which reaches all of the points on the surface of the pit at exactly the same time. If all of the surrounding forces are equal, the pit will be uniformly compressed.

One of the biggest problems of any implosion device using high explosives is to ensure a symmetric collapse occurs, and to avoid any jet formation and instabilities that, provoking a predetonation, would result in a low energy release (yield).

A further strategy to penetrate or perforate a target is the use of so-called **kinetic energy (KE) munition**. In this context the term "kinetic energy" indicates that the destructive energy of the munition originates from its high kinetic energy ($T = \frac{1}{2}mv^2$). For this reason, a high mass (m) and a high velocity (v) of the penetrator are desirable. The Bernoulli equation shows the relationship between the penetration depth (P) and the length of the penetrator (L) with the densities of the penetrator (ρ_P) and the target (ρ_T):

$$P \sim L \sqrt{\frac{\rho_P}{\rho_T}}$$

The constant η depends on the velocity of the penetrator (v_P). For a typical penetrator velocity of 1600 m s^{-1} the constant η is about 0.66:

$$P = \eta L \sqrt{\frac{\rho_P}{\rho_T}} = 0.66 \times L \sqrt{\frac{\rho_P}{\rho_T}} \quad \text{(for } v_P = 1600 \text{ m s}^{-1}\text{)}.$$

Since the density of the target material is usually 7.85 g cm^{-3} (armor grade steel) one can only influence the penetration depth by the velocity, density and length of the penetrator. Table 7.2 summarizes the penetration depths depending on the penetrator material used (penetrator density). It is apparent why from a strategical point of view penetrators made of highly dense (and pyrophoric) depleted uranium are very suitable.

It should also be noted that increasing the length of the penetrator is not always easy because long penetrators need to be stabilized by tail fins to avoid spinning and to guarantee a hitting angle of exactly 0° (Fig. 7.10). (N.B. The NATO hitting angle is defined as the angle between the penetrator's trajectory and a line perpendicular to the target surface. This means that with a NATO angle of 0° the penetrator hits the target directly perpendicular to the target surface, as desired.)

Tab. 7.2: Penetration depth as a function of the penetrator material used (density) for a typical penetrator of 80 cm lengthwith a velocity of 1600 m s^{-1}.

penetrator material	penetrator density / g cm^{-3}	Penetration depth / cm
armor steel	7.85	53
tungsten	19.3	83
depleted uranium (DU)	19.0	82
tantalum	16.7	77

Fig. 7.10: Typical kinetic energy munition (KE) which is stabilized by tail fins. (This diagram is reproduced with slight modification from the original of Prof. Dr. Manfred Held, who is herewith thanked for his permission to reproduce this.)

While a relatively small shaped charge (see above) travels with relatively low velocity and therefore requires only a small launch platform, kinetic energy munitions (KE) usually need large and therefore less mobile cannons to achieve the high velocities required. This makes shaped charges particularly useful for launching from highly mobile platforms such as helicopters. But the lower velocity of a shaped charge device compared to kinetic energy munition makes the shaped charge device more vulnerable during the duration of its flight.

7.2 Detonation Velocities

In the discussion of shaped charges (see above) it was shown that the brisance (see Ch. 2.2) of the main charge used is of utmost importance. The brisance value (B) is defined as the product of the loading density (ρ), the specific energy (F for "force of an explosive") and the detonation velocity D:

$$\text{brisance: } B = \rho F D$$

The specific energy ("force") of an explosive (F) on the other hand can be calculated according to the general equation of state for gases:

$$\text{specific energy: } F = p_e V = n\,RT$$

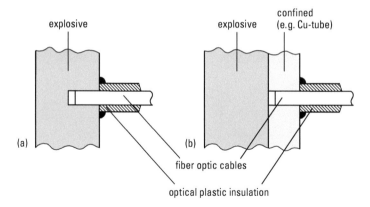

Fig. 7.11: Position of an optical fiber in case of an unconfined (a) and a confined (b) explosive charge.

Therefore, we can conclude that for shaped charges the loading density of the high explosive and its detonation velocity (or to be more precise the Gurney velocity, see Ch. 7.3) are relevant performance parameters.

In Chapter 4.2.2 we already discussed methods for the theoretical calculation of the detonation velocity and detonation pressure. In this chapter we now want to focus on the experimental determination of the detonation velocity. Bearing in mind that detonation velocities of known high explosives may reach up to 10,000 m s^{-1}, the experimental determination of the detonation velocity is not easily achieved. There are several methods which are suitable to measure the detonation velocity [38]. Most of these methods are based on the fact that the detonation process is accompanied by the emission of light (autoluminous process). Depending on the measuring equipment selected, the methods for the detonation velocity determination can be divided into,

1. optical methods, which are based on the use of different types of high-speed cameras (distance-time curves) and
2. electrical methods, which are based on the use of different types of velocity probes combined with an electronic counter or an oscilloscope.

In recent years, the fiberoptic technique has widely been used for the determination of the detonation velocity. Optical fibers are capable of detecting and transmitting a light signal accompanying the detonation wave front. This light signal may be recorded by optical methods (using a high-speed streak camera), or it may be transformed into an electric signal (by a fast photodiode) which is then recorded by a suitable ultrafast signal recording technique (fast-storage oscilloscope or multichannel analyzer). The optical fibers also serve as a convenient means of transporting the signal from the experimental assembly in the firing area to the recording shed. The length of the optical fiber may exceed 20 m. Usually highly flexible optical fibers with a relatively low attenuation factor, 1 mm core diameter and 2.2 mm

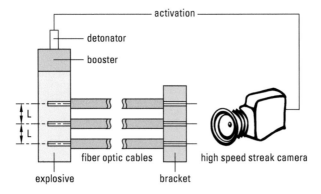

Fig. 7.12: Schematic representation of the determination of the detonation velocity using the optical fiber/streak camera technique.

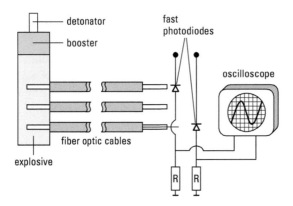

Fig. 7.13: Schematic representation of the test arrangement for the determination of the detonation velocity using the optical fiber/fast photodiode/oscilloscope technique.

outside diameter of the black plastic jacket may be used. If the explosive charge is unconfined, the optical fibers are placed directly into the explosive charge to a depth of 2/3 of the charge diameter. If the explosive charge is confined, the fiber is placed through a hole in the wall of the metal or plastic confinement right to the inner surface of the tube (Figs. 7.11 and 7.12).

Figures 7.12 and 7.13 show the experimental set-up for using a high-speed streak camera or a fast photodiode/oscilloscope for the determination of the detonation velocity.

As discussed above, a convenient way of measuring the detonation velocity is to convert the light signal using a fast photodiode that has a rising time of about 10 ns into an electric signal that may be recorded by either a fast oscilloscope (Fig. 7.13) or a multi-channel analyzer (Fig. 7.14).

Fig. 7.14: Recording the detonation velocity (VOD, velocity of detonation) using a multichannel analyzer.

Fig. 7.15: Schematic representation (left) and photo of a KV-250 detonation chamber (up to 250 g TNT or equivalent).

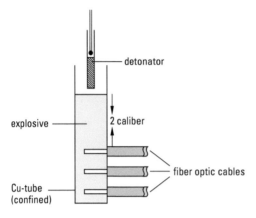

Fig. 7.16: Schematic set-up of a test for the determination of the detonation velocity using the optical fiber method.

Fig. 7.17: Typical set-up for a simple determination of the detonation velocity using the optical fiber method.

For the experimental measurement of the detonation velocity in a chemical laboratory (indoors), it is advisable to carry out the detonation experiment in a so-called detonation chamber (Fig. 7.15). Usually, the explosive is filled (pressed or melt-cast) into a plastic or metal tube (sealed on one end) which is wider than the critical diameter. This tube contains at least two, but preferably more (for mean value calculation) holes for the optical fibers a distance of approx. 1 in. An experimental set-up for measuring the detonation velocity using the optical fiber method is shown in Figures 7.16 and 7.17. For the measurement it is important to address the following:

1. The density of the explosive under investigation in the tube has to be constant and known exactly.
2. The diameter of the confinement tube has to be above the critical diameter (see Ch. 3, Fig. 3.2). While for many secondary explosives 1 inch is a "good"

diameter to start with, for primary explosives usually much smaller diameters (ca. 5 mm) are sufficient. In any case measurements with different tube diameters are recommended in order to ensure convergence of the detonation velocity on increasing tube diameter (to make sure to be above the critical diameter).
3. The closest optical fiber should not be closer to the detonator than one, preferably two calibers, to ensure that the detonation wave can stabilize and that it originates exclusively from the explosive under investigation (and not from the detonating device).

Table 7.3 and Figure 7.18 show the dependency of the calculated (EXPLO5, see Ch. 4.2.2) and experimentally measured (optical fiber method, multi channel analyzer) detonation velocities of a new high explosive (NG-A) on the loading density. The good agreement between the calculated and measured detonation velocity values reflects the quality of the measurement (time accuracy: ± 0.1 μs, accuracy of the measured detonation velocity ± 0.2%, highest measurable velocity 10,000 m s^{-1}) and gives credence to the theoretically calculated parameters.

Tab. 7.3: Measured and calculated detonation velocities of NG-A as a function of the loading density.

loading density / g cm^{-3}	D / m s^{-1} (measured)	D / m s^{-1} (calculated)
0.61	4181	4812
1.00	6250	6257

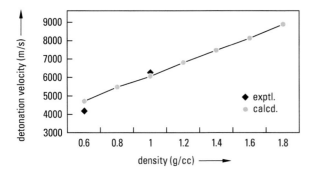

Fig. 7.18: Measured (exptl.) and calculated (calcd.) detonation velocity (det. vel.) of NG-A as a function of the loading density (density).

7.3 Gurney Model

As was already discussed in Chapter 7.1 (shaped charges), it is not only the detonation velocity, but also the so-called Gurney velocity, that determines how quickly metal fragments are ejected from an explosive charge with a specific shape (bombs, grenades). This question was researched by Ronald W. Gurney in 1943. Gurney suggested that there is a simple dependence releating to the mass of the metal confinement (M) and the explosive (C) on the fragment velocity (V). The simple Gurney model developed in Aberdeen (MD) assumes the following:
- The energy released on detonation is essentially completely changed into the kinetic energy of the detonation gases *and* kinetic energy of the metal fragments.
- The energy used for the deformation or fragmentation of the material used for confinement can essentially be ignored.
- During detonation, the explosive is spontaneously ($\Delta t = 0$) transformed into homogeneous and chemically completely changed, gaseous products under high pressure.
- The gaseous detonation products expand with uniform density and a linear velocity gradient.
- The chemical detonation energy of the explosive is changed into kinetic energy, until the fragments have a steady-state velocity (Fig. 7.19), from which the Gurney velocity can be calculated.

Fig. 7.19: Dependence of the normalized fragment velocity of 16 explosives (Cu cylinder) on the expansion volume. Steady-state at velocity/final Velocity = 1.

Fig. 7.20: Construction of a cylindrical (a), a spherical (b) and a sandwich configuration (c).

The fragment velocity is largely dependent on the shape of the charge. For cylindrical charges (which are a good approximation for most bomb and missile (rocket) warheads) (Fig. 7.20):

$$\frac{V}{\sqrt{2E}} = \left(\frac{M}{C} + \frac{1}{2}\right)^{-0.5}$$

The constant $\sqrt{2E}$ is the so-called Gurney velocity (in km s^{-1}), which is dependent on the nature of the explosive.

Spherical charges (Fig. 7.20) which have initiating charges in the centre and are used as an approximation for grenades (and many cluster bombs), have a similar relationship:

$$\frac{V}{\sqrt{2E}} = \left(2\frac{M}{C} + \frac{3}{5}\right)^{-0.5}$$

Charges in the shape of a symmetrical sandwich (Fig. 7.20) (e.g. reactive armor) obey the following relationship:

$$\frac{V}{\sqrt{2E}} = \left(2\frac{M}{C} + \frac{1}{3}\right)^{-0.5}$$

The Gurney velocity $\sqrt{2E}$ is decisive for the performance of the explosive used. As we already saw in Ch. 3, Kamlet and Jacobs introduced the Φ parameter which relates the detonation velocity and the detonation pressure for $C_aH_bN_cO_d$ explosives with the heat of detonation Q (in cal g^{-1}), where M is the molecular weights of the gaseous detonation products (in g mol^{-1}) and N is the number of moles of gaseous detonation products per gram of explosive (for charge densities of $\rho > 1$ g cm^{-3}):

$$\Phi = N\sqrt{MQ}$$

To approximate the Gurney velocity, Hardesty, Kamlet et al. have suggested the following relationship:

$$\sqrt{2E} = 0.6 + 0.54\sqrt{1.44\,\Phi\rho}$$

therefore

$$\sqrt{2E} = 0.877\,\Phi^{0.5}\rho^{0.4}$$

It was only in 2002 that A. Koch et al. could show that the Gurney velocity $\sqrt{2E}$ and the detonation velocity (D = VoD) of an explosive can roughly be described using the following simple relationship:

$$\sqrt{2E} = \frac{3\sqrt{3}}{16}D \approx \frac{D}{3.08}$$

Therefore, there is a simple approximation for the Gurney velocity $\sqrt{2E}$ for either a pure explosive or for an explosive formulation, when the charge density ρ and detonation velocity are known (Tab. 7.4).

Tab. 7.4: Dependence of the Gurney velocity $\sqrt{2E}$ on the detonation velocity ($D = V_0 D$).

explosive	charge density ρ / g cm^{-3}	D / km s^{-1}	$\sqrt{2E}$ / km s^{-1}
Composition A-3[a]	1.59	8.14	2.64
Composition B[b]	1.71	7.89	2.56
HMX	1.835	8.83	2.87
octol (75/25)[c]	1.81	8.48	2.75
PETN	1.76	8.26	2.68
RDX	1.77	8.70	2.82
tetryl	1.62	7.57	2.46
TNT	1.63	6.86	2.23
tritonal[d]	1.72	6.70	2.18

[a] 88 % RDX, 12 % binder and plasticizer
[b] 60 % RDX, 39 % TNT, 1 % binder
[c] 75 % HMX, 25 % RDX
[d] 80 % TNT, 20 % Al

Tab. 7.5: Gurney velocity $\sqrt{2E}$ obtained according to method A: $\sqrt{2E} = D/3.08$ (see Tab. 1.1) and Method B: $\sqrt{2E} = 0.404 + 1.020\,\rho - 0.021\,c + 0.184\,(b/d) + 0.303\,(d/a)$.

explosive	charge density ρ / g cm^{-3}	D / km s^{-1}	$\sqrt{2E}$ / km s^{-1} method A	$\sqrt{2E}$ / km s^{-1} method B
HMX	1.835	8.83	2.87	2.90
PETN	1.76	8.26	2.68	2.97
RDX	1.77	8.70	2.82	2.87
tetryl	1.62	7.57	2.46	2.41
TNT	1.63	6.86	2.23	2.42

The most recent approximation to estimate the Gurney velocity has been suggested by Locking (BAE) which has the following form [see: P. M. Locking, 29th Int. Symp. On Ballistics, Edinburgh, Scotland, May 9–13, **2016**]:

$$EG = \sqrt{2E} = \frac{D}{f_x} \quad \text{with} \quad f_x = 18.0467\,(1 + 1.3\,\rho_x/1000)/\rho_x^{0.4}$$

A very simple approach to approximate the Gurney velocity for $C_aH_bN_cO_d$ explosives (density ρ, in g cm^{-3}) was introduced in 2008 and is based on the following equation:

$$\sqrt{2E}\,[\text{km s}^{-1}] = 0.404 + 1.020\,\rho - 0.021\,c + 0.184\,(b/d) + 0.303\,(d/a)$$

Some results are summarized in Table 7.5. However, this method fails for explosives which do not contain hydrogen atoms and is also only recommended for preliminary categorization.

Of course, the Gurney velocity can also be determined experimentally by using the measured fragment velocity V. It can be concluded that the relationship discussed above provides a good approximation, but that the Gurney velocity is also dependent on other factors including, for example, the material of the confinement used (in particular its density). This can be seen in Table 7.6.

Figure 7.21 also shows that the Gurney velocity for a specific explosive, with a particular charge density, is only constant when the surface density (in g cm^{-3}) of the cylinder walls is above a critical value of approx. 1 g cm^{-2}.

The most recent work, which has mainly been reported by Joe Backofen, shows that the Gurney velocity is not only dependent on the explosive, the mass of the confinement, and the shape of the charge, but that it is also dependent on the properties of the material (see Table 1.2 and Fig. 7.20). Subsequently he proposed a two-step model to describe this: a brisant shock-dominated first step, followed by a gas-dynamic second (propulsion) step.

Tab. 7.6: Dependence of the experimentally determined Gurney velocities $\sqrt{2E}$ on the cylinder material used.

explosive	$\sqrt{2E}$ (steel) / m s^{-1}	$\sqrt{2E}$ (copper) / m s^{-1}
Composition A-3	2416	2630
Composition B	2320	2790
octol (75/25)	2310	2700
TNT	2040	2370
tetryl	2209	2500

Fig. 7.21: Dependence of the Gurney velocity $\sqrt{2E}$ on the surface density (in g cm^{-2}) of the cylinder walls for Composition B ($\rho = 1.71$ g cm^{-3}).

Short et al. [*Combustion and Flame* **1981**, *43*, 99–109] used these detonation products to derive wall velocities in cylinder tests of $C_aH_bN_cO_d$ explosives by least squares fitting of the experimental data as follows:

$$V_{\text{cylinder wall}} = 0.368 \times \varphi^{0.54} \times \rho^{0.84} \, (R - R_0)^{(0.212 - 0.065 \times \rho)}$$

$$\varphi = N \times M^{1/2} \times Q^{1/2}$$

where $V_{\text{cylinder wall}}$ is the cylinder wall velocity in km/s, $(R-R_0)$ is the actual radial expansion in cm, ρ is the loading density in g/cm³, N is the number of moles of gaseous detonation products per gram of explosive, M is the average molecular weight of these gases, and Q is the heat of detonation in calories per gram.

7.3.1 Example: Calculation of the Gurney Velocity for a General Purpose Bomb

The MK80 GBP series is the abbreviation for a series of flying bombs whose effect is based on pressure and splitter effect. All variations of this series show an optimized aerodynamic shape so that they can be used in quick-flying fighter planes. The MK80 series consists of bombs which are predominantly composed of the explosive payload (TNT, Composition B, tritonal ...), one or more fuses (nose and tail) and the tail unit with stabilizing fins or breaking facility. Typically, the mass of the explosive corresponds to approximately half of the total mass.

The MK84 (Fig. 7.22) is the largest bomb in the MK80 series. With a length of 3.8 m and a diameter of 46 cm they can be considered for the calculation of the Gurney velocity as being cylindrical. The total mass of the MK84 is 907 kg and the mass of the HE is approximately 430 kg.

Fig. 7.22: MK84 general purpose bomb.

Assuming that the payload of 430 kg of HE is only made up of Tritonal (80 % TNT, 20 % Al), the detonation velocity can be calculated to be $D = 6.70$ km s^{-1}, based on a density of $\rho = 1.72$ g cm^{-3}. Following the relationship discussed above:

$$\sqrt{2E} = \frac{3\sqrt{3}}{16} D \approx \frac{D}{3.08}$$

The Gurney velocity can be calculated to be

$$\sqrt{2E} = 2.175 \text{ km s}^{-1}.$$

Therefore, if it is assumed to be a cylindrical charge, the fragment velocity can be approximated as follows:

$$\frac{V}{\sqrt{2E}} = \left(\frac{M}{C} + \frac{1}{2}\right)^{-0.5}$$

and

$$V = \sqrt{2E}\left(\frac{M}{C} + \frac{1}{2}\right)^{-0.5}$$

$$V = 2.175 \text{ km s}^{-1}\left(\frac{477 \text{ kg}}{430 \text{ kg}} + \frac{1}{2}\right)^{-0.5} = 3.5 \text{ km s}^{-1}$$

7.4 Plate Dent Tests vs. Fragment Velocities

A standard test procedure for determining the output of a detonator is the plate dent test. In the plate dent test, a detonator is assembled on a metallic plate of specified hardness and thickness. The depth of the dent produced after firing of the detonator is used as a measure of the strength of the detonator.

The Small Scale Shock Reactivity Test (SSRT test) follows a similar principle. In this test, the explosive under investigation ("sample" in Fig 7.23) is initiated by a commercial detonator and the depth in a solid aluminum block is measured for example by filling it with fine sand (SiO_2) and correlating the dent volume with the power of the explosive (Fig. 7.23, Tab. 7.7).

Fig. 7.23: Small Scale Shock Reactivity Test (SSRT test): schematic drawing (top left), photograph of test set-up (top right) and dents in solid aluminum disks after the test.

Tab. 7.7: SSRT test results for three high explosives (RDX. CL-20 and TKX-50).

explosive	mass [mg]	dent [mg SiO$_2$]
RDX	504	589
CL20	550	947
TKX-50	509	857

The depth of the dent is somewhat proportional to $\int p\, dt$ [N m^{-2} s = kg m^{-1} s^{-1}] (Figs. 7.24 and 7.25), which corresponds to the impulse density I_D (defined as: impuls I [kg m s^{-1}] per area (A [m^2])):

depth of dent	$\propto$	$\int p\, dt$	[N m^{-2} s = kg m^{-1} s^{-1}]
impulse:		$I = m\, v$	[kg m s^{-1}]
impulse density:		$I_D = \dfrac{I}{A}$	[kg m^{-1} s^{-1}]

However, often the detonation of the next assembly (upon which the detonator acts in the explosive train) is governed, not by the time integral of a pressure pulse (Figs. 7.24 and 7.25), but by the instantaneous shock pulse. In other words, while the depth of the dent is somewhat proportional to $\int p\, dt$ [N m^{-2} s = kg m^{-1} s^{-1}]

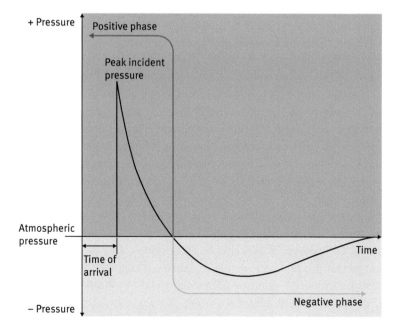

Fig. 7.24: Typical pressure-time history.

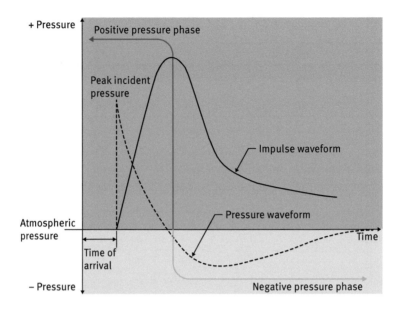

Fig. 7.25: Typical impulse waveform.

(Figs. 7.24 and 7.25), the strength of a detonator often corresponds better to an instantaneous shock pulse.

N.B.: A shock pulse (shock or pressure wave) develops when two pieces of moving metal contact each other in an initial impact. This shock pulse is in the ultrasonic frequency range and typically occurs at around 36 kHz. The amplitude of the shock pulse is proportional to the velocity of the impact.

It has been shown that increasing the length of the secondary (high) explosive column above a certain point does not change the velocity of the flyer plate, but increases the depth of the dent. This is due to the fact that the dent block responds to the time integral of the pressure pulse (Figs. 7.24 and 7.25). However, a longer column of an explosive will have the same detonation characteristics and thus throws the flyer plate with the same velocity. Therefore, a better method for determining the output of a detonator may be the calculation or measurement of fragment velocities, since the instantaneous shock pulse is proportional to the acceleration of a metal plate.

Calculated results on the acceleration of metal plates resulting from different explosive charges were obtained using the one-dimensional computer hydrocode EP [N. F. Gavrilov, G. G. Ivanova, V. N. Selin, V. N. Sofronov, UP-OK program for solving continuum mechanics problems in one-dimensional complex, VANT. Procedures and programs for numerical solution of mathematical physics problems, (3), p. 11–21, **1982**.] (Figs. 7.26, 7.27 and 7.28). As a result of such calculations, a rather complete comparative picture of explosion properties and characteristics of shock wave action onto barriers was obtained for several new and standard explosive

7.4 Plate Dent Tests vs. Fragment Velocities

Fig. 7.26: Acceleration of plates by free charges of HMX (1), TKX-50 (2), MAD-X1 (3) and RDX (4).

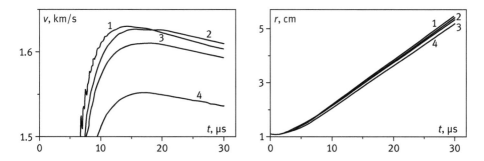

Fig. 7.27: Acceleration of Tantalum cylindrical layers by charges of HMX (1), TKX-50 (2), MAD-X1 (3) and RDX (4).

Fig. 7.28: Acceleration of a cylinder (1–v, 2–r, 3 – E_k) by a porous charge of TKX-50 (left) and cylinders by porous charges of HMX (1), TKX-50 (2), MAD-X1 (3) and RDX (4).

materials. The results which were obtained indicate that in a wide range of their initial states (porosity, inert binders), the new explosive materials TKX-50 and MAD-X1 possess higher characteristics of explosion and shock wave action on practically every compact barrier in comparison with many standard explosive materials, and in particular in comparison with the most used military explosive, RDX.

Fig. 7.29: Schematic design of the DAX assembly. The PDV velocity-probe that tracks the witness foil acceleration is fixed to the end of a 100 mm long acrylic extension and fiber coupled to the PDV detection hardware.

Recently, Lorenz et al. developed a relatively simple test that experimentally obtains accurate measurements of the detonation velocity, pressure and expansion energy of new high explosives (*PEP* **2015**, *40*, 95–108). The so-called Disc Acceleration eXperiment (DAX test) provides an initial condition of steady detonation and a charge-geometry amenable to 2D hydrodynamic simulations.

Using a combination of timing pins and a one laser velocimetry probe, the disc Acceleration eXperiment, or DAX, generates data from which detonation velocities, C-J pressures and the detonation energy (at relative volume expansions of 2–3) can be obtained. Figure 7.29 shows a schematic design of the DAX assembly. Using a PBX9407 booster (94 % RDX, 6 % VCTFE), most new and existing explosives obtain steady-state propagation well before the front reaches the end of the rate stick.

The PDV diagnostic measures the accelerating witness plate, which serves as the primary measure of explosive performance for DAX. A very detailed analysis of the plate velocity allows accurate estimates of the C-J pressure, product gas energies and JWL EOS's.

Figure 7.30 shows the disc trajectory (velocity vs. time) data for a typical single DAX shot. The initial motion of the foil's rear surface (away from the sample) is a free surface "jump-off" that is approximately twice the in-material particle velocity. The disc continues to accelerate in steps, due to the reverberating pressure wave inside the disc. This is a common condition for foils and thin plates driven by explosives. The time-dependent pressure wave (Taylor wave) behind the detonation front, results in a series of compression and release waves in the disc, which are observed as accelerations and decelerations of the back surface of the witness disc. A generalized position-time description of the reverberating waves is shown in Fig. 7.31.

Figure 7.32 shows velocity-time profiles of the accelerating witness disc using two different witness disc materials and thicknesses. In the "coast" region (around 2 µs), the copper disc has 33 % more kinetic energy than the aluminum disc. This

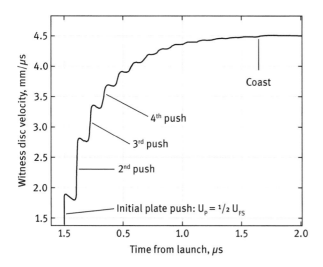

Fig. 7.30: Disc trajectory (velocity vs. time) data for a typical single DAX shot.

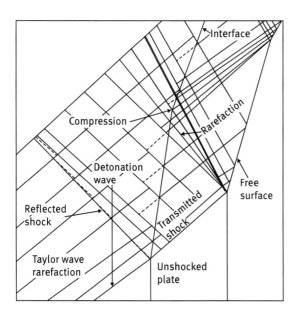

Fig. 7.31: Position-time trajectories for the compression and release waves.

indicates the improved coupling efficiency for the heavier witness disc. The disc velocity corresponds to the internal energy change in the expanding product gases, but the coupling to the plates varies with plate mass.

The detonation velocity can be directly obtained from the piezoelectric timing pin data (cf. Fig. 7.29), and the difference (in mm) where the pins are placed into

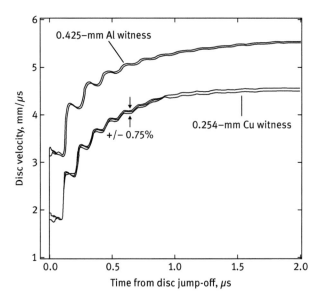

Fig. 7.32: DAX data for a PETN-based explosive using two different witness disc materials (Cu and Al) and disc thicknesses.

the detonation tube. However, the detonation pressure (C-J pressure) has always been more difficult to determine. This is because the end of the reaction zone is not always obvious – even if the pressure of the detonation could be measured with high precision. One solution is to simply estimate the C-J pressure from the density and detonation velocity:

$$P_{C\text{-}J} \approx \frac{1}{4} \rho_0 D_V^2$$

However, this approximation is not very accurate, and more precise methods have been described in the literature (see for example: *PEP* **2015**, *40*, 95–108).

7.5 Underwater explosions

Depending on the medium in which the pressure-wave generated by an explosive expands into its surroundings, one can distinguish between
(i) underwater explosions (UNDEX) and
(ii) air explosions (AIREX).

The unique properties of underwater explosions are due to the high velocity of sound in water meaning that the pressure-wave travels approximately four times

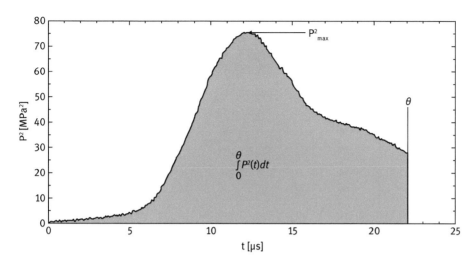

Fig. 7.33: Calculation of the shock energy equivalent E_s.

faster in water than it does in air. Furthermore, due to the high density and low compressibility of water, the destructive energy (from the explosion) can be efficiently transferred over relatively large distances. The most important effects caused by an underwater explosion are the corresponding shock-wave and the gas bubble pulsations.

During detonation, an explosive undergoes rapid conversion into gaseous products with high temperature and pressure. If detonation occurs underwater, complex physicochemical processes take place. Nevertheless, taking into account the damage which is caused by the detonation, this process (underwater explosion) can be considered as resulting from a shock wave (Fig. 7.33) and the subsequent gas bubble pulsations of the detonation products (Figs. 7.34 and 7.35). The shock wave which is generated propagates radially outward from its source and acts on the surrounding water. At the front of the shock wave, the pressure increases violently to the peak pressure point and then decays exponentially to hydrostatic pressure. The pressure pulses of the detonation gas products are much slower in comparison with the shock wave. Initially, the gases which have been compressed to high pressure cause rapid expansion of the gas bubble. Due to the inertia of outwardly moving water, the bubble expands to pressures lower than the point of pressure equilibrium and this continues until the pressure difference becomes large enough to stop the outward flow of water. Subsequently, the higher pressure of the surrounded water reverses the process causing contraction of the gas bubble. Similarly, the reverse process does not stop at the point of pressure equilibrium, but recompresses the detonation products to higher pressure values. Afterwards, the second pulsation occurs (an acoustic pulse without generation of the shock wave). The bubble expansion-contraction oscillations referred to as *bubble pulses*

Fig. 7.34: Schematic presentation of a pulsating and migrating bubble.

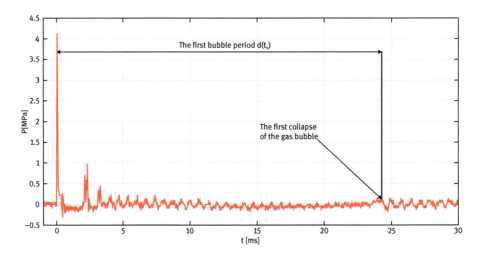

Fig. 7.35: Plot of the bubble energy equivalent $E_B = T_b^3$.

(*bubble collapse*) take place several times. Each subsequent pulse is weaker due to loss of energy. Because of high buoyancy, the bubble of gases moves upward in a jumping manner.

The underwater explosion test is a widely used method to compare the effectiveness of various explosives and is a convenient method for measuring the brisance and heaving power. This approach is based on following hypothesis:

- the shock energy from an underwater explosion measures the explosive's shattering action in other materials (μs time-scale);
- the bubble energy from the underwater explosion measures the heaving action of the explosive (ms time-scale).

The shock energy is the compressional energy radiated from an underwater detonation. In the tests, it is obtained by measuring the area under the $P^2(t)$ curve at a known distance. The total explosive energy is postulated to be the sum of the shock wave and bubble energies. Additionally, underwater tests also can measure the shock wave impulse – the rate of change of pressure of the shock wave – which is another indicator of explosive strength. The shock wave impulse is derived by measuring the area under the $P(t)$ curve for a selected integrated time interval at a known distance from the explosion.

Key characteristics of underwater explosion phenomena which are used to evaluate the performance of explosives are:

a) shock wave:
- the peak overpressure (P_m);
- the time constant (θ; the time required for the pressure to fall to the value $P_m \cdot e^{-1}$);
- the overpressure (above the hydrostatic pressure) change with time ($P(t)$);
- the impulse of the shock wave, (I);

The impulse, I is calculated using following equation:

$$I = \int_0^t P(t)\,dt \tag{1}$$

where:

$P(t)$ – overpressure (above the hydrostatic) change with time;
P_0 – hydrostatic pressure;

the upper limit of the integral is usually 5 θ.
- energy flux density (energy);

The energy flux density of shock wave (E, value of work done on a surface or the energy behind the shock front per unit area) corresponds to

$$E = \frac{1}{\rho_0 C_0} (1 - 2.422 \cdot 10^{-4} P_m - 1.031 \cdot 10^{-8} P_m^2 I = \int_0^t P^2(t)\,dt \tag{2}$$

where:

$\rho_0 C_0$ – the acoustic impedance of the water (ρ_0 – density of water; C_0 – sound velocity in water);
the coefficients –2.422 and –1.031 take into account the correction for after-flow.
t – usually 5 θ.
b) oscillation of the first bubble:
- the maximum pressure of the first bubble pulse (P_B);
- the first bubble phase duration (T_1);
- the maximum bubble radius (A_{max}).

A consequence of the many advantages (such as the huge amount of information which can be obtained during the test as well as the low costs, and the easy performance of the underwater explosion), this method has been standardized and is used to compare the initiating capability of detonators [5]. The primary shock wave energy and the bubble gas energy are determined by measuring the overpressure of the shock wave energy generated in water. The methodology which is described in the European Standard (Explosives for civil uses – Detonators and relays – Part 15: Determination of equivalent initiating capability, EN 13763-15) is based on the following equations which are used for describing the primary shock-wave energy and bubble energy [4, 6, 7].

The full description of shock wave energy generated in water
The primary shock-wave energy (E_{SW}) can be determined from the following equation

$$E_{BW} = \frac{4\pi R^2}{\rho_w C_w} \int_{t_o}^{t_o+\theta} P^2(t)dt \qquad (3)$$

where:

R – distance between detonator and pressure sensor;
ρ_w – water density;
C_w – sound velocity in water;
t_o – moment the shock wave starts;
θ – time at which the sensor output has decreased to $P_{max} \cdot e^{-1}$.

The full description of the bubble energy generated in water
The time which is measured between the moment of the primary shock wave registration and the consecutive pressure wave is used to calculate the bubble gas energy (E_{BW}):

$$E_{SW} = \frac{A \cdot \sqrt{(BH+C)^5}}{M_{HE}} t_b^3 \qquad (4)$$

where:

A, B, C – constants which depend on the experimental conditions;
H – depth the detonator is immersed;
M_{HE} – mass of explosive;
t_b – time interval between the shock-wave pressure peak and the first collapse of the gas bubble.

Assuming there are no boundary effects, the so-called Willis formula can be used to calculate the bubble energy from the first period of pulsation of the gas bubble (t_b) [7]:

$$t_b = \frac{1.135 \cdot \sqrt[3]{E_{BW}} \cdot \sqrt{\rho_w}}{\sqrt[6]{p_h^5}} = \frac{1.135 \cdot \sqrt[3]{E_{BW}} \cdot \sqrt{\rho_w}}{\sqrt[6]{(\rho_w \cdot g \cdot h + 101325)^5}} \qquad (5)$$

where:

p_h – total hydrostatic pressure at the charge depth
g – gravitational acceleration.

The total explosive energy (E) is assumed to be the sum of the shock wave and bubble energies.

$$E = E_{SW} + E_{BW} \qquad (6)$$

– Shock energy equivalent, E_s (Figure 7.33):

$$E_s = \int_{t_o}^{t_o+\theta} P^2(t)\,dt \qquad (7)$$

– Bubble energy equivalent, E_B (Figure 7.35):

$$E_B = t_b^3 \qquad (8)$$

Check your knowledge: See Chapter 14 for Study Questions.

8 Correlation between the Electrostatic Potential and the Impact Sensitivity

8.1 Electrostatic Potentials

In Chapters 4.2.2 and 4.2.3 we have seen that it is often possible to predict performance parameters using quantum mechanics without requiring experimental data. There are also huge efforts being made towards being able to theoretically predict, not only the performance but also the sensitivity parameters (if possible using quantum mechanical ab initio or DFT calculations), and particularly to allow a prediction of the sensitivity of new molecules which have not been synthesized previously. This would mean that the preparation of sensitive molecules which would have little chance of ever being used as energetic materials could be avoided. In addition, such sensitivity predictions would help to increase safety and also improve financial issues related to for synthetic projects so that they could be used more effectively for more specific targets. In comparison to computer calculations, the synthetic work in chemical laboratories requires a lot more effort and is much more time consuming and therefore considerably more expensive. As discussed previously in Chapter 6.1, the impact sensitivity is one of the most important parameters for the estimation of the sensitivity of highly energetic compounds. At this point it is worthwhile to repeat that although the drop-hammer method for the determination of the impact sensitivity uses a mechanical method, the initiation most likely also occurs as a result of hotspots. Thus it occurs thermally. It is thanks to the work of Peter Politzer, Jane Murray and Betsy Rice [39–44] in particular, that it was realized that for covalently bonded molecules, the electrostatic potential (ESP) of an isolated molecule can be correlated with the (bulk) material properties of the compound in the condensed phase.

The electrostatic potential $V(r)$ for a particular molecular hypersurface (normally 0.001 electron bohr^{-3} surface) is defined as follows:

$$V(r) = \sum_i \frac{Z_i}{|R_i - r|} - \int \frac{\rho(r')}{|r' - r|} dr'$$

Here Z_i and R_i are the charge and coordinates (position) of atom i and $\rho(r)$ is the electron density. The electrostatic potential can be obtained either from X-ray diffraction investigations, but more commonly it is obtained using quantum chemical calculations. Regions in which the ESP is positive are electron deficient and their electron density is low in this area. In contrast, regions which have negative potentials are electron rich and have a higher electron density. The relative magnitude and extension of positive and negative potentials on a given hypersurface of a mol-

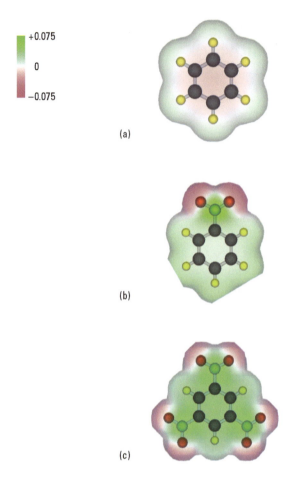

Fig. 8.1: Electrostatic potential for benzene (a), nitrobenzene (b) and trinitrobenzene (c) on the 0.001 electron bohr^{-3} hypersurface with the color code from +0.075 a.u. (green) to −0.075 a.u. (red).

ecule are of decisive importance for the impact sensitivity of the compound. In the following section, the 0.001 electron bohr^{-3} hypersurface will always be considered. For "typical", non-impact sensitive organic molecules, the areas of the negative potentials are smaller but significantly stronger than the positive ones. In contrast to this, in impact sensitive compounds the regions of positive potentials are still larger, but as strong as or stronger than the negative regions (Fig. 8.1). Politzer and Rice were able to show that the impact sensitivity can be described as a function of this anomalous behavior. Figure 8.1 shows the calculated electrostatic potentials of benzene, nitrobenzene (both insensitive) and trinitrobenzene (impact sensitive). It can also be said that the less evenly the electron density is distributed

Tab. 8.1: Conversion of $h_{50\%}$ values into the SI consistent impact sensitivity in J.

compound	$h_{50\%}$ / cm	impact sensitivity / J
PETN	13	3
RDX	28	7
HMX	32	8
HNS	54	13
FOX-7	126	31
NTO	291	71

(without taking into account extreme values due to charges above atoms of the electron donating or electron withdrawing substituents) the more sensitive a compound will behave. Moreover, it could be determined that for aromatic compounds (see Fig. 8.1) the impact sensitive species show a stronger electron deficient region above the aromatic ring (more positive potential) in comparison with the less sensitive species. It appears to be the case that the more impact sensitive compounds are significantly more electron deficient in the centre of their structures (more positive potential) than their insensitive analogues. Above the C—NO$_2$ bond in the more impact sensitive trinitrobenzene there also is a region of positive potential.

In addition to the qualitative conclusions (see Fig. 8.1), attempts have also been made to be able to quantitatively predict the impact sensitivity, in particular for the important organo-nitro compounds. In this, the average electrostatic potential V_{Mid} (kcal mol^{-1}) at the "mid point" of the C—N bond, or the averaged potential $V_{\text{Mid}}^{\star}$ (kcal mol^{-1}) for all C—N bonds in the molecule, plays an important role:

$$V_{\text{Mid}} = \frac{Q_C}{0.5\,R} + \frac{Q_N}{0.5\,R}$$

$$V_{\text{Mid}}^{\star} = \frac{1}{N}\sum_{i=1}^{N}\left(\frac{Q_C}{0.5\,R} + \frac{Q_N}{0.5\,R}\right)$$

For organo-nitro compounds, the following correlation between $V_{\text{Mid}}^{\star}$ and the (widely used in the USA) $h_{50\%}$ value has been proposed by Rice et al. [44]:

$$h_{50\%} = y_0 + a\,\exp.(-b\,V_{\text{Mid}}^{\star}) + c\,V_{\text{Mid}}^{\star}$$

The $h_{50\%}$ value is directly correlated to the impact sensitivity and corresponds to the drop-height in cm, for which 50% of the samples explode in a drop hammer test using a 2.5 kg mass (Tab. 8.1). In the equation above, y_0, a, b and c are constants with the following values:

$a = 18922.7503$ cm,
$b = 0.0879$ kcal mol^{-1},
$c = -0.3675$ cm kcal^{-1} mol^{-1},
$y_0 = 63.6485$ cm.

On this basis, it should also be possible for other classes of compounds to establish similar correlations, in order to predict the impact sensitivities at least semiquantitatively.

8.2 Volume-Based Sensitivities

Politzer, Murray et al. have also researched the possibility of a link between the impact sensitivities of energetic compounds and the space available for their molecules in the crystal lattices [39f, g]. To measure this space, they used the equation

$$\Delta V = V_{\text{eff.}} - V(0.002)$$

where $V_{\text{eff.}}$ is the effective molecular volume obtained from the crystal density and $V(0.002)$ is that enclosed by the 0.002 a.u. contour of the molecule's gas phase density, determined computationally. When the experimental impact sensitivity (h_{50} values) was plotted against ΔV for a series of 20 compounds (Fig. 8.2, a), the nitramines (▲) formed a separate group showing little dependence on ΔV. Their impact sensitivities correlate well with an anomalous imbalance in the electrostatic potentials on their molecular surfaces, which is characteristic of energetic com-

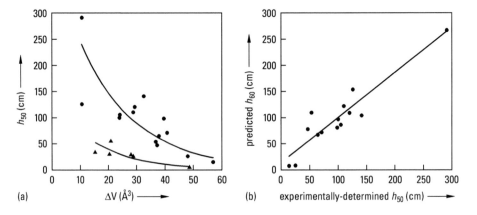

Fig. 8.2: Measured impact sensitivities (h_{50} values) plotted against computed ΔV for various energetic compounds (▲ = nitramines, ● = non-nitramine energetic compounds) (a). Comparison of impact sensitivities predicted on an electrostatically corrected volume-based model [39f,g] (b).

pounds in general (see above, Ch. 8.1). The imbalance is symptomatic for the weakness of the N—NO$_2$ bonds, caused by depletion of electronic charge. Thus, for compounds (e.g. nitramines) in which trigger-linkage-rupture is a key step in detonation initiation, the surface potential imbalance can be symptomatic of sensitivity. The impact sensitivities of non-nitramines (Fig. 8.2, a), on the other hand, depend on ΔV a lot more, and can be quite effectively related to it if an electrostatically-based correction term is added (Fig. 8.2, b). Why might this be a factor? Perhaps the availability of more space enhances the molecule's ability to absorb and localize the external energy coming from the impact, for example either vibrationally or translationally.

The volume-based (ΔV) h_{50} values for the non-nitramine energetic compounds were predicted using the equation

$$h_{50} = a\,(\Delta V)^{\frac{1}{3}} + \beta v \sigma_{tot}^2 + y$$

where α (−234.83), β (−3.197) and y (962.0) are constants, $\sigma_{tot}^2 = \sigma_+^2 + \sigma_-^2$ (with σ_+^2, σ_-^2 and σ_{tot}^2 being the positive, negative and total variances of the electrostatic potential) and v being an electrostatic balance parameter defined as $v = (\sigma_+^2 \sigma_-^2) / (\sigma_+^2 + \sigma_-^2)^2$.

Check your knowledge: See Chapter 14 for Study Questions.

9 Design of Novel Energetic Materials

9.1 Classification

As we have already discussed above, secondary explosives (or high explosives) are homogeneous substances which can either be covalent or ionic. However, in both cases it is important that the oxidizer and fuel are combined within *one* molecule (i.e. in one covalent molecule or in one ionic formula unit). For example, this is the case for TNT, in which the three nitro groups correspond to the oxidizer, and the C–H backbone acts as the fuel. Such molecules belong to the "**oxidation of the carbon backbone**" group of substances (Tab. 9.1).

In the TNT molecule, all of the angles involving the C atoms are approximately 120° (C_6 ring) or 109° (CH_3 group) and they therefore correspond almost to the ideal angles for a sp^2 and sp^3 hybridized atom. If one now considers a strained ring or cage structure, for example in CL-20, the energy of the molecule is increased as a result of the ring or **cage strain**. Consequently, in the explosive decomposition forming CO, CO_2, H_2O and N_2, more energy is released than would be in an unstrained system (Tab. 9.1).

Principally, members of both of the above mentioned classes of compounds (oxidation of the molecule back-bone; ring and cage strain) can be exothermic or endothermic compounds. An example of the former is TNT with $\Delta_f H° = -295.5$ kJ kg^{-1}, while RDX shows a positive standard enthalpy of formation of $\Delta_f H° = +299.7$ kJ kg^{-1}.

A third class of highly energetic materials is always endothermic and is called **nitrogen-rich molecules** (Tab. 9.1). The reasons why nitrogen-rich molecules are particularly interesting as highly energetic compounds will be considered in more detail in the next section.

Tab. 9.1: Overview of different approaches for the design of highly energetic compounds.

type	example	comments	structure
oxidation of the C back-bone,	TNT, PETN, RDX	compound can be exothermic or endothermic, compounds can be covalent or ionic	Fig. 1.3
ring or cage strain	CL-20, ONC	most endothermic, most covalent	Fig. 1.6
nitrogen-rich molecules	TAGzT, Hy-At	always endothermic, compounds can be covalent or ionic	Fig. 2.2

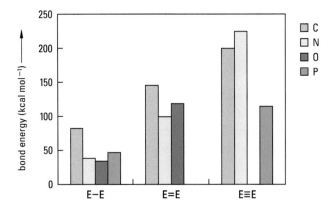

Fig. 9.1: Approximate bond energies for C—C, N—N, O—O and P—P single, double and triple bonds (in kcal mol⁻¹).

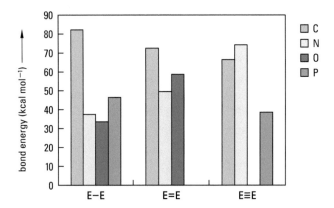

Fig. 9.2: Average element-element bond energies per 2-electron bond (in kcal mol⁻¹).

If we first consider the bond energies for typical C—C, N—N and O—O single, double and triple bonds, we can see that the bond strength increases from single to double to triple bonds (which is unsurprising) (Fig. 9.1).

However even more important is the question whether an element-element double bond is twice as strong as a single bond, and a whether a triple bond is three times as stable as a single bond. This means that we have to average the bond energy per 2-electron bond. This situation is illustrated in Figure 9.2.

Now we can see that the element nitrogen is unique as the bond energy per 2-electron bond increases from a single to a double to a triple bond. The surprising enormously high bond energy of the N≡N triple bond is responsible for the fact that all nitrogen-rich and polynitrogen compounds show a strongly exothermic decomposition when N_2 is formed. Furthermore, we can see that unlike nitrogen, carbon prefers single bonds over double or triple bonds. This means that thermodynamically, acetylene H—C≡C—H trimerizes in an exothermic reaction forming benzene, while thermodynamically, N≡N would never oligomerize or polymerize to N_x (x = 4, 6, 8). The reason for the particular stability of the N≡N triple bond in comparison to carbon can be found in the small size of the N atoms (better overlapping) (covalent radii: N = 0.70 Å, C = 0.77 Å), and also as a result of the different hybridization of N in N_2 in comparison with C in HCCH, in which the N—N σ bond in N_2 has a considerably higher p-character than the C—C bond in acetylene does:

N≡N:	σ-LP(N)	64 % s	36 % p
	σ-(N—N)	34 % s	66 % p
H—C≡C—H	σ(C—C)	49 % s	51 % p
	σ(C—H)C	45 % s	55 % p
	H	100 % s	

The large jump in the average 2-electron bond energies between the double and the triple bonds is particularly valuable for nitrogen. This means that during the development of new energetic substances with a very high nitrogen content (N > 60 %) and a N—N bond order smaller than or equal to 2, large amounts of energy (for explosions or rocket propulsion) will be released on decomposition to molecular dinitrogen.

9.2 Polynitrogen Compounds

It would be particularly interesting to take polynitrogen compounds into consideration (100 % N). However, no other modification of nitrogen except N_2 is known which is metastable under normal conditions (1 atm, 298 K) in the condensed phase.

In recent years, however, the search for further energy-rich forms of nitrogen has intensified using both theoretical and experimental methods. Apart from their purely theoretical interest, the possible potential uses of metastable forms of N_x as highly energetic materials, which would only produce hot air (hot dinitrogen) on decomposition, has made N_x molecules particularly interesting. Generally, calculations show that the decomposition of highly energetic forms of poly-nitrogen to N_2 should release over 10 kJ/g energy. This energy release per gram of condensed material is much higher than that of any of the state-of-the-art propellants or explosives used today. Some of the forms of nitrogen we will consider in the following section have already been identified as being vibrationally stable, thermodynamically metastable N_x molecules. Particularly interesting in this context is the N_6 molecule, which has been most intensively investigated theoretically. The nitrogen triazide molecule N $(N_3)_3$, has also been investigated by using theoretical calculations and its synthesis has been attempted, but only the formation of five equivalents of

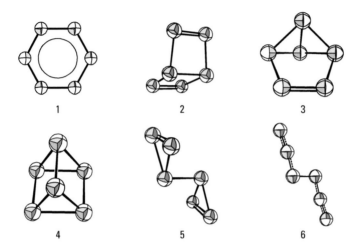

Fig. 9.3: Structures of six different N_6 isomers [45–46].

Tab. 9.2: Compilation of N_6 isomers and their isoelectronic organic CH analogues [45–46].

N_6 form (see Fig. 9.3)	point group	stability	rel. energy (in kJ mol^{-1})	CH analogue	point group
1	D_{6h}	unstable	899	Benzene	D_{6h}
2	C_{2v}	stable	1037	Dewar benzene	C_{2v}
3	C_{2v}	stable	890	Benzvalene	C_{2v}
4	D_{3h}	stable	1384	Prismane	D_{3h}
5a	C_{2h}	transition state	1020	Bicyclopropene	C_{2h}
5b	C_{2v}	transition state	1041	Bicyclopropene	C_{2v}
6a	C_1	transition state	769	–	–
6b	C_2	transition state	769	–	–
N_2	$D_{\infty h}$	stable	0	Acetylene	$D_{\infty h}$

N_2 and large quantities of energy were observed. The six most important possible structures of N_6 are summarized in Figure 9.3, while Table 9.2 shows a comparison of the N_6 molecules with some known, isoelectronic organic C–H compounds. For the different N_6 structures (**1–6**) five have (**1–5**) classic analogues in organic chemistry, however only the N_6 analogues **2**, **3** and **4** represent stable minimum structures. The diazide structure **6** is the only isomer which doesn't have a classic analogue in organic chemistry and represents a transition state.

Since only isomers **2**, **3** and **4** are stable minima, we will concentrate on them in the following discussion. Structure **3** contains an almost pre-formed N_2 unit (N1–N2) and in-depth studies show that this compound is therefore not kinetically stabilized enough for it to be of practical interest. The relative energies of **2** and **4** are 1037 and 1384 kJ mol^{-1} higher than that of N_2. These values correspond to spe-

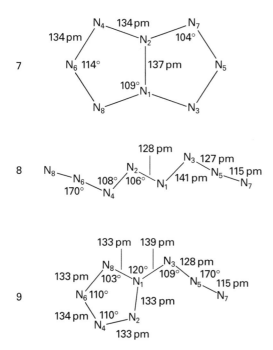

Fig. 9.4: Structures of three different N$_8$ isomers (**7–9**).

cific energies of 14 and 19 kJ/g. These energies are noticeably high, considering that the high-performance explosives used today have typical values of 6 kJ/g.

Furthermore, both molecules **2** and **4**, have relatively high values for the lowest frequency normal vibration which is found at approximately 450 cm^{-1}. These values suggest, that the structures of these two isomers are relatively rigid and that the highly thermodynamically favorable decomposition reaction (see above) to N$_2$ should have a significant activation energy. Therefore, we can summarize that the N$_6$ analogues of Dewar benzene (**2**) and prismane (**4**) are the best candidates for N$_6$ for real high-energy materials. Structure **4** appears to be the most promising preparatively, since the unimolecular dissociation into three N$_2$ molecules is symmetry forbidden (4 + 4 + 4) and would possess considerable activation energy.

On the basis of quantum mechanical calculations, it has also recently been shown that azidopentazole (**9**, Fig. 9.4) is possibly the global minimum on the N$_8$ energy hypersurface. Azidopentazole has a significant energy barrier with respect to the ring closing reaction (**9 → 7**) and it can be expected that azidopentazole is also stable, also with respect to reverse cyclization. Therefore azidopentazole probably represents not only the global minimum on the N$_8$ energy hypersurface, but should also be a synthetically realistic target. But how could one attempt to synthesize azidopentazole? The most promising route is probably the reaction of phenylpentazole (prepared from the corresponding diazonium salt and the azide ion) with

Tab. 9.3: Summary of N_8 isomers.

N_8 isomer	point group	stability	rel. energy (in kJ mol^{-1})
Pentalene bicycle (7)	D_{2h}	stable	245.1
N_8 chain (8)	C_s	stable	246.9
Azidopentazole (9)	C_s	stable	229.7
N_2	$D_{\infty h}$	stable	0.0

a covalent azide compound. The structures and relative energies of the three most probable N_8 isomers are summarized in Figure 9.4 and Table 9.3.

$$C_6H_5-NH_2^+ \xrightarrow{NaNO_2,\ HCl,\ 0\ °C} C_6H_5-N_2^+Cl^-$$

$$C_6H_5-N_2^+Cl^- + AgPF_6 \longrightarrow C_6H_5-N_2^+PF_6^- + AgCl$$

$$C_6H_5-N_2^+PF_6^- + NaN_3 \longrightarrow C_6H_5-N_5 + NaPF_6$$

$$C_6H_5-N_5 + R-N_3 \longrightarrow N_5-N_3\ (9) + C_6H_5-R$$

It is interesting to note at this point that polynitrogen compounds are not only of great interest as explosives, but also as rocket propellants. For example, a specific impulse of 408 s and a combustion chamber temperature of 7500 K have been predicted for N_4 (T_d) (and for N_8 an I_{sp} of 500 s). From experience, since an increase in the specific impulse of 20 s results in a doubling of the payload, this would be a massive improvement. Furthermore, since polynitrogen compounds exhibit smoke-free combustion, they are also of big interest for tactical rockets.

Although currently some very nitrogen-rich compounds are known, there is no all-nitrogen/polynitrogen compound in use or even under investigation. Nonetheless, the (calculated) advantages of this class of compounds is summarized again in the following:
- only gaseous reaction products (N_2)
- large positive standard enthalpies of formation ($\Delta_f H°$)
- large propulsive or explosive power
- very high specific impulse (200 % > value for hydrazine)
- very high (adiabatically calculated) reaction temperature (up to 7500 K)
- smoke-free combustion
- minimal signature of a rocket engine
- low erosion of gun barrels (no formation of iron carbide)

In recent years it has been possible to increase the static pressure during material synthesis and investigations by more than a factor of 50. This means that it is possible to access pressures which were previously unthinkable. This break-

Tab. 9.4: Correlation of the applied pressure and material changes.

p / bar	pΔV / kcal mol^{-1}	effect
5 000	2	pressing the substance
50 000	20	deformation of bonds
500 000	200	formation of new bonds with new electronic states

through was made possible by the huge advances in the diamond-cell technique. The pressure at the centre of the earth corresponds to a value of approximately 3 500 000 bar = 3 500 kbar = 3.5 Mbar. The highest pressure that has been reached using the diamond-cell is 5.5 Mbar. It has been assumed, that the highest possible pressure accessible using this technique should be approximately 7.5 Mbar. At this pressure, the metallization of diamond begins, which is linked to a large decrease in volume, so that as a result the diamonds crack. The change of the inner energy (U) of a substance is dependent on the variables temperature (T) and pressure (p):

$$\Delta U = T \Delta S - p \Delta V$$

Due to the temperature dependent term $T \Delta S$, the state (solid, liquid, gas) can be changed, e.g. by melting or evaporation. However, the pressure-dependent term $p \Delta V$ is more dominant, because the pressure can be increased to a much greater extent than the temperature. Table 9.4 summarizes the effects at static pressures of 5 000 bar, 50 000 bar or 500 000 bar, that can be achieved for a compound which shows a decrease in volume of $\Delta V = 20$ cm^3 mol^{-1}.

The application of static pressures in the range of 5 000 to 50 000 bar has relatively speaking only the small effect of pressing the substance together and bending of the bonds. These effects can also be achieved by changing the temperature at ambient pressure. When the pressure reaches 500 000 bar, a completely new situation arises: old bonds can be broken, new bonds made and new electronic states occupied, which result in drastic changes to the physical properties of the substance.

As a highlight of such investigations, the results of the high-pressure research on nitrogen should be mentioned. As a result of the triple bond in the dinitrogen molecule, N_2, it is one of the most stable diatomic molecules. At low temperatures and pressures nitrogen can be condensed, but this solid still contains diatomic N_2 molecules and is an insulator with a large band gap.

In 1985 McMahan and LeSar predicted that the triple bond in molecular nitrogen should be breakable under very high pressures and a solid should be formed which consists of trivalent (i.e. three-coordinate) nitrogen atoms (pressure-coordination rule). Such structures already exist at normal pressures for the other group 15 elements phosphorus, arsenic, antimony and bismuth. The transformation pres-

Fig. 9.5: Singly bonded polymeric nitrogen (cubic modification). (Reproduced with permission from Prof. Dr. J. Evers, http://www.cup.uni-muenchen.de/ac/evers/index.html)

sure for nitrogen should lie in a range between 500 and 940 kbar. An estimation of the bond energies predicts 38 kcal mol^{-1} for a single bond and 226 kcal mol^{-1} for a triple bond. This predicts a difference of 188 kcal mol^{-1}, and corresponds to the value of 200 kcal mol^{-1} (cf. Tab. 9.4).

The preparation of trivalent nitrogen was achieved in 2004 by Eremets et al. in a diamond cell at 1 150 000 bar and 2 000 K [47–49]. The crystallographic data for the trivalent nitrogen is: cubic, lattice parameter a = 3.4542(9) Å. A three-dimensional structure which consisted of trivalent nitrogen atoms (Fig. 9.5) was found. The N—N bond length at 1.1 Mbar is 1.346 Å, and the NNN angle is 108.8°. The nitrogen atoms form screws of trivalent atoms which are connected to form a three-dimensional network.

Unfortunately, this cubic polynitrogen is not stable under normal conditions and already decomposes back to molecular dinitrogen at a pressure of approximately 42 GPa.

The pentazolates are the final, all-nitrogen members of the azole series. Pentazolate salts have been notoriously elusive, despite enormous efforts to make them in either the gaseous or condensed phases over the last hundred years. Quite recently, Oleynik et al. [*Chem. Mater.* **2017**, *29*, 735–741] reported the successful synthesis of a solid state compound consisting of isolated pentazolate anions N_5^-, which involved compressing and laser heating cesium azide (CsN_3) mixed with cryogenic liquid N_2 in a diamond anvil cell. The experiment was suggested by theory, which predicted transformation of the mixture would occur at high pressures to form the new compound, cesium pentazolate (CsN_5). This work provides critical insight into the role of extreme conditions in probing and exploring unusual bonding types that ultimately leads to the formation of novel, high-nitrogen and also high-energy density species. Fig. 9.5a shows the X-ray structures of two CsN_5 polymorphs at 60 GPa.

In this context, it should also be mentioned that very recently Zhang et al. [**Science** *355*, 374–376.] were able to finally report the synthesis and characteriza-

Fig. 9.5a: Crystal structures of two energetically very similar CsN$_5$ polymorphs at 60 GPa.

tion of a salt containing the pentazolate anion which is stable at room temperature and can be prepared and isolated under standard conditions: $(N_5)_6(H_3O)_3(NH_4)_4Cl$.

9.3 High-Nitrogen Compounds

One synthetic approach is to synthesize nitrogen-rich compounds (N > 60%), which also contain oxidizing groups (e.g. nitro: $-NO_2$, nitrimino: $=N-NO_2$, Fig. 9.6), in order to achieve the best possible oxygen balance (Ω). This is important because in principal, it is always better to have the oxidizer and fuel combined in *one* molecule (in contrast to mixtures), because this generally results in superior detonation parameters such as a higher detonation velocity and detonation pressure. Of course, the new compounds synthesized also should not be worse in any of the positive properties RDX shows (e.g. thermal stability, low sensitivity etc.). Table 9.5 shows a summary of the desired properties for new, nitrogen-rich highly energetic compounds.

$$-\overset{|}{\underset{|}{C}}-NO_2 \qquad -\overset{|}{\underset{|}{C}}-O-NO_2 \qquad -\overset{|}{\underset{|}{C}}-\underset{H}{\overset{|}{N}}-NO_2 \qquad \diagdown C=N-NO_2$$

(a) \qquad\qquad (b) \qquad\qquad (c) \qquad\qquad (d)

Fig. 9.6: Oxidizing groups which can facilitate a positive oxygen balance; Nitro (a), Nitrato (b), Nitramino (c), Nitrimino (d).

Tab. 9.5: Desired properties of new nitrogen-rich highly energetic compounds.

performance	detonation velocity	$D > 8500$ m s^{-1}
	detonation pressure	$p > 340$ kbar
	heat of explosion	$Q > 6000$ kJ kg^{-1}
stability	thermal stability	$T_{dec.} \geq 180$ °C
	impact sensitivity	IS > 7 J
	friction sensitivity	FS > 120 N
	electrostatic sensitivity	ESD > 0.2 J
chemical properties	hydrolytically stable,	
	compatible with binder and plasticizer,	
	low water solubility[a] (or non-toxic),	
	smoke-free combustion,	
	long-term stable (> 15 years under normal conditions)	

[a] low octanol-water partition coefficient

Of course, it is difficult to fulfill all these criteria, in particular as further practical aspects also have to be taken into account, including: simplest possible synthesis, a cheap and sustainable synthesis (no chlorinated solvents), as well as being able to automate the individual synthetic steps [50, 51]. So far it has been possible to identify two classes of compounds which show many of the properties mentioned above:
- tetrazole compounds [52, 53],
- trinitroethyl compounds [54].

9.3.1 Tetrazole and Dinitramide Chemistry

While triazole compounds are not often energetic enough and pentazole species are kinetically too labile, the tetrazole derivatives often combine the desired endothermicity with kinetic stability (Fig. 9.7). Aminotetrazole (AT) and diaminotetrazole (DAT) are good starting materials for the synthesis of new highly energetic, nitrogen-rich compounds. AT and DAT can be synthesized in accordance with the synthetic schemes outlined in Figure 9.8. For DAT, the synthesis starts from thiosemicarbazide (easily obtained from KSCN and hydrazine) in acidic medium.

Fig. 9.7: The basic structures of 1,2,4-triazole (a), tetrazole (b) and pentazole compounds (c).

9.3 High-Nitrogen Compounds

Fig. 9.8: Two synthetic routes for the synthesis of aminotetrazole (AT, a) and the synthetic route for diaminotetrazole (DAT, b).

Often, energetic compounds composed of salts show advantages over their covalent analogues; they show a lower vapor pressure for example, essentially eliminating the risk of inhalation of toxic compounds. Furthermore ionic compounds usually have higher densities, higher thermal stabilities and larger critical diameters (IM) [55]. On the other hand, salts are often water soluble and hygroscopic, which is negative. The solid state density (ρ) is a particularly important parameter since not only the detonation pressure (p) but also the detonation velocity (D) are directly dependent on the density, which we have already seen using the simple relationship of Kamlet and Jacobs which was discussed in Chapter 3 [10].

$$p_{C\text{-}J} \text{ [kbar]} = K\rho_0^2 \Phi$$

$$D\,[\text{mm } \mu\text{s}^{-1}] = A\,\Phi^{0.5}(1 + B\rho_0)$$

The constants K, A and B are defined as follows:

$$K = 15.88$$
$$A = 1.01$$
$$B = 1.30$$

The parameter Φ is therefore

$$\Phi = N(M)^{0.5}(Q)^{0.5}$$

286 — 9 Design of Novel Energetic Materials

Fig. 9.9: Synthesis of diaminotetrazolium dinitramide (HDAT⁺DN⁻).

Fig. 9.10: Structures of aminotetrazolium dinitramide (HAT⁺DN⁻, left) and diaminotetrazolium dinitramide (HDAT⁺DN⁻, right).

where
- N is the no. of moles of gas released per gram of explosive,
- M the mass of gas in gram per mole of gas and
- Q is the heat of explosion in cal per gram.

Aminotetrazole (AT) and diaminotetrazole (DAT) can easily be converted into the perchlorate salts, which can subsequently be converted into the highly energetic salts HAT⁺DN⁻ and HDAT⁺DN⁻ by reaction with potassium dinitramide (KDN). Figure 9.9 shows the synthesis of diaminotetrazolium dinitramide. The structures of both high-energy salts was determined using single crystal X-ray diffraction (Fig. 9.10).

The most important parameters for the characterization of the performance of secondary explosives for use in energetic formulations in war heads and general purpose bombs (GPBs) are the heat of detonation (Q), the detonation pressure (p) and the detonation velocity (D). For comparison, Figure 9.11 shows a summary of the important values for aminotetrazolium dinitramide (HAT⁺DN⁻), diaminotetrazolium dinitramide (HDAT⁺DN⁻) and RDX.

Tetrazole compounds are also good starting materials for the synthesis of energetic materials with substituted side-chains attached to the tetrazole skeleton. For example, tetrazole can be alkylated by reaction with 1-chloro-2-nitro-2-azapropane and converted into various energetic compounds (Figs. 9.12 and 9.13).

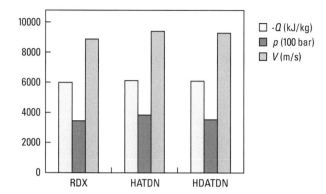

Fig. 9.11: Comparison of the performance data for aminotetrazolium dinitramide (HAT⁺DN⁻), diaminotetrazolium dinitramide (HDAT⁺DN⁻) and hexogen (RDX).

Fig. 9.12: Synthesis of 1-chloro-2-nitro-2-azapropane.

Compound **8** for example is stable up to 184 °C (melting point of 150 °C), is not impact sensitive (IS > 100 J) and is only slightly friction (FS = 120 N) and electrostatic sensitive (ESD = 0.22 J). The performance parameters such as the detonation velocity (8467 m s⁻¹), detonation pressure (273 kbar) and heat of explosion (5368 kJ kg⁻¹) are significantly higher than the values for TNT and only marginally lower than those of RDX.

Out of the different possible derivatives of tetrazoles, we have so far only discussed (**A**) and nitriminotetrazole (**D**) (Fig. 9.14) in more detail. Although the azide-substituted tetrazole compound of type **C** is very energetic, it is still not useful for practical application since it is extremely sensitive. Additionally of interest are the neutral and ionic nitrotetrazole derivates of type **F**, which are more stable (Fig. 9.14).

While free nitrotetrazole is extremely sensitive and only of limited thermal stability, the salts of nitrotetrazole are, at least thermally, often significantly more stable. The best synthetic route starts from 5-aminotetrazole via the corresponding copper salt (M = alkali metal):

288 — 9 Design of Novel Energetic Materials

Fig. 9.13: Synthesis of tetarazole compounds with energetic side-chains.

Fig. 9.14: Tetrazole derivatives with different substituents.

Fig. 9.15: Synthesis of various energetic nitrotetrazolate salts.

Although AgNT can be obtained from NaNT and AgNO$_3$, ammonium nitrotetrazolate is best prepared as the hemihydrate (Fig. 9.15) by the acidification of sodium nitrotetrazolate with sulfuric acid, followed by extraction and reaction with ammonia gas. This compound is particularly suited to the synthesis of further energetic nitrotetrazolate salts such as, for example the hydrazinium, guanidinium, aminoguanidinium, diamonoguanidinium and triaminoguanidinium salts (Fig. 9.15).

Even though they are mainly of academic interest, the nitrotetrazole compounds are very suitable species to study using ^{15}N NMR spectroscopy. Figure 9.16 shows an example of very nicely resolved ^{15}N NMR spectra of free nitrotetrazole, as well as of both the methylated isomers.

A very interesting approach to the synthesis of new, ethyl-bridged, energetic tetrazole compounds, both neutral and ionic has been reported recently by Shreeve and co-workers [56a, b]. They react *in situ* generated cyanazide, N$_3$CN, with 1,2-diaminoethane to form the bridged tetrazole derivative **A** (Fig. 9.17), which can sub-

Fig. 9.16: ^{15}N NMR spectra of 2-methyl- (a), 1-methyl-nitrotetrazole (b) and nitrotetrazole (c).

Fig. 9.17: Synthetic scheme for the preparation of briged nitriminotetrazole compounds.

sequently be converted with HNO_3 to the corresponding neutral nitrimino derivative **B**. Finally, using hydrazine, compound **B** can be converted to the energetic hydrazinium salt **C**, which is surprisingly thermally stable (m. p. = 223 °C).

Another very promising tetrazole compound, bistetrazolylamine (H_2BTA), can be obtained directly from sodium dicyanamide and sodium azide [56c]:

$$\text{NaN(CN)}_2 + 2\text{ NaN}_3 \xrightarrow{\text{HCl}} \text{H}_2\text{BTA} \cdot \text{H}_2\text{O} \xrightarrow{-\text{H}_2\text{O}} \text{H}_2\text{BTA}$$

The anhydrous compound has a density of 1.86 g cm^{-3}, good performance data (V_0D = 9120 m s^{-1}, p_{C-J} = 343 kbar) but a very low sensitivity (impact > 30 J, friction > 360 N). Energetic salts of the type [Cu(H$_2$BTA)$_2$][NO$_3$]$_2$ are also known.

A new synthetic approach for high-nitrogen compounds has recently been suggested by Charles H. Winter et al. which uses poly(pyrazolyl)borate complexes [56d]. Since a borate anion [BR]$_4^-$ carries only one negative charge but can include four heterocyclic ring systems R, theoretically, up to 8 N$_2$ molecules can get liberated if R is a tetrazole ring system, i. e. if we are talking about a tetrakis(tetrazolyl)borate anion.

9.3.2 Tetrazole, Tetrazine and Trinitroethyl Chemistry

Through the combination of very oxygen-rich trinitroethyl groups with the nitrogen-rich and endothermic tetrazole and tetrazine ring systems, the syntheses of three highly energetic trinitroethyl derivatives were achieved:
1. trinitroethyl-tetrazole-1,5-diamine (TTD),
2. bis(trinitroethyl)-tetrazole-1,5-diamine (BTTD),
3. bis(trinitroethyl)-1,2,4,5-tetrazine-3,6-diamine (BTAT).

The synthesis of these compounds is relatively facile using trinitroethanol and the corresponding amines, while for the synthesis of TTD diaminotetrazole, BTTD TTD and for BTAT, diamino-tetrazine was used (Fig. 9.18). At pH values above 6, trinitroethanol behaves as an acid, so that for the Mannich reaction shown in Figure 9.18, two different mechanisms can be discussed.

An important characteristic of these compounds is that due to the dipolar nitro groups, the molecules form strong intra and intermolecular interactions including hydrogen bonds, which contribute not only to an increase in the thermal stability, but also to a higher density. This is particularly clear in trinitroethanol due to its intermolecular dipolar nitro group interactions (Fig. 9.19). The molecular structures of TTD, BTTD and BTAT which were obtained using single crystal X-ray diffraction are shown in Figures 9.20–9.22. The high densities, combined with the endothermic character of the tetrazole or tetrazine moieties and the good oxygen balance of the trinitroethyl group, all result in very good detonation parameters (Tab. 9.6).

As can be observed in Table 9.6, TTD and BTAT have good to very good performance parameters with acceptable sensitivities, however, since TTD decomposes at 126 °C, its thermal stability is too low. In contrast, BTAT is stable to 184 °C (DSC, 5 °C/min). A long-term study at 140 °C for 48 h also showed no decomposition of BTAT.

At this point it should be pointed out, that BTAT has the same CHNO composition as CL-20, but it is significantly less sensitive than the energetic ε modification of CL-20. Figure 9.23 shows a photo of a steel-sleeve test (Koenen test) performed using BTAT with a critical diameter of 10 mm (see Ch. 6.1).

Fig. 9.18: Mannich reaction for the synthesis of trinitroethyl-tetrazole-1,5-diamine (TTD), bis(trinitroethyl)-tetrazole-1,5-diamine (BTTD) and bis(trinitroethyl)-1,2,4,5-tetrazine-3,6-diamine (BTAT).

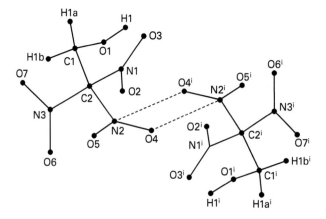

Fig. 9.19: Intermolecular dipolar nitro group interactions in trinitroethanol.

Tab. 9.6: Properties of TTD and BTAT in comparison with RDX.

	TTD	BTAT	RDX
formula	$C_3H_5N_9O_6$	$C_6H_6N_{12}O_{12}$	$C_3H_6N_6O_6$
Ω / %	−15.2	−10.9	−21.6
$\Delta_f H°(s)$ / kJ mol^{-1}	+356	+336	+85
$\Delta_f U°(s)$ / kJ kg^{-1}	1443	852	280
$T_{dec.}$ / °C	126	184	202
ρ / g cm^{-3}	1.831	1.886	1.80
D / m s^{-1}	9194	9261	8894
p_{C-J} / kbar	370	389	338
T_{ex} / K	4650	4867	4290
Q_{C-J} / kJ kg^{-1}	6018	6135	5875
V_0 / l kg^{-1}	788	743	797
impact sensitivity / J	30	7	7
friction sensitivity / N	40	160	120

As we have already seen, trinitromethane and trinitroethanol are important starting materials in the synthetic chemistry of highly nitrated compounds. The precursor potassium trinitromethanide can be prepared best from tetranitromethane (TNM):

$$O_2N-\underset{NO_2}{\underset{|}{C}}-NO_2 \xrightarrow{\text{KOH, Glycerol, H}_2\text{O}} [O_2N-C(NO_2)_2]^{\ominus} K^{\oplus}$$

Free trinitromethane can be obtained by acidification with phosphoric acid:

$$K^+[C(NO_2)_3]^- \xrightarrow{H_3PO_4} H-C(NO_2)_3$$

Tetranitromethane can itself be prepared in good yield by the nitration of acetic anhydride or isopropanol using conc. HNO_3. The synthesis requires very exact control of the reaction conditions and is not without danger.

$$4 \ (CH_3CO)_2O + 4 \ HNO_3 \rightarrow C(NO_2)_4 + 7 \ CH_3COOH + CO_2$$

Finally, trinitroethanol (TNE) can be obtained by the condensation of trinitromethane with formaldehyde (paraformaldehyde) in acidic conditions:

$$H-C(NO_2)_3 + H_2CO \rightarrow HO-CH_2-C(NO_2)_3$$

Fig. 9.20: Molecular structure of trinitroethyl-tetrazole-1,5-diamine (TTD) from single crystal X-ray diffraction data.

Fig. 9.21: Molecular structure of bis(trinitroethyl)-tetrazole-1,5-diamine (BTTD) from single crystal X-ray diffraction data.

The corresponding chlorocarbonate can be obtained by the reaction of free TNE with phosgene, and is also an important building block in the chemistry of polynitro-compounds (Fig. 9.24).

9.3 High-Nitrogen Compounds — 295

Fig. 9.22: Molecular structure of bis(trinitroethyl)-1,2,4,5-tetrazine-3,6-diamine (BTAT) from single crystal X-ray diffraction data.

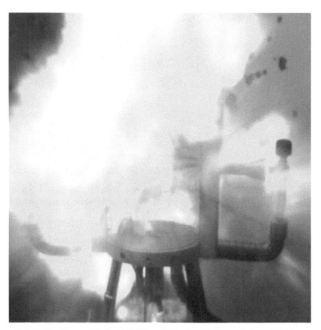

$C_6H_6N_{12}O_{12}$, $T_{dec} > 180\,°C$, $\Sigma\,(O,N) > 82\,\%$, $D = 9261$ m s^{-1}, $P_{CJ} = 389$ kbar

Fig. 9.23: Explosion of bis(trinitroethyl)-1,2,4,5-tetrazine-3,6-diamine (BTAT) in Koenen test.

A better method for obtaining the chlorocarbonate is by stirring a mixture of TNE, FeCl$_3$, CCl$_4$ and a few drops of water under reflux for an hour.

Fig. 9.24: Synthesis of trinitroethanol and trinitroethylchlorocarbonate.

The synthesis of hydrazinium nitroformate (HNF) occurs best through the reaction of anhydrous hydrazine with trinitromethane (nitroform) in methanol or diethyl ether:

$$H-C(NO_2)_3 + N_2H_4 \rightarrow \underset{\text{HNF}}{[N_2H_5]^+[C(NO_2)_3]^-}$$

9.3.3 Ionic Liquids

The term ionic liquids refers to liquids that exclusively contain ions. They are liquid salts which do not need to be dissolved in a solvent such as water. As a rule, the term ionic liquids is used when the corresponding salts are liquid at temperatures lower than 100 °C. Examples of suitable cations are alkylated imidazolium, pyridinium, ammonium or phosphonium ions. The size and the symmetry of the participating ions prevents the formation of a strong crystal lattice, meaning that only small amounts of thermal energy are needed to overcome the lattice energy and break up the solid crystal structure.

Current research is focused on finding suitable nitrogen-rich anions, so that ionic liquids could also be used as fuels in bipropellants (see Ch. 2.2 and Ch. 2.4). In comparison with the currently used fuels such as MMH and UDMH, ionic liquids would possess a negligible low vapor pressure and therefore considerably reduce the inhalation toxicity. Initial investigations have shown that the dicyanamide anion, $[N(CN)_2]^-$, in combination with the 1-propargyl-3-methyl-imidazolium cation may be a suitable ionic liquid for rocket propulsion. However, using WFNA as the oxidizer, the ignition delay is 15 ms, comparative to 1 ms for the MMH / NTO system and 9 ms for MMH / WFNA [57].

Tab. 9.7: Influence of the water content on the specific impulse (in s) of oxidizing and reducing salt mixtures.

oxidizer	fuel	I_{sp}^* (0 % water)	I_{sp}^* (20 % water)	I_{sp}^* (40 % water)
HAN	HAA	372	356	319
HAN	AA	366	342	303
HAN	HA	378	355	317
ADN	HAA	376	352	317
ADN	AA	362	335	303
ADN	HA	373	347	314
AN	AA	349	324	281
AN	HAA	366	342	303
AN	HA	359	334	293
HN	AA	374	349	309
HN	HAA	401	354	316
HN	HA	379	354	314
	Hydrazine	224		

Ionic liquids should also be very suitable for use as monopropellants, but the salts used must contain either the oxidizer and fuel combined, or salt mixtures which contain both oxidizing and reducing salts. Since these mixtures are homogeneous systems which contain both the oxidizer and fuel, they can be labeled as monopropellants, just as hydrazine is. Particularly interesting are salt mixtures which are less toxic and have a lower vapor pressure than hydrazine. Such mixtures are also known as "green propellants". Suitable anions are the nitrate or dinitramide ions [58]. A combination which has already been studied intensively as an oxidizer is the HAN, hydroxylammonium nitrate system. ADN, ammonium nitrate (AN) and hydrazinium nitrate (HN) have also been investigated. As fuels, hydroxylammonium azide (HAA), ammonium azide (AA) or hydrazinium azide (HA) may be appropriate. As a rule, these salt mixtures are not used as pure substances on safety grounds, but with 20 or 40 % water added; they then decompose catalytically in an exothermic reaction. Table 9.7 shows the dependence of the calculated specific impulses on the water content for such salt mixtures.

Quite recently several new energetic ionic liquids (EILs) have been synthesized either consisting of imidazole salts or 1,2,4-triazole salts but until now only few EILs from derivatives of tetrazole have been described in the literature. Tetrazoles are built of 5-membered heterocycles with four nitrogen atoms in the ring. The heats of formation ($\Delta_f H°$) get more and more positive with increasing number of nitrogen atoms (imidazole $\Delta_f H° = +58.5$ kJ mol^{-1}, 1,2,4-triazole $\Delta_f H° = +109.0$ kJ mol^{-1}, 1,2,3,4-tetrazole $\Delta_f H° = +237.2$ kJ mol^{-1}). Furthermore the use of tetrazoles is of great interest since the generation of molecular nitrogen as an end-product of propulsion is highly desirable. Also the formation of carbon residues can be minimized. Tetrazoles can be synthesized by many different synthetic routes, which are mostly low in price and also portable in larger scales.

R¹, R² = Me, Et, CH$_2$CH$_2$-N$_3$, NH$_2$
R³ = H, Me, NH$_2$, CH$_2$-N$_3$
X = NO$_3^-$, N(NO$_2$)$_2^-$, ClO$_4^-$, N$_3^-$

Fig. 9.25: 1,4,5-Substituted tetrazolium cation with different energetic anions X.

X = NO$_3^-$ (**1**), N(NO$_2$)$_2^-$ (**2**), ClO$_4^-$ (**3**), N$_3^-$ (**4**)

Fig. 9.26: Energetic Ionic Liquids based on the 1,4,5-trimethyltetrazolium (TMTz⁺) cation.

Tab. 9.8: Calculated gas phase heats of formation of **1–4**.

	1	2	3	4
$\Delta H°f(g)$ / kcal mol^{-1}	128	173	136	248

In the future, 1,4,5-substituted tetrazolium salts with the general formula depicted in Fig. 9.25 should be synthesized and evaluated. Due to the introduction of hydrophobic side chains (except for NH$_2$) one would expect these compounds to have melting points below 100 °C.

It should be advantageous to introduce methyl-, ethyl-, ethylazide- and aminogroups, on the 1- and 4-position of the tetrazole ring system as well as hydrogen, methyl-, amino- and methylazido groups to the 5-position. Theoretical calculations have shown that these tetrazolium salts in combination with classic energetic and oxidizing anions like nitrate (NO$_3^-$), dinitramide (N (NO$_2$)$_2^-$), perchlorate (ClO$_4^-$) or azide (N$_3^-$) should have high heats of formation and, consequently, high specific impulses. For example, the 1,4,5-trimethyltetrazolium cation (Fig. 9.26), which is not able to form classical hydrogen bonds, has a calculated gas phase enthalpy of formation of 127.6 kcal mol^{-1}. In combination with the energetic anions X = nitrate, dinitramide, perchlorate, and azide (Fig. 9.25) four new EILs, with high heats of formation can be discussed. Table 9.8 summarize the calculated thermodynamic parameters.

A significant step to a lower-toxicity bipropulsion system would be the demonstration of hypergolicity (spontaneous ignition) between an ionic liquid, which is a paragon of low vapor toxicity, and a **safer oxidizer**. Apart from cryogens, hydrogen peroxide seems to be especially promising because of its high performance, less-

Fig. 9.27: Borohydride based ionic liquids [THTDP][BH$_4$] (**1**) and [THTDP][Al(BH$_4$)$_4$] (**2**).

Tab. 9.9a: Drop test results of **2** on four oxidizers.

	90% H$_2$O$_2$	98% H$_2$O$_2$	N$_2$O$_4$ (NTO)	WFNA[b]
reaction	ignition	ignition	ignition	explosion
ignition delay / ms	< 30	< 30	ignition in vapor phase before the liquids combined	–

[a] WFNA = white fuming nitric acid

toxic vapor and corrosivity, and its environmentally benign decomposition products, which make handling this oxidizer considerably less difficult than N$_2$O$_4$ or HNO$_3$. A high fuel performance can be fostered by light metals with large combustion energies and relatively light products. Elements with considerable performance advantages and nontoxic products are aluminum and boron. The need for light combustion products through the production of hydrogen gas and water vapor is fulfilled by a high hydrogen content. Aluminum and boron are well known for their ability to serve as hydrogen carriers in neutral and ionic molecules.

Quite recently, Schneider and other researchers from the Air Force Research Laboratory at Edwards AFB tested various ILs with 90% and 98% H$_2$O$_2$ [57b]. In view of the advantages of high hydrogen content, ILs containing [Al(BH$_4$)$_4$]$^-$ ions may be viewed as a densified form of hydrogen stabilized by metal atoms. In this recently published study, Schneider et al. investigated the new compounds trihexyltetradecylphosphonium tetrahydridoborate [THTDP][BH$_4$] (**1**) and [THTDP][Al(BH$_4$)$_4$] (**2**) (Fig. 9.27).

Compounds **1** and **2** were subjected to drop tests to determine their reactivity with common propulsion oxidizers, including 90% and 98% H$_2$O$_2$ (Table 9.9a). While **1** lights only 3 s after dropping onto H$_2$O$_2$, the ignition delay of **2** was quite short. The hypergolic dicyanamide ILs display changes in delay times (with white

Tab. 9.9b: Computed equilibrium I_{sp} values[b].

O/F (mass %)	NTO / MMH	N_2H_4 / H_2O_2	(THTDP)[Al(BH_4)$_4$] (2)
50 / 50	287	292	218
60 / 40	311	**304**	226
70 / 30	**321**	303	257
80 / 20	286	269	291
85 / 15			**302**
90 / 10			286

[a] chamber pressure = 15 bar, expansion pressure = 0.05 bar

fuming nitric acid, WFNA) from 30 ms to 1000 ms upon reversal of the order of addition of oxidizer and fuel. In contrast, the ignition of **2** is equally fast regardless of the order of addition. These simple drop tests place only upper limits on the ignition delays because ignition may be initiated by hydrogen, which burns with an almost invisible flame. However, these tests do demonstrate that an IL with a complex [Al(BH_4)$_4$]⁻ anion is universally reactive with traditional rocket oxidizers including lower hazard H_2O_2. The computed specific impulses for various H_2O_2/**2** ratios in comparison with the systems NTO(N_2O_4) / MMH and N_2H_4 / H_2O_2 are summarized in Tab. 9.9b. Furthermore, this new class of ILs holds the potential for enabling high-performing, noncryogenic, green bipropulsion for the first time.

9.4 Dinitroguanidine Derivatives

Exceeding the detonation performance of known explosives such as RDX (1,3,5-trinitro-1,3,5-hexahydrotriazine) is a long term research goal in the field of new energetic compounds. The development of strained cages like CL-20 (HNIW: 2,4,6,8,10,12-hexanitro-2,4,6,8,10,12-hexaazaiso-wurzitane) or ONC (octanitrocubane) has attracted lots of interest, however both materials have several drawbacks. The existence of polymorphs as well as a cost intensive preparation makes both compounds disputable for contemporary application. As an aside, highly nitrated compounds like HNB (hexanitrobenzene) show the most promising properties, but their chemical stability towards acids or bases is often not good enough. As the material density is the key for well performing high explosives, a different approach has been started with the concept of zwitterionic structures. A molecule capable of fulfilling these requirements is 1,2-dinitroguanidine (DNQ), which has been reported and well described in the literature previously. The crystallographic characterization of DNQ shows different nitro group surroundings, contrary to dissolved DNQ which shows rapid prototropic tautomerism. The presence of an acidic nitramine (pK_a = 1.11) in DNQ makes it possible to produce monoanionic dinitro-

Fig. 9.28: Synthesis of ammonium dinitroguanidine (ADNQ) and 1,7-Diamino-1,7- dinitrimino-2,4,6-trinitro-2,4,6-triaza-heptane (APX).

guanidine salts. Nucleophilic substitution reactions involving the dinitroguanidine anion are reported in literature as well.

1,2-Dinitroguanidine (**1**) can easily be synthesized as reported by Astrat'yev. The synthetic route is shown in Figure 9.28. Since it is known that strong nucleophiles like hydrazine or ammonia can substitute nitramine groups at elevated temperatures, **ADNQ** (ammonium dinitroguanidine) was produced from a stoechiometric amount of ammonium carbonate to avoid the free base. 1,7-Diamino-1,7-dinitrimino-2,4,6-trinitro-2,4,6-triaza-heptane (**APX**) was also synthesized according to the method of Astrat'yev, starting from 1,3-bis(chloromethyl)nitramine (**2**) and potassium dinitroguanidine (**3**). The former compound is readily available from the nitrolysis of hexamine with subsequent chlorination.

Differential scanning calorimetry (DSC) measurements were used to determine the decomposition temperatures of APX and ADNQ, and indicate that decomposition of APX starts with an onset temperature of 174 °C. In contrast to this behavior the decomposition temperature of the ionic compound ADNQ is 197 °C. In addition, both compounds were tested according to the UN3c standard in a Systag, FlexyTSC Radex oven at 75 °C for 48 hours with the result, that no weight loss or decomposition products were detected.

The detonation parameters calculated with the EXPLO5 program using the experimentally determined densities (X-ray) as well as some sensitivity values are summarized in Table 9.10. Especially the detonation parameters of APX are particularly promising exceeding those of TNT and RDX and in part even those of HMX. The most important criteria of high explosives are the detonation velocity (D = APX: 9540, ADNQ: 9066, TNT: ~7178, RDX: 8906, HMX: 9324 m s^{-1}), the detonation pressure (p_{C-J} = APX: 398, ADNQ: 327, TNT: 205, RDX: 346, HMX: 393 kbar) and the

Tab. 9.10: Detonation parameters.

	APX	APX + 5 % PVAA	ADNQ
ρ / g cm^{-3}	1.911	1.875	1.735
Ω / %	−8.33	−16.28	−9.63
Q_v / kJ kg^{-1}	−5935	−5878	−5193
T_{ex} / K	4489	4377	3828
p_{C-J} / kbar	398	373	327
D / m s^{-1}	9540	8211	9066
V_0 / l kg^{-1}	816	784	934
IS / J	3	5	10
FS / N	80	160	252
ESD / J	0.1	–	0.4

energy of explosion (Q_v = APX: −5943, ADNQ: −5193, TNT: 5112, RDX: −6043, HMX: −6049 kJ kg^{-1}) [59a].

The sensitivity and performance values (Tab. 9.10) revealed that ADNQ may be a high-performing but still relatively insensitive high explosive, while APX needs to be classified as a primary explosive due to its high impact sensitivity.

9.5 Co-Crystallization

Until recently, co-crystallization methods have only been used in pharmaceuticals. However, the hope in the energetics area is that due to strong intermolecular interactions (π stacking, H-bonding) one could reduce the sensitivity of a particular good performer which is too sensitive (e.g. CL-20, APX etc.) when co-crystallized with another compound. Unfortunately, when a high explosive is co-crystallized with a non-energetic compound, the energetic component inevitably sees its explosive power diluted. Quite recently, Matzger et al. reported the first energetic-energetic co-crystal composed of CL-20 and TNT in a 1:1 molar ratio [59c]. The co-crystal CL-20 · TNT also exemplifies the capabilities of aliphatic, highly nitrated compounds to co-crystallize despite their lack of strong, predictable interactions. With no available π-π stacking, this co-crystal forms based on a series of CH hydrogen bonds between nitro group oxygens and aliphatic hydrogens as well as interactions between the electron-deficient ring of TNT and nitro groups of CL-20 (Fig. 9.29). The co-crystal CL-20 · TNT has a density of 1.91 g cm^{-3}, which is somewhat lower than those of CL-20 polymorphs (1.95–2.08 g cm^{-3}), but still substantially higher than that of TNT (1.70–1.71 g cm^{-3}). The impact sensitivity of CL-20 · TNT was measured using a drop test and determined to be approximately half as sensitive as CL-20. Therefore, incorporating insensitive TNT into a co-crystal with sensitive CL-20 greatly reduces its impact sensitivity, potentially improving the viability of CL-20

Fig. 9.29: Molecular assembly in the co-crystal CL-20·TNT.

in explosive applications. This initial study holds promise that more energetic–energetic co-crystals may be discovered from among the nitro-rich, non-aromatic compounds that dominate the field.

9.6 Future Energetics

Great strides have been made in increasing performance and decreasing sensitivity in energetic materials since the first commercialization of nitroglycerine (NG) in the form of dynamite in 1867 by Alfred Nobel. However, the DoD continues rely on traditional, half-century old energetics to meet their combat needs. New energetic materials must be developed to enable DoD to advance beyond traditional thought on energetic capabilities, to transition to the next stage in combat, and to maintain a lethality overmatch with all enemies.

Energetic materials are most commonly used in either high explosives or propellant formulations. Certain parameters are important in determining the effectiveness of new molecules in these formulations, including high densities (ρ), good oxygen balance (Ω) and high detonation/combustion temperatures and high specific impulses (I_{sp}) for rocket propellant formulations and lower combustion temperatures combined with a high force and pressure and a high N_2/CO ratio of the reaction gases for gun propellants.

Using the heat of explosion (Q), the detonation velocity (D) and the detonation pressure (P) as a measure for the performance of a high explosive, one can clearly see from Fig. 9.30 that since the time NG was taken into service the performance of chemical explosives has improved substantially.

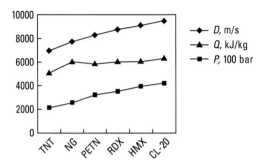

Fig. 9.30: Performance of chemical explosives.

Tab. 9.11: Nitrogen content (N) and oxygen balance (Ω) of conventional HEs.

	N / %	Ω / %
TNT	18.5	−73.9
PETN	17.7	−10.1
RDX, HMX	37.8	−21.6

In spite of the many years of research, there are limited possibilities to realize a substantial increase in performance from conventional C-H-N-O explosives. Recent advances in energetics energy output have come in improved processing or inclusion of energetic binders to increase overall formulation energy, but limited success has been realized in the development of novel energetics. One reason for this is that conventional nitramine and nitroaromatic explosives such as TNT, RDX, HMX and other similar molecules share the same three limitations (Table 9.11):
(i) they are not nitrogen-rich (N ≤ 50 %);
(ii) the oxygen-balance is not close to zero;
(iii) formulations (mixtures of various HEs) are required to achieve a good oxygen balance.

Nitrogen rich molecules are desired as energetics because of the high energy content in N—N bonds. Oxygen balance is defined as the percentage of used oxygen that remains or is needed after an oxidation reaction and can therefore be positive or negative. Materials with an oxygen balance close to zero are typically, but not always, more effective energetics since all of the oxygen is used up in reaction. An oxygen balance can be modified through formulation additions to bring the overall formulation oxygen balance close to zero.

Researchers have already realized the energy content limit for CHNO based molecules. Research needs to expand beyond this way of thinking and increase efforts to explore different molecular structures and molecular make-up in order to

Fig. 9.31: Chemical Structure for HZT and TAGZT.

Tab. 9.12: N content and Ω of high-N compounds.

		N / %	Ω / %
1st generation	HZT	85	−63
	TAGZT	82	−73
2nd generation	TKX-50	59	−27

realize the substantial increase in performance that will be required for future combat systems. Early research has shown that materials with a high nitrogen content offer many advantages to those with carbon backbones, including the potential for vastly increased energy content. Research into molecules with high nitrogen content (> 50 %) has shown potential for a substantial increase in available energy. The first generation of high-nitrogen compounds, such as hydrazinium azotetrazolate (HZT) and triamino-guanidinium azotetrazolate (TAGZT) (Fig. 9.31), did indeed meet the criteria for being nitrogen rich and proved to be very desirable ingredients in erosion-reduced gun propellants (see Fig. 4.4a), however, due to the unfavorable oxygen balance such compounds are not suitable as energetic fillers in high explosive compositions (Tab. 9.12).

The second generation of high-nitrogen compounds which have improved oxygen balances such as TKX-50 and ABTOX (Fig. 9.32), combines desirable high nitrogen content with a good oxygen balance (Tab. 9.12). These compounds are therefore more suitable for use in high-explosive formulations. Moreover, materials with an oxygen balance close to zero are also suitable as powerful ingredients in solid rocket propellants. An increase of the I_{sp} of only 20 s would be expected to increase the payload or range by ca. 100 %. Related to this, Fig. 4.4a shows the computed performance parameters for conventional and high-N gun propellants.

Fig. 9.32: Synthesis of 1-BTO from DAGL (top) and synthesis of TKX-50 and ABTOX (bottom).

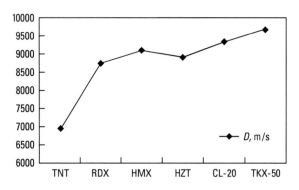

Fig. 9.33: Performance of conventional and high-nitrogen explosives (D = VoD).

The computed and predicted performance values not only exceed the first generation of high-nitrogen compounds (*e.g.* HZT) but, in case of TKX-50, also the performance values of RDX and HMX (Fig. 9.33).

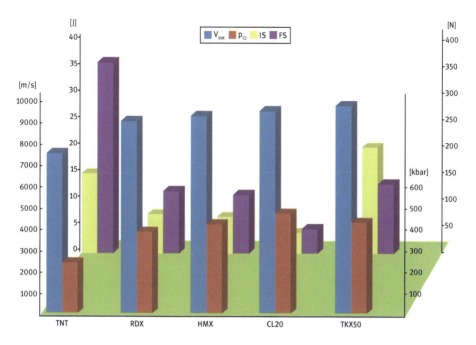

Fig. 9.34: Performance (detonation velocity and C-J pressure) and sensitivity data (impact and friction) of TKX-50 in comparison with other high explosives.

TKX-50 (bis(hydroxylammonium) 5,5'-bis(tetrazolate-1N-oxide)) is one of the most promising ionic salts being developed as a possible replacement for RDX [59d–f]. It can be prepared on a > 100 gram-scale scale by the reaction of 5,5'-(1-hydroxytetrazole) with dimethyl amine which forms the bis(dimethylammonium) 5,5'-(tetrazolate-1N-oxide) salt which is then isolated, purified and subsequently reacted in boiling water with two equivalents of hydroxylammonium chloride to form TKX-50, dimethylammonium chloride and HCl (Fig. 9.32).

TKX-50 shows high performance with respect to its detonation velocity and C-J pressure, but a desirable low sensitivity towards impact and friction (Fig. 9.34).

The thermal behavior of TKX-50 (bis(hydroxylammonium) 5,5'-(tetrazolate-1N-oxide)) and the kinetics of its thermal decomposition were studied using differential scanning calorimetry (DSC) and thermogravimetric analysis (TGA). The thermal decomposition of TKX-50 starts to occur within the range 210–250 °C depending on the heating rate used.

By applying multiple heating rate DSC measurements and Ozawa's isoconversional model free method, an activation energy of 34.2 kcal mol^{-1} and pre-exponential factor of $1.99 \cdot 10^{12}$ s^{-1} were calculated from the DSC peak maximum temperature – heating rate relationship.

By applying nonisothermal TGA experiments and the Flynn–Wall isoconversional model free method, it was found that the activation energy changes with

Fig. 9.35: The initial pressure on the interfaces of the explosive-barrier from different materials (from polystyrene to tungsten).

conversion and lies between 34.7 and 43.3 kcal mol^{-1}, while the pre-exponential factor ranges from $9.81 \cdot 10^{11}$ to $1.79 \cdot 10^{16}$ s^{-1}.

The enthalpy of formation of TKX-50 was calculated as ΔH_f° (TKX-50) = +109 kcal mol^{-1}. The experimentally determined value based on bomb calorimetry measurements is ΔH_f° (TKX-50) = +113 ± 2 kcal mol^{-1}.

The calculated detonation parameters as well as the equations of state for the detonation products (EOS DP) of the explosive materials TKX-50 and MAD-X1 (and also for several of their derivatives) were obtained using the computer program EXPLO5 V.6.01. These values were also calculated for standard explosive materials which are commonly used such as TNT, PETN, RDX, HMX, as well as for the more powerful explosive material CL-20 for comparison. The determination of the detonation parameters and EOS DP was conducted both for explosive materials having the maximum crystalline density, and for porous materials of up to 50 % in volume. The influence of the content of the plastic binder which was used (polyisobutylene up to 20 % in volume) on all of the investigated properties was also examined.

Calculated results on shock wave loading of different inert barriers in a wide range of their dynamic properties under explosion on their surfaces of concrete size charges of different explosive materials in various initial states were obtained with the use of the one-dimensional computer hydrocode EP. Barriers due to materials such as polystyrene, textolite, magnesium, aluminum, zinc, copper, tantalum or tungsten were examined (Fig. 9.35). Initial values of pressure and other parameters of loading on the interface explosive-barrier were determined in the process of conducted calculations. Phenomena of propagation and attenuation of shock waves in barrier materials were considered too for all possible situations.

From these calculations, an essentially complete overview of the explosion properties and characteristics of shock wave action on barriers was obtained for several new and also for several standard explosive materials for comparison

Fig. 9.36: Acceleration of copper plates by free charges of HMX (1), TKX-50 (2), MAD-X1 (3) and RDX (4).

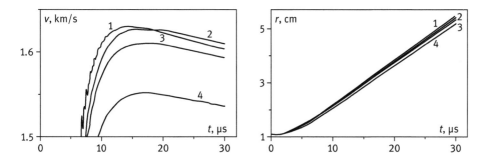

Fig. 9.37: Acceleration of Tantalum cylindrical layers by charges of HMX (1), TKX-50 (2), MAD-X1 (3) and RDX (4).

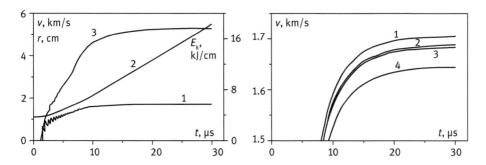

Fig. 9.38: Acceleration of a cylinder (1 = v; 2 = r; 3 = E_k) by a porous charge of TKX-50 (left) and by porous charges of HMX (1), TKX-50 (2), MAD-X1 (3) and RDX (4).

(Figs. 9.36, 9.37 and 9.38). The results which were obtained suggest that in a wide range of their initial states (porosity, inert binders), the new explosive materials TKX-50 and MAD-X1 possess better explosive properties and shock wave action on practically every compact barrier which was considered, in comparison with stan-

Tab. 9.13: Results of a small scale shock reactivity test (SSRT).

explosive	mass [mg]	dent [mg SiO$_2$]
RDX	504	589
CL20	550	947
TKX50	509	857

Tab. 9.14: Calculated cylinder energies for TKX-50 and RDX.

compound V/V$_0$	E_C / kJ cm^{-3}	% of standard			
		TATB	PETN	HMX	Cl-20
TKX-50 2.2	−8.16	168	128	109	90
RDX 2.2	−6.94	143	109	93	77

Tab. 9.14a: Detonation velocities (VoD) of TKX-50 depending on the tube diameter.

sample	column diameter/mm	density/g cm^{-3}	VoD/m s^{-1}
TKX-50 / binder = 95.5 / 4.5	20 × 20	1.79	8699
TKX-50 / binder = 95.5 / 4.5	30 × 30	1.80	8774
TKX-50 / binder = 95.5 / 4.5	40 × 40	1.80	8994
TKX-50 / binder = 95.5 / 4.5	50 × 50	1.81	8996
TKX-50 / binder = 95.5 / 4.5	60 × 60	1.80	9037

dard explosive materials including the military explosive RDX. The large number of the calculated results which were obtained can be used to study the influence of different factors on the explosion and shock wave action of new explosive materials. Furthermore, some of these results can then be used to plan experiments to confirm the predicted properties.

An experimental small scale shock reactivity test (SSRT) (Tab. 9.13) was in good agreement with the calculated cylinder energies (Tab. 9.14).

The detonation velocity was measured by the electrical method using an electronic timer and fiber optic probes. Apparently TKX-50 has a large critical diameter (> 3 cm), but shows a high detonation velocity of 9460 m s^{-1} in a cylindrical 40 mm tube.

Quite recently, researchers from China's Xi'an Modern Chemistry Research Institute investigated the detonation velocities of TKX-50 depending on the tube diameter (Tab. 9.14a) (*Propellants Explos. Pyrotech.* **2019**, *44*, 408–412).

Tab. 9.14b: The thickness of the organic glass gap and the corresponding shock-wave pressure for TKX-50, HMX and CL-20 based explosives.

explosive	thickness / mm	pressure / GPa
TKX-50	18.45	6.24
RDX	40.50	3.02
CL-20	49.51	2.64

Tab. 9.15: Compatibility of TKX-50 performed according to a vacuum stability test (STANAG 4556 ED.1), 40 hrs, 100 °C (compatible 0.0–3.0 mL).

Material	Reactivity of mix (mL)	Result
TKX-50	0.067	pass
TKX-50 & CuO	0.62	pass
TKX-50 & steel shavings	negligible	pass
TKX-50 & Al powder	0.13	pass
TKX-50 & R8002	2.4	pass
TKX-50 & BDNPA/F	0.94	pass
TKX-50 & chlorez wax	negligible	pass
TKX-50 & Viton	negligible	pass
TKX-50 & DNAN	negligible	pass
TKX-50 & TNT	negligible	pass
TKX-50 & NTO	negligible	pass
TKX-50 & HMX	negligible	pass
TKX-50 & PrNQ	1.2	pass

The detonation velocities were obtained for TKX-50-based explosive columns with different diameters, and an obvious scale effect was observed. Steady detonation can perhaps not be achieved when the column diameters are less than 30 mm.

The shock sensitivity was determined by the same authors using the gap test (Tab. 9.14b).

The critical initiation pressures of the TKX-50-based explosives that were examined are much higher than those for HMX and CL-20-based explosives.

While TKX-50 shows good compatibility with other energetic fillers, binders and plasticizers as well as with most metals and metal oxides (Tab. 9.15), its purity is crucial in order for it to exhibit good compatibility with all of the above for use in formulations.

The vacuum stability test (VST, Tab. 9.15) is frequently used for the determination ofthe compatibility of energetic materials with contact materials as defined in STANAG 4147 and for quality tests of energetic ingredients (other STANAGs). A

Fig. 9.39: Structure of carbonic dihydrazidinium bis[3-(5-nitroimino-1,2,4-triazolate)] (CBNT).

dried and accurately weighed sample of an energetic material (or mixture of materials) is placed in the heating tube followed by its assembly and evacuation. The heating tube is immersed in a constant temperature bath (usually 100 °C or 120 °C) for a period of 40 hrs and the volume of evolved gas is recorded. Most explosives or explosive formulations yield less than 1 cm³ of gas per gram of an explosive during 40 hrs at 120 °C. Vacuum stability test finds its wide application in qualification, surveillance, manufacture, quality control and R&D of a whole range of energetic materials.

For many years, the synthesis of tetrazino-tetrazine 1,3,6,8-tetraoxide (TTTO) (see figure below) had been one of the holy grails of N–O-chemistry within the area of energetic materials. Recently, a ten-step synthesis of TTTO has been accomplished by Tartakokovsky et al. [*Angew. Chem. Int. Ed.* **2016**, *55*, 11472–11475]. The synthetic strategy was based on the sequential closure of two 1,2,3,4-tetrazine 1,3-dioxide rings by the generation of oxodiazonium ions and their intramolecular coupling with tertbutyl-NNO-azoxy groups. TTTO is considered as being an example of a new high-energy compound.

Another very promising ionic energetic material is carbonic dihydrazidinium bis[3-(5-nitroimino-1,2,4-triazolate)] (CBNT) which was first synthesized in 2010 by Shreeve et al. (*J. Am. Chem. Soc.* **2010**, *132*, 11904) (Fig. 9.39). The salt is thermally stable up to 220 °C and has a high density of 1.95 g cm^{-3} and very good calculated detonation parameters of VoD = 9399 m s^{-1} and p$_{C-J}$ = 360 kbar.

Chemists at LMU have developed two new secondary explosives based on bispyrazolylmethane: one of which (**A, BDNAPM**) is stable up to 310 °C, and the other has an exceptional detonation velocity of over 9300 m s^{-1} (**B, BTNPM**)

Tab. 9.16a: Sensitivity and performance data for BTNAPM and BTNPM.

	BDNAPM	BTNPM	BTNPM + Al (85/15%)	HMX
Formula	$C_7H_6N_{10}O_8$	$C_7H_2N_{10}O_{12}$		$C_4H_8N_8O_8$
FW [g mol^{-1}]	358.19	418.15		296.16
IS [J]	11	4		7 [1]
FS [N]	> 360	144		112 [1]
ESD-test [J]	> 1	0.6		0.2
N [%]	39.10	33.50		37.84
Ω [%]	−40.20	−11.48	−23.1	−21.61
$T_{dec.}$ [°C]	310	200		320
Density [g cm^{-3}]	1.836 (173 K) 1.802 (298 K)	1.934 (298 K)	202 (298 K)	1.905 (298 K)
$\Delta_f H°$ / kJ mol^{-1}	205.1	378.6		116.1
$\Delta_f U°$ / kJ kg^{-1}	655.8	976.8		492.5
EXPLO6.02 values:				
$-\Delta_E U°$ [kJ kg^{-1}]	5052	6254	8003	5794
T_E [K]	3580	4570	5527	3687
p_{C-J} [kbar]	295	393	360	389
D [m s^{-1}]	8372	9300	8583	9235
Gas vol. [L kg^{-1}]	706	704	524	767

(Tab. 9.16a). In collaboration with the US Army Research Laboratory (ARL), the detonation velocities were experimentally determined using the LASEM method (Laboratory-Scale *Method* for Estimating *Explosive* Performance, see chapter 9.7).

A B

Prior to this work, methylene-bridged bispyrazolylmethane derivatives had not been described as high-energy materials. Compound **A** is thermally stable up to 310 °C and shows a detonation velocity of 8332 m s^{-1}. HNS (VoD = 7629 m s^{-1}) is currently the most commonly used explosive in oil exploration. However, compound A has a much better performance than HNS.

The detonation parameters of the hexanitro species **B** are even better, with a detonation velocity of 9300 m s^{-1} and a detonation pressure of 30 GPa. Currently, the most commonly used explosive in military applications is RDX (hexogen), which has a significantly lower performance (8803 m ss^{-1} and 33.8 GPa) than compound **B**.

The synthesis of both compounds is relatively straightforward and is shown in the following equation:

Some energetic pyrazoles (most notably dinitropyrazole and trinitro pyrazole, as well as the corresponding N-methylated derivatives) have proven to be sensitizing agents, with problematic levels of toxicity causing rashes, blisters, and very bad headaches over time. These symptoms have been found to be especially prevalent when larger scales of the energetic compounds are being handled.

Nitrate esters (e.g. NG) generally have a bad reputation because they show high sensitivities and often low thermal stabilities. However, they are essential ingredients in NC based propellant formulations.

A new and promising nitrate ester based on the bi-isooxazole unit has recently been reported by Sabatini et al. [*Eur. J. Org. Chem.* **2017**, 1765–1768]. The compound, bi-isoxazoletetrakis(methyl nitrate), can easily be synthesized in a metal-free [3+2] cycloaddition reaction as shown in Fig. 9.40.

The sensitivities of the new tetranitrate ester are comparable to those of PETN (IS = 3 J, FS = 60 N, ESD = 0.0625 J), and the compound has a calculated detonation velocity and detonation pressure between those of NG and PETN (Tab. 9.16b).

Bi-isoxazoletetrakis(methyl nitrate) has the potential to be used in multiple functions. The characteristics and performance properties of the new nitrate ester suggest that this material has the potential to serve as: (a) a potential ingredient in melt-castable eutectic formulations; (b) a potential energetic plasticizing ingredient with nitrocellulose-based propellant formulations, in an effort to reduce the volatility/migration issues that arise during cook-off; (c) a potential ingredient in percussion primer compositions.

Another very promising melt-cast explosive is the compound bis(1,2,4-oxadiazole)bis(methylene) dinitrate (Tab. 9.15a). An improved synthesis for this compound has already been recently published by Sabatini, Chavez et al. (OPR&D, in press: 10.1021/acs.oprd.8b00076) and is shown in Fig. 9.40a.

Fig. 9.40: Synthesis of bis(1,2,4-oxadiazole)-bis(methylene) dinitrate.

Tab. 9.16b: Energetic properties of bi-isoxazoletetrakis(methyl nitrate).

	bi-isoxazoletetrakis (methyl nitrate)	PETN	NG
m.p. / °C	121.9	141.3	14
$T_{dec.}$ / °C	193.7	170.0	50.0
$\Omega(CO_2)$ / %	−36.7	−10.1	+3.5
$\Omega(CO)$ / %	0	+15.2	+24.7
ρ / g cm^{-3}	1.786	1.760	1.60
p_{C-J} / kbar	271	335	253
VoD / m s^{-1}	7837	8400	7700
$\Delta H°_f$ / kJ mol^{-1}	−395	−539	−370

The physical properties and energetic performance of bis(1,2,4-oxadiazole)bis(methylene) dinitrate (**2**) are summarized in Tab. 9.15a.

To summarize this chapter, it can be concluded that most of the presently used high explosives are nitramine compounds, e.g. RDX and HMX (also CL-20). Nitrate esters are often too sensitive and thermally not stable enough, but are needed for double-base propellant formulations. Whereas nitromethyl and especially trinitromethyl compounds suffer from low thermal stability, the related fluoro-nitro derivatives $R_2CF(NO_2)$ are expensive and usually very toxic. One area of high-energy materials that needs to be focused on in the future is that of polycyclic aliphatic nitramines, including cages and fused rings. One promising example of such a compound, and a step in the right direction, is bicycle-HMX (BCHMX [cis-1,3,4,6-tetranitrooctahydroimidazo-[4,5-d]imidazole]), which was first synthesized by Zeman and Elbeih [*PEP* **2013**, *38*, 238–243 & 379–385).

Fig. 9.40a: Synthesis of bis(1,2,4-oxadiazole)-bis(methylene) dinitrate.

Tab. 9.15a: Calculated physical properties of bis(1,2,4-oxadiazole)bis(methylene) dinitrate (2) in comparison with those of TNT and Composition B.

property	2	TNT	Comp. B
m.p./°C	84.5	80.4	78–80
$T_{dec.}$/°C	183	295	200
$\Omega(CO_2)$/%	−33	−74	–
ρ/g cm^{-3}	1.83	1.65	1.68–1.74
p_{C-J}/GPa	29.4	20.5	26–28
VoD/m s^{-1}	8180	6950	7800–8000

BCHMX

BCHMX (1,3,4,6-tetranitrooctahydroimidazo[4,5-d]imidazole) is usually prepared from KTACOS (potassium imidazo[4,5-d]imidazole-1,3,4,6-tetrasulfonate) which can easily be obtained from potassium sulfamate, formaldehyde, and glyoxal.

Future research will lead to even more powerful, high-nitrogen high-oxygen explosives with enhanced and superior detonation parameters to fulfil the lethality requirements of all forces. New energetics will provide vastly increased energy content over RDX, up to several times the energetic performance. In addition, these materials will have a high energy density with a high activation energy.

With these capabilities, DoD will be able to develop new applications for energetic materials and vastly improve performance in current munitions. The increased performance of the next generation of energetics will enable DoD to meet the same strategic goals with fewer munitions and less energetic material. The increase in energy output can be harnessed in munitions that require a fraction the amount of energetics to deliver the same payload or thrust profile. This will allow engineers to put energetics in munitions that have never had them before.

9.7 Detonation Velocities from Laser-induced Air Shock

Laser-induced air shock from energetic materials (LASEM) is a laboratory-scale technique for estimating the detonation velocity of energetic materials [see Gottfried et al.: *Angew Chem Int. Ed.*, **2016**, *55*, 16132–16135; *Prop. Expl. Pyrotech.*, in press: DOI: 10.1002/prep.201600257 and *Phys. Chem. Chem. Phys.* **2014**, *16*, 21452]. A focused nanosecond laser pulse is used to ablate and excite an energetic material residue, forming a laser-induced plasma (Fig. 9.41). Exothermic chemical reactions in the plasma accelerate the laser-induced shock wave generated by the plasma formation – thus, energetic materials produce faster shock waves than inert materials (as measured on the microsecond timescale by a high-speed camera). The measured laser-induced air shock velocity can be correlated to measured detonation velocities from large-scale testing, and used to estimate the detonation velocity of the material. However, the energetic material is not detonated by the laser, and therefore, LASEM only provides an estimate of the maximum attainable detonation velocity for an energetic material. Despite this, it has the distinct advantage of requiring only milligram quantities of the energetic material.

Fig. 9.41: Comparison between the LSDW (laser-supported detonation wave) and the blast wave, which are distinct when the laser interrogates the sample off-angle (a) and combined when the propagation of the laser is perpendicular to the sample (b).

New energetic material candidates are often initially synthesized in only small quantities for safety reasons, and the subsequent scale-up to multiple gram quantities can be costly. While the shot-to-shot variation in the measured laser-induced shock velocities can be significant due to the stochastic nature of the laser-material interaction, many laser shots can be acquired to obtain a reproducible average at minimal expense (especially compared to large-scale detonation testing, where significant safety hazards increase the cost of testing). The usefulness of the method was originally demonstrated by comparing the laser-induced air shock velocities of inert and energetic materials, followed by calibration of these values by correlating the measured laser-induced air shock velocities from conventional military explosives to the measured detonation velocities from large-scale detonation testing. The calibration fit was validated using conventional energetic materials (both monomolecular and composite) which were not used to develop the correlation. Estimated detonation velocities for materials where measured detonation velocities have not been reported were also calculated for nanoscale cyclotrimethylene trinitramine (nano-RDX) and three high-nitrogen explosives. This method has also recently been extended to mixtures of military explosives with metal additives (aluminum or boron).

In this chapter, the estimated detonation velocities of six new explosives, dihydroxylammonium 5,5′-bistetrazole-1,1′-diolate (TKX-50), dihydroxylammonium 5,5′-bis(3-nitro-1,2,4-tri-azolate-1N-oxide) (MAD-X1), bis(4-amino-3,5-dinitropyrazol-1-yl)methane (BDNAPM), bis(3,4,5-trinitropyrazol-1-yl)methane (BTNPM), 5,5′-bis(2,4,6-trinitrophenyl)-2,2′-bi- (1,3,4-oxadiazole) (TKX-55), and 3,3′-diamino-4,4′-azoxyfurazan (DAAF), are discussed and compared to the theoretically predicted detonation velocities obtained using two different thermochemical codes (EXPLO5 V6.01 and CHEETAH V8.0). TKX-50, MAD-X1, BDNAPM, BTNPM, TKX-55, and DAAF, are members of the group of recently synthesized explosives which are amongst the most promising prospective candidates for use in practical applications (Fig. 9.42).

Figs. 9.43-I and 9.43-II show snapshots from the high-speed video of the laser excitation of the six explosives mentioned above. The selected photos represent

Fig. 9.42: Chemical structures of TKX-50, MAD-X1, BDNAPM, BTNPM, TKX-55, and DAAF.

the typical laser shot for each material. The brightness was increased (+40%) for all of the images in Fig. 9.43-I to enable enhanced visualization. Adjustment of the brightness and/or contrast for the first few frames was necessary during measurement of the shock wave position, in order to improve visualization of the shock wave in the presence of the defocused plasma emission. The few individual shots where the plasma emission completely obscured the shock wave position in the first frame were discarded (however, a total of 15 videos with measurable shock waves in the first frame were obtained for each sample). Since the camera was focused in front of the laser-induced plasma to enhance visualization of the shock wave, the plasma emission is slightly out of focus. The purple light in the snapshots is a result of the strong CN emission, while the white light is residual continuum emission from the laser-induced plasma. It was found that TKX-55 (Fig. 9.43-I-a) and DAAF (Fig. 9.43-I-d) produce the most intense plasma emission, followed by TKX-50 (Fig 9.43-I-b), BDNAPM (Fig. 9.43-I-e), and MAD-X1 (Fig. 9.43-I-f). BTNPM (Fig. 9.43-I-c) produced very little visible plasma emission following laser excitation. It has been previously observed that, in general, the most powerful energetic materials produce the smallest amount of visible plasma emission.

In Figure 9.43-II-d–f, a thin, dark string of excited material coincident with the laser beam (center of images) is particularly visible in the later frames. It also appears, to a lesser extent, in Figure 9.43-I-b (TKX-50) and in later frames of TKX-55, once the plasma emission has decreased (not shown). The structure of the plasma plume region is also distinct for these energetic materials (DAAF, BDNAPM, MAD-X1). The only conventional military explosive known to show similar features is TATB, which could indicate that these effects are related to the low thermal sensitivity of these energetic materials. BTNPM is the only explosive in this group that does not show a similar interaction with the laser beam, and it has the lowest decomposition temperature and a higher impact and friction sensitivity compared to most of the other materials.

Fig. 9.43-I: Selected snapshots from the high-speed video of the laser excitation of (a) TKX-55, (b) TKX-50, (c) BTNPM, (d) DAAF, (e) BDNAPM), and (f) MAD-X1.

Fig. 9.43-II: Selected snapshots from the high-speed videos of the laser excitation of (a) TKX-50, (b) BTNPM, and (c) MAD-X1.

Occasionally, unusual features appeared in the high-speed video for random laser shots of a few of the energetic materials. For example, in Figure 9.43-II, distortion of the leading edge of the laser-induced shock wave in air is shown for (a) TKX-50, (b) BTNPM, and (c) MAD-X1 (in order of increasing degree of severity). For these, a second, much smaller plume of reacting material appeared to form near the top of main laser-induced plasma (coincident with the laser beam). This distortion is so severe in the MAD-X1 shot, that a bubble was formed in the main shock wave, which persisted for tens of microseconds. The cause of this behavior is currently unknown, but has so far never been observed in the high-speed videos of conventional military explosives.

The measured characteristic laser-induced air shock velocities (with 95 % confidence intervals), along with the estimated and theoretically predicted detonation velocities calculated using two thermochemical codes (EXPLO5 V6.01 and CHEETAH V8.0), are shown in Table 9.17 for comparison. The average measured detonation velocities from the large-scale detonation testing of some conventional military explosives are also listed in Table 9.17. The percentage difference between the predicted or measured detonation velocities and the estimated detonation velocities from LASEM are given in parentheses. The average difference between the estimated detonation velocities and EXPLO5 predictions is 3.1 % (3.6 % for the six explosives mentioned above and 2.5 % for conventional military explosives). For CHEETAH predictions, the average difference is 3.9 % (4.8 % for the six explosives mentioned above and 2.8 % for conventional military explosives). Using the values in the CHEETAH database, the average difference between

Tab. 9.17: LASEM results in comparison with theoretical predictions and large-scale detonation test results (% difference from estimated detonation velocity indicated in parentheses).

Sample	Laser-induced shock velocity $[m \cdot s^{-1}]$	Estimated V_{LASEM} $[km \cdot s^{-1}]$	EXPLO5 v6.01 V_{C-J} $[km \cdot s^{-1}]$	CHEETAH v8.0 V_{C-J} $[km \cdot s^{-1}]$	Measured $V_{det.}$ at TMD $[km \cdot s^{-1}]$
TKX-50	835 ± 11	9.56 ± 0.28	9.767 (2.1 %)	9.735 (1.8 %)	9.432 (−1.4 %)
MAD-X1	807 ± 9	8.86 ± 0.22	9.195 (3.6 %)	9.267 (4.4 %)	n/a
BDNAPM	798 ± 9	8.63 ± 0.21	8.332 (−3.6 %)	8.171 (−5.6 %)	n/a
BTNPM	850 ± 13	9.91 ± 0.31	9.304 (−6.5 %)	9.276 (−6.8 %)	n/a
TKX-55	782 ± 11	8.23 ± 0.26	8.030 (−2.5 %)	7.548 (−9.0 %)	n/a
DAAF	774 ± 11	8.05 ± 0.26	8.316 (3.2 %)	8.124 (0.9 %)	8.11 ± 0.03 (1.0 %)
TNT	731 ± 9	6.99 ± 0.23	7.286 (4.1 %)	7.192 (2.8 %)	7.026 ± 0.119 (0.5 %)
HNS	740 ± 8	7.20 ± 0.21	7.629 (5.6 %)	7.499 (4.0 %)	7.200 ± 0.071 (0.0 %)
NTO	784 ± 10	8.30 ± 0.25	8.420 (1.4 %)	8.656 (4.1 %)	8.335 ± 0.120 (0.4 %)
RDX	807 ± 8	8.85 ± 0.19	8.834 (−0.2 %)	8.803 (−0.5 %)	8.833 ± 0.064 (−0.2 %)
CL-20	835 ± 10	9.56 ± 0.24	9.673 (1.2 %)	9.833 (2.8 %)	9.57 (0.1 %)

the estimated and predicted values for CHEETAH was 2.4 % (1.0 % for TXK-50 and DAAF; 3.0 % for conventional military explosives).

In contrast to the predicted detonation velocities, the difference between the estimated detonation velocities and measured large-scale detonation velocities for conventional military explosives is only 0.2 %. The accuracy of LASEM for estimating the detonation velocities of high-nitrogen energetic materials is not known, since few large-scale results have been reported in the literature, other than for DAAF and TKX-50. The estimated detonation values agree with the reported detonation velocities for DAAF and TKX-50 to within less than 1.5 %. The LASEM results suggest that TKX-55, BDNAPM, and BTNPM have higher detonation velocities than those predicted by the thermochemical codes, while the estimated detonation velocities for MAD-X1 and TKX-50 are slightly lower than those predicted by EXPLO5 or CHEETAH. A comparison of the two thermochemical codes shows that while the magnitude of the difference between the estimated detonation velocities and predicted detonation velocities may differ (especially for TKX-55), the predicted detonation velocities are typically either both higher or both lower than the estimated (or measured) detonation velocities.

High-speed video investigations confirm that the most powerful explosive among those investigated in this study is BTNPM (since it produced the least intense plasma emission); the weakest are TKX-55 and DAAF. Moreover, the structure of the plasma plume regions of DAAF, BDNAPM, MAD-X1, and TKX-55 appear to reflect the low thermal sensitivity of these energetic materials and the high thermal sensitivity of BTNPM.

The estimated detonation velocities of the six explosives which were mentioned above, which were based on the measured characteristic laser-induced air shock velocities, are in very good agreement with the detonation velocities calculated using the EXPLO5 V6.01 and CHEETAH V8.0 thermochemical codes. The average difference between the estimated detonation velocities and those calculated using EXPLO5 is 3.6 %, while using CHEETAH it is 4.8 %. The LASEM results show that TKX-55, BDNAPM, and BTNPM have higher detonation velocities than those calculated using EXPLO5 or CHEETAH, while the estimated detonation velocities for MAD-X1 and TKX-50 are somewhat lower than those calculated using the thermochemical codes. The detonation velocities calculated using EXPLO5 and CHEETAH differ slightly from each other (especially for TKX-55), nevertheless they show the same tendency (in general, the values are either both higher or both lower than the estimated – or measured – detonation velocities). In comparison to the predicted detonation velocities (EXPLO5: 2.5 %; CHEETAH: 2.8 %), the difference between the estimated detonation velocities and measured large-scale detonation velocities for conventional military explosives is only 0.2 %. Moreover, for DAAF and TKX-50, the estimated detonation values are in agreement with the reported detonation velocities to within less than 1.5 %.

9.8 Thermally Stable Explosives

A new field of secondary explosives was opened up and developed by the investigation of HNS by SHIPP, and also with the investigation of TATB by JACKSON and WING, and is commonly referred to as "thermally stable explosives".

These explosives generally have lower energetic performances in comparison with RDX and would never be suitable replacements for RDX. However, they are exceptional since they show very high decomposition points and very good long-term stabilities. Therefore, such properties open-up the possibility for their use in special applications such as (i) developing new oil wells or renewable energy (geothermal energy) (Fig. 9.44), (ii) space applications and (iii) nuclear weapons.

Since it has a high decomposition temperature of 350 °C, TATB (Fig. 9.45) is a very interesting compound for these applications, but due to the high costs involved in its synthesis and its extreme insensitivity, it is mainly used in military (especially nuclear) applications. Although HNS has a lower decomposition temperature of 320 °C, it also has the advantage of lower production costs, which is why it is the commonly used explosive for civil applications. Two interesting, highly thermally stable explosives were synthesized by COBURN, namely N(2,4,6-trinitrophenyl)-1H-1,2,4-triazol-3-amine (PATO) in 1969, and the 3,5-dinitro-$N2,N6$-bis(2,4,6-trinitrophenyl)pyridine-2,6-diamine (PYX) in 1972 (Fig. 9.45).

The structures in Fig. 9.45 show that thermally stable explosives commonly combine aromatic systems – normally benzene – with nitro groups. The combination of both systems in one molecule can result in an increase in the thermal stabil-

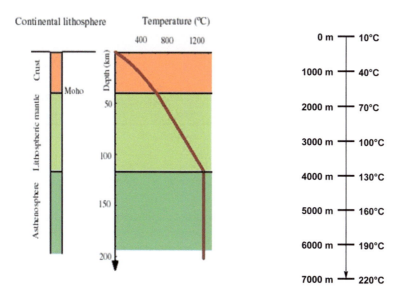

Fig. 9.44: Effect of the variation of temperature depending on the depth (in the ground).

Fig. 9.45: Explosives which show high thermal stabilities a) HNS; b) TATB; c) PATO; d) PYX.

Tab. 9.18: Energetic performance of RDX in comparison with those of HNS, TATB, PYX, and PATO.

	RDX	HNS	TATB	PYX	PATO
IS / J	7.5	5	50	25	8
FS / N	120	240	353	250	–
ESD / J	–	–	–	–	–
$T_{Dec.}$ / °C	210	320	350	350	310
ρ / g cm^{-3}	1.806	1.74	1.80	1.757	1.94
EXPLO5 6.01 values:					
T_{det} / K	3800	3676	3526	3613	3185
P_{C-J} / kbar	352	243	296	252	313
$V_{det.}$ / m s^{-1}	8815	7612	8310	7757	8477
V_0 / L kg^{-1}	792	602	700	633	624

ity. For example, HNS shows a decomposition temperature which is approximately 111 °C higher than that observed for RDX.

A short overview of the energetic performances of explosives with high thermal stabilities in comparison with that of RDX is given in Tab. 9.18.

The connection of two picryl moieties through a 1,3,4-oxadiazole bridge was reported by Dacons and Sitzmann, who described the synthesis of 2,5-bis(2,4,6-trinitrophenyl)-1,3,4-oxadiazole (DPO, Fig. 9.46), which is structurally related to the below mentioned TKX-55.

DPO

Fig. 9.46: Structure of DPO.

Due to the need for thermally highly stable secondary explosives for the oil industry, several possible HNS replacements have been suggested. However, HNS replacements should have the following properties:
1. Show an extremely high thermal stability – which means a decomposition point over 300 °C, as well as long-term stability with no decomposition at 260 °C over a period of 100 h.
2. Detonation velocity greater than 7500 m · s^{-1}.
3. Specific energy greater than 975 kJ/kg. The specific energy is calculated according to Equation:
$$F = p_e \cdot V = n \cdot R \cdot T$$
(n is the number of products of decomposition, R is the gas constant, and T is the detonation temperature).

The specific energies for RDX, HNS, TATB, PYX, and PATO are given below.

	RDX	HNS	TATB	PYX	PATO
Specific energy	1026	752	839	777	675

4. Impact sensitivities > 7.4 J and friction sensitivities > 235 N. These values are slightly higher than those expected for RDX replacements (IS > 7 J; FS > 120 N).
5. Since these explosives are mostly for intended use in civil applications, the total costs should not be greater than 500 Euro/kg.
6. Finally, the critical diameter should be smaller than for HNS. The critical diameter indicates the minimum diameter of an explosive charge at which detonation can still take place. It also depends on the texture of a compound. In cast charges, it is larger than in pressed ones.

To obtain compounds with properties which correspond to the strict requirements outlined above for improved heat-resistant explosives, four general approaches have been outlined by J. P. AGRAWAL over the past few years which can be used to impart and increase the thermal stability:

- Salt formation
  The thermal stability can be enhanced via salt formation. The following example, in Fig. 9.47, is a good illustration of this approach.
- Introduction of amino groups
  Introducing amino groups into nitrated benzene rings in the *ortho* position influences the thermal stability. This is one of the simplest and oldest approaches for enhancing the thermal stability of explosives. The "push-pull" effect generated by the combination of electron withdrawing nitro groups next to electron donating amino groups, as well as the formation of hydrogen bonds, is responsible for the high stability towards heat or other external stimuli like friction or impact which is observed for such compounds. A good illustration is given by the introduction of amino groups to 1,3,5-trinitrobenzene in Fig. 9.48
- Introduction of conjugation
  A very good example for the introduction of conjugation in explosive molecules, and its ability to impart higher thermal stability, is given by hexanitrostilbene (HNS). Another interesting explosive with a similar structure to HNS is hexanitroazobenzene (HNAB). In HNAB, a conjugated system is also present, which increases the thermal stability in comparison with its monomer. In Fig. 9.49, these two examples are shown.
- Condensation with triazole rings
  Various studies on the synthesis of picryl- and picrylamino substituted were reported in detail by Coburn & Jackson. In these investigations, 1,2,4-triazoles, 1,2,4-triazole or amino-1,2,4-triazole was reacted with picryl chloride (1-chloro-2,4,6-trinitrobenzene). One of the most interesting of these molecules is PATO (3-picrylamino-1,2,4-triazole), a well known, thermally stable explosive, which is obtained by the condensation of picryl chloride with 3-amino-1,2,4-triazole (Fig. 9.50). A similar compound was prepared by Agrawal and co-workers, namely 1,3-bis(1',2',4'-triazol-3'-ylamino)-2,4,6-trinitrobenzene (BTATNB), which

$T_{dec.} = 237°C$ $T_{dec.} = 333°C$

Fig. 9.47: Salt formation based on 3,3'-diamino-2,2'4,4',6,6'-hexanitrophenylamine.

9.8 Thermally Stable Explosives

1,3,5-Trinitrobenzene
T_{dec} 122 °C

2,4,6-trinitroaniline (TNA)
T_{dec} 192 °C

1,3-Diamino-2,4,6-trinitrobenzene
T_{dec} 286 °C

1,3,5-Triamino-2,4,6-trinitrobenzene (TATB)
T_{dec} >320 °C

Fig. 9.48: The introduction of amino groups into 1,3,5-trinitrobenzene results in an increase in the thermal stability.

2,4,6-trinitrotoluene (TNT)
T_{dec} 295 °C

Hexanitrostilbene (HNS)
T_{dec} >320 °C

2,4,6-trinitroaniline (TNA)
T_{dec} 192 °C

Hexanitroazobenzene (HNAB)
T_{dec} 215 °C

Fig. 9.49: Introduction of conjugation monomer vs. dimer. On formation of the dimer, the thermal stability increases.

Fig. 9.50: Condensation with a triazole rings PATO (left) and BTATNB (right).

was synthesized by the reaction of styphnyl chloride (1,5-dichloro-2,4,6-trinitrobenzene) with two equivalents of 3-amino-1,2,4-triazole. BTATNB exhibits a slightly higher thermal stability than PATO and is safer with respect to its impact and friction sensitivity.

PATO can be obtained in almost quantitative yields from 3-amino-1,2,4-triazole – which is a common herbicide (Amitrole) – and picryl chloride which can be obtained from conversion of picric acid (*PEP* **2019**, *44*, 203–206):

Agrawal et al. also reported the synthesis of BTDAONAB (Fig. 9.51), which does not melt below 550 °C and was proposed to be a better example of a thermally stable explosive than TATB. According to the authors, this material has a very low impact (21 J), and no friction sensitivity (> 360 N) and is thermally stable up to 550 °C. With these reported properties, BTDAONAB is superior to all of the nitro-aromatic compounds discussed above. BTDAONAB has a VoD of 8300 m/s while TATB is about 8000 m/s.

Moreover, another nitro-aromatic compound (BeTDAONAB), similar to Agrawal's BTDAONAB has recently been published by Keshavarz et al. and is also reported to be very insensitive. In this compound, the terminal triazole moieties have

Fig. 9.51: Molecular structure of BTDAONAB.

Fig. 9.52: Molecular structure of BeTDAONAB.

Tab. 9.19: Comparison of the thermal and explosive properties of TATB, HNS, BTDAONAB, and BeTDAONAB.

Property	TATB	HNS	BTDAONAB	BeTDAONAB
density / g cm^{-3}	1.94	1.74	1.97	1.98
DTA (exo) / °C	360	353	550	275
DSC (exo) / °C	371	350		268
Ω_{CO} / %	−18.6	−17.8	−6.8	−5.9
IS / J	50	5	21	21
FS / N	>353	240	353	362
VoD / m s^{-1}	7900	7600	8600	8700
p_{C-J} / kbar	273	244	341	354

been replaced by two energetic (more endothermic) tetrazole units (Fig. 9.52, Tab. 9.19).

The drawback of these two thermally extremely stable compounds (BTDAONAB and BeTDAONAB, Tab. 9.19) is that they can't be initiated with presently used primary explosives such as lead azide (LA) or lead styphnate (LS).

New Developments:

Quite recently, the synthesis of the new heat-resistant explosive (T_{dec} = 335 °C) 5,5′-bis(2,4,6-trinitrophenyl)-2,2′-bi(1,3,4-oxadiazole) (TKX-55) has been reported. The compound can be prepared from (demilitarized) TNT and shows very promising detonation parameters (VoD = 8030 m s^{-1}, p_{C-J} = 273 kbar (Fig. 9.53).

Even more thermally stable ($T_{dec.}$ = 360 °C) than TKX-55 is a new calixarene explosive (Fig 9.54) which can be prepared by a facile synthetic procedure. However, the performance (VoD = 7865 m s^{-1}, p_{C-J} = 289 kbar) is somewhat lower than that of TKX-55.

Tetrazolyl triazolotriazine (Fig 9.54) was also recently suggested as a possible candidate for use as a TATB replacement. However, its decomposition temperature of 305 °C is somewhat lower than that of TATB (360 °C).

There is clearly demand for new thermally highly stable explosives, and the race is still on to find better and cheaper examples for use in oil and gas exploration, space applications, and nuclear weapons.

Fig. 9.53: Synthesis of TKX-55 (5) from TNT.

Fig. 9.54: Molecular structures of a new calixarene explosive (left; R = R$_1$ = H) and of tetrazolyl triazolotriazine (right).

---

i Check your knowledge: See Chapter 14 for Study Questions.

---

10 Synthesis of Energetic Materials

10.1 Molecular Building Blocks

Most inorganic or organic energetic materials contain at least one of the building blocks shown in Table 10.1 and can therefore be classified by their functional groups.

Tab. 10.1: Molecular building blocks for energetic materials.

group	energetic substance	example
$-O-O-$	organic peroxo compounds	TATP
ClO_3^-	chlorate	$KClO_3$
ClO_4^-	perchlorate	AP
$-NF_2$	organic difluoramine derivatives	HNFX-1
$-NO_2$	organic nitro compounds	TNT
$-O-NO_2$	organic nitrato- or nitroesters	NG, PETN, NG-A
NO_3^-	nitrate	AN
$-N=N-$	organic or inorganic diazo compounds	TAGzT
$-N_3$	covalent azides	GAP
N_3^-	ionic azides	$[N_2H_5][N_3]$, hydrazinium azide
$-NH-NO_2$, $-NR-NO_2$	nitramino compounds	RDX, HMX
$N(NO_2)_2^-$	dinitramide	ADN
$=N-NO_2$	nitrimino compounds	nitriminotetrazole
$-CNO$	fulminates	$Hg(CNO)_2$, mercury fulminate
$-C\equiv C-$	acetylides	Ag_2C_2, silver acetylide

The molecular structures of nitriminotetrazole and tetrakis(difluoramino)octahydro-dinitro-diazocine (HNFX) have not been discussed so far and are shown in Figure 10.1.

Fig. 10.1: Molecular structures of nitriminotetrazole (a) and tetrakis(difluoramino)octahydro-dinitro-diazocine (HNFX) (b).

It is interesting that nitriminotetrazole contains an oxidizing nitrimino group in addition to the endothermic ring, which helps to create a more positive oxygen balance ($\Omega = -12.3\,\%$). HNFX works similarly as it corresponds to a HMX molecule in which two of the nitramine groups are exchanged for $C(NF_2)_2$ groups. In this example, the two remaining nitramine groups and the introduced difluoroamine groups act as strong oxidizers.

10.2 Nitration Reactions

As we have seen in Table 1.2, practically all formulations used by the military contain TNT, as well as RDX and/or HMX. Additionally, PETN and nitroglycerine (in dynamite) play an important role. All these substances are obtained by nitration reactions. An overview of different types of nitration is given in Table 10.2.

Tab. 10.2: Possible nitration reactions.

reaction	C nitration	O nitration	N nitration
product	nitro compounds, $R-NO_2$	nitrate ester, $R-O-NO_2$	nitramine, $R-NH-NO_2$
examples	TNT, PA, tetryl, TATB, HNS	NC, NG, PETN, NG-A	RDX, HMX, NQ
nitrating agent	mixed acid: HNO_3/H_2SO_4	mixed acid: HNO_3/H_2SO_4	mixed acid: HNO_3/H_2SO_4, HNO_3 (100%)

The nitration of organic compounds still occurs almost exclusively using mixed acid or 65% to 100% HNO_3. Mixed acid is a mixture of concentrated nitric acid and concentrated sulfuric acid, which acts as a very strong nitrating agent due to the presence of the NO_2^+ ion.

$$HNO_3 + H_2SO_4 \longrightarrow H_2ONO_2^+ + HSO_4^-$$

$$H_2ONO_2^+ + HSO_4^- + H_2SO_4 \longrightarrow NO_2^+ + H_3O^+ + 2\,HSO_4^-$$

Aromatic hydrocarbons can be converted into nitro compounds via an electrophilic substitution using mixed acid. For example, trinitrotoluene is obtained by the direct nitration of toluene using mixed acid. Other good nitrating agents are $NO_2^+ BF_4^-$ and $NO_2^+OSO_2CF_3^-$ (nitronium triflate) in CH_2Cl_2 with ultrasonic activation.

10.2 Nitration Reactions

A good example of an N-nitration which can be undertaken using mixed acid or 100 % HNO$_3$, is the nitration of aminotetrazole to nitriminotetrazole (Fig. 10.2) [59b]:

Fig. 10.2: Nitration of aminotetrazole to nitriminotetrazole.

Nitriminotetrazole can also be obtained in good yield by the diazotization of nitroaminoguanidine to nitroguanylazide and subsequent cyclization under alkaline conditions (Fig. 10.3).

Fig. 10.3: Synthesis of nitriminotetrazole from nitroaminoguanidine.

A further example of a nitration using nitric acid is the preparation of RDX from hexamethylentetramine (HMTA) via triacetyltriazine (TART) in a two-step reaction (Fig. 10.4):

Fig. 10.4: Synthesis of hexogen (RDX) from hexamethylentetramine (HMTA) via triacetyltriazine (TART).

In modern synthetic chemistry N_2O_5 can also be used as a nitrating agent; it is advantageous because anhydrous conditions can be used and, if pure N_2O_5 is used, there are no acidic impurities present. If acidic impurities do not matter, a mixture of N_2O_5 and HNO_3 is also a good and strong nitrating agent.

Previously, technical N_2O_5 was mainly obtained by the dehydration of nitric acid at −10 °C, as N_2O_5 is the anhydride of nitric acid. It is an easily sublimed solid (subl. 32 °C, 1 bar).

$$12\,HNO_3 + P_4O_{10} \xrightarrow{-10\,°C} 6\,N_2O_5 + 4\,H_3PO_4$$

Since 1983, the technical synthesis usually used follows that developed by Lawrence Livermore National Laboratory, in which the electrolysis of nitric acid in the presence of N_2O_4 results in the formation of a approx. 15–20 % solution of N_2O_5 in anhydrous nitric acid.

$$2\,HNO_3 \xrightarrow{N_2O_4,\ 2e^-} N_2O_5 + H_2O$$

In the following, we want to look at a further way to produce pure and almost acid-free N_2O_5. This was developed into a semi-industrial process (pilot plant) as late as 1992 in the DRA Laboratories (Defence Research Agency). This method involves the gas-phase ozonization of N_2O_4 using an ozone-oxygen mixture with an approx. 5–10 % ozone content [60, 61].

$$N_2O_4 + O_3 \longrightarrow N_2O_5 + O_2$$

Solutions of pure N_2O_5 in chlorinated organic solvents (CH_2Cl_2, $CFCl_3$) are mild nitrating agents, which have recently found a wide range of applications (Tab. 10.3).

An example of a ring-opening nitration reaction is shown below.

$$\text{cyclo-}(CH_2)_nX \xrightarrow{N_2O_5,\ CH_2CL_2,\ 0-10\,°C} O_2NO-(CH_2)_n-X-NO_2$$
$$(n = 2\ \text{oder}\ 3;\ X = O\ \text{oder}\ NR\ (R = \text{Alkyl}))$$

Tab. 10.3: Synthetic application of CH_2Cl_2 / N_2O_5 solutions.

reaction type	product
aromatic nitration	C–NO_2
nitrolysis	N–NO_2
ring opening reactions	N–NO_2 or O–NO_2
selective nitration	O–NO_2 (rarer N–NO_2)

If the substituent on nitrogen is an H atom and not an alkyl group, the nitration occurs with preservation of the cyclic, four-membered ring structure.

$$\text{cyclo-}(CH_2)_3\,NH \xrightarrow{N_2O_5,\ CH_2CL_2,\ 0\text{–}10\,°C} \text{cyclo-}(CH_2)_3\,N(NO_2)$$

In a similar synthetic route, the strong oxidizers TNAZ (trinitroazetidine) and CL-20 were prepared (Fig. 1.7).

ADN is a unique highly energetic material, because it does not contain carbon, chlorine or fluorine. ADN is an environmentally benign high energy inorganic oxidizer salt, first synthesized in 1971 at the Zelinsky Institute of Organic Chemistry in Moscow. It is claimed that ADN-based solid propellants are in operational use in Russian TOPOL intercontinental ballistic missiles and that ADN was previously produced in ton-scale quantities in the former USSR. ADN forms colorless crystals and is a very oxygen-rich nitrogen oxide, which possesses excellent properties for use as an explosive (in combination with strong reducing agents such as aluminum, aluminum hydride or organic compounds) or solid rocket propellant. Additionally, the absence of halogens makes ADN an environmentally friendly solid rocket propellant and makes RADAR detection of the rocket emission more difficult. ADN is currently being tested as one of the most promising replacements for ammonium perchlorate [62].

Ab initio calculations predict a structure with C_2 symmetry for the free dinitramide ion, $N(NO_2)_2^-$ (Fig. 10.5), while in solution and in the solid state the local symmetry is essentially C_1. This can be explained on the basis of weak cation-anion interactions or interactions with the solvent, since the dinitramide ion is very easy to deform because of the very small barrier to rotation of the NNO_2 moiety (< 13 kJ mol^{-1}) [63].

Fig. 10.5: Two different views of the dinitramide ion with C_2 symmetry.

Although ADN is already industrially prepared, the lab synthesis is not trivial and depends on the precise control of the reaction conditions. Normally, ADN is synthesized by the nitration of ammonia using N_2O_5 (prepared by the ozonization of NO_2) in a chlorinated solvent (Fig. 10.6):

Fig. 10.6: Schematic diagram of the apparatus used for the small-scale lab-preparation of N_2O_5 (right) and ADN (left).

synthesis of N_2O_5:
$$2\,NO_2 + O_3 \rightarrow N_2O_5 + O_2$$

synthesis of ADN:
$$NH_3 + N_2O_5 \rightarrow (O_2N)NH_2 + HNO_3$$
$$(O_2N)NH_2 + N_2O_5 \rightarrow (O_2N)_2NH + HNO_3$$
$$(O_2N)_2NH + NH_3 \rightarrow [NH_4^+][N(NO_2)_2]^-$$
$$2\,HNO_3 + 2\,NH_3 \rightarrow 2\,[NH_4^+][NO_3]^-$$
$$\overline{4\,NH_3 + 2\,N_2O_5 \rightarrow [NH_4^+][N(NO_2)_2]^- + 2\,[NH_4^+][NO_3]^-}$$

Note: Free dinitraminic acid is a very strong acid with a pK_A value of −5.6. It is unstable however and decomposes quickly to N_2O and HNO_3.

Industrially, ADN is commonly prepared by the nitration of urea. In the first stage, urea is converted into urea nitrate using dilute nitric acid, which is then dehydrated to form nitrourea by reaction with concentrated sulfuric acid. The 1,1-dinitrourea isomer can subsequently be obtained by nitrating nitrourea with a strong nitrating agent ($NO_2^+\,BF_4^-$, N_2O_5); the dinitrourea then reacts with gaseous ammonia to form ADN and urea:

$$H_2N-CO-NH_2 \xrightarrow{\text{dil. } HNO_3} [H_2N-CO-NH_3]^+[NO_3]^-$$

$$[H_2N-CO-NH_3]^+[NO_3]^- \xrightarrow{\text{konz. } H_2SO_4} H_2N-CO-NH(NO_2)$$

$$(O_2N)HN-CO-NH_2 \xrightarrow{N_2O_5} (O_2N)_2N-CO-NH_2$$

$$(O_2N)_2N-CO-NH_2 + 2\ NH_3 \longrightarrow [NH_4^+][N(NO_2)_2]^- + H_2N-CO-NH_2$$

Using a new process, ADN can be prepared in a very environmentally friendly manner (without the use of chlorinated organic solvents) by the direct nitration of salts (potassium or ammonium) of sulfamic acid (amidosulfonic acid, H_2N-SO_3H) using mixed acid (HNO_3/H_2SO_4):

$$H_2N-SO_3 \xrightarrow{HNO_3/H_2SO_4,\ T<-25\,°C} N(NO_2)_2^-$$

Sulfamic acid (amidosulfonic acid) itself can be obtained simply from urea and pyrosulfuric acid (disulfuric acid, $H_2S_2O_7$):

$$CO(NH_2)_2 + H_2S_2O_7 \rightarrow 2\ H_2N-SO_3H + CO_2$$

Conversion of ADN into different metal salts of ADN is facile by reaction with the corresponding metal hydroxide:

$$MOH + [NH_4][N(NO_2)_2] \rightarrow [M][N(NO_2)_2] + NH_3 + H_2O$$

10.3 Processing

Most explosives which are used by the military are solids under normal conditions and are usually obtained as granules on a technical scale. These granules are then mixed with other explosives and energetic (e.g. binder, plastiziser) or non-energetic additives (anti-oxidants, wetters, wax) and made into their final form using one of the following technical processes:
1. melting, mostly under vacuum,
2. pressing, often under vacuum,
3. extruding at high pressures.

Melting (melt-casting) is one of the oldest technical processes and is normally used for formulations which contain TNT as the melt-castable component and RDX or HMX as the energetic filler. This is essentially the reason why TNT is still often used today. TNT melts at approx. 80 °C and has a very high ignition temperature of 240 °C (Fig. 10.7). The melting and decomposition points of RDX and HMX lie close together, while other energetic compounds such as NQ or TATB decompose at much lower temperatures than their estimated melting points (Fig. 10.7). The main disadvantage of the melt process is, that during cooling cracks can form in the explosive, which has a negative influence on its performance and sensitivity.
To overcome the shortcomings of melt-casting, liquid polymers (e.g. HTPB, GAP etc.) are used as binders while explosives such as RDX or HMX are used as fillers for PBXs. In this so-called **normal-casting** process, the explosive (RDX, HMX) is mixed with a binder (HTPB) which may be crosslinked with a diisocyanate to pro-

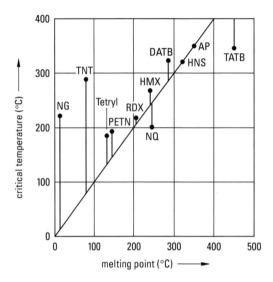

Fig. 10.7: Melting points and critical temperatures of some energetic materials. (This diagram is reproduced in slightly modified form from the original with permission from Prof. Dr. Manfred Held.)

vide a cured and dimensionally stable matrix. This process is often performed under vacuum, and filling can also be agitated by vibration in order to remove any gas bubbles trapped in the matrix. This type of casting is generally adopted for large caliber warheads.

A powder is generally used in **pressing**, which is mainly undertaken under vacuum. The advantage in comparison with the melt process is the low temperature. However, this process is also more hazardous which is why it is always automated or remote controlled. The subsequent mechanical processing of the finished explosive block is also relatively dangerous and should always be carried out using remote-control.

Extruding is usually carried out using a screw-type extruding machine as it is known from the plastic industry. In this process, the explosive and polymer binder are pre-mixed and then mixed, compressed and extruded through a small hole under high pressure using the extruder. This process is particularly suitable for the preparation of polymer bound explosives (PBXs).

10.4 Safe Handling of Spent Acid

A number of severe accidents in the explosives industry have occurred in connection with the chemical instability of spent acids which are generated in industrial-scale nitration units. In order to counter this hazard, safety charts have been constructed by several authors, including BIAZZI.

Nitration reactions are not only exothermic (energy releasing), they also produce under controlled reaction conditions nitrated organic compounds which can be highly explosive. This makes the nitration of organic compounds a potentially hazardous process, and it is therefore perhaps not surprising that the accidental explosion of many industrial nitration plants has been reported in the literature over the years. In addition to the above-mentioned aspects, it also has to be taken into account that unlike common explosives such as TNT, compounds such as nitroglycerine and nitroglycerol (both nitric esters) are known to become very unstable if they are subjected to prolonged contact with acids. During the production of each of these compounds, a considerable number of accidents have occurred predominantly during handling or storing of the spent acid. Spent acid is used to denote the residual acid which remains after the nitration process. The main problematic steps in the production of nitric esters on an industrial scale are the separation of the desired nitric ester from the spent acid and the undesired accumulation of the nitric esters in points such as transfer lines and collecting areas, as well as the build-up of dangerous quantities of nitric ester over periods of time. It is paramount that the unintentional accumulation of nitric acid esters is avoided at all costs, since acidic nitric esters are chemically unstable and sensitive mixtures.

Acidic solutions hydrolyze nitric esters which results in the formation of the corresponding alcohol and nitric acid according to the following equation:

$$RCH_2NO_3 + H_2O \rightarrow RCH_2OH + HNO_3 \qquad (1)$$

Alcohols show a higher sensitivity to wet oxidation than esters. The alcohol group is the most sensitive part and importantly, the alcohol functional group doesn't exist in a completely esterified molecule. Oxidation of the alcohols occurs as follows:

$$RCH_2OH + O \rightarrow RCHO + H_2O \qquad (2)$$

$$RCH_2NO_3 + O \rightarrow RCHO + HNO_3 \qquad (3)$$

Based on reaction (2), it can be concluded that an increase in the number of OH groups (dinitroglycerine or mononitroglycol content) results in a decrease in the stability. At this stage it must be pointed out that equation (3) is probably of importance at elevated temperatures, where the decomposition is a combination of not only thermal but also of oxidative aspects. The latter involves the breaking-up of radicals in the dinitroglycerine or mononitroglycol molecule.

Decomposition may also be affected by the following:

$$RCH_2ONO_2 + NO^+ + 3 H_2O \rightarrow RCH_2ONO + NO_3^- + 2 H_3O^+ \qquad (4)$$

$$RCH_2ONO_2 + HNO_2 \rightarrow RCH_2NO_2 + HNO_3 \qquad (5)$$

Organic nitrites may also be formed by the reaction of nitrogen dioxide with the organic alcohol:

$$RCH_2OH + 2NO_2 \rightarrow RCH_2NO_2 + HNO_3 \quad (6)$$

Organic nitrites are more unstable than nitric acid esters and undergo oxidation as follows:

$$RCH_2NO_2 + O \rightarrow RCHO + HNO_2 \quad (7)$$

Considering reactions (4)–(7), it can be summarized that it is the formation of nitrites that explains the strong decomposition effect of HNO_2 and lower nitrogen oxides, in addition to the autocatalytic character of the decomposition reactions.

Since the separation of nitric esters from spent acid as well as the unwanted accumulation of nitric esters have been shown to be the main causes of accidents at nitroglycerine and nitroglycol units, clearly the safety of industrial nitration plants depends hugely on the composition of spent acid – assuming that the raw materials are in accordance with the specifications. The content of dissolved ester in the acid phase, the content of HNO_3 dissolved in the oil phase, as well as the content of hydroxyl groups arising from mononitroglycol or dinitroglycol in the acid phase are all determined by the composition and temperature of the spent acid.

As a result, it can be concluded that in the manufacture of nitroglycerine and nitroglycol it is imperative that the spent acid composition regions which are safe (or dangerous) in contact with nitric esters are known. The so-called "high safety

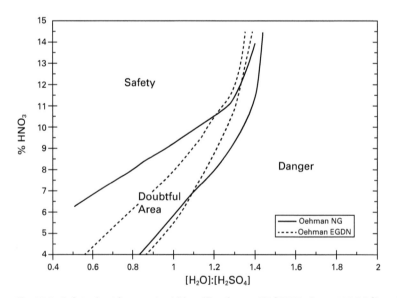

Fig. 10.8: Safety chart for spent acid handling from a NG/EGDN plant at 70 °C (from BIAZZI).

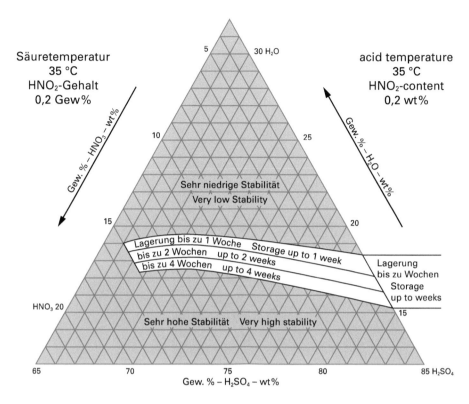

Fig. 10.9: Safety chart for spent acid handling from a NG/EGDN plant at 35 °C (from BIAZZI).

area" in Figure 10.8 corresponds to the area in which a maximum of 0.2% HNO_2 is formed after heating for 2 h at 70 °C. The region designated the "normal safety area" is where 0.2–1% HNO_2 is formed under the same conditions. These statements are only valid when pure materials are used. Systems which contain a high HNO_2 content widen the danger area. Figure 10.9 has been specifically constructed for a system of 35 °C and a HNO_2 content of 0.2 wt %. Figure 10.10 depicts a safety chart for spent acid handling from a NG plant at 35 °C.

Check your knowledge: See Chapter 14 for Study Questions.

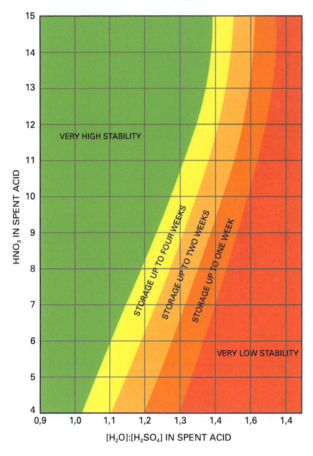

Fig. 10.10: Safety chart for spent acid handling from a NG plant at 35 °C (from BIAZZI).

11 Safe Handling of Energetic Materials in the Laboratory

11.1 General

A chemical research laboratory is different from an industrial production facility for many reasons. First of all, only much smaller quantities of energetic materials are handled in a chemical research laboratory and secondly, during research, the properties of new, energetic materials are not known and therefore particular care must be taken. One of the most important safety rules can be worked out from the "rule of thumb", which relates the distance D (in m), which offers a chance of survival, to the mass w (in kg) of the explosive. For a typical secondary explosive at large distances, the proportionality constant is approximately 2:

$$D = cw^{0.33} \approx 2\,w^{0.33}$$

It is important to note that this approximation is based only on the pressure/impulse of the shock wave and does not take into account the fragment formation (e.g. from confined charges, fume-cupboard front shields in the laboratory, reaction vessel glass).

In order to work safely with highly energetic materials in the chemical laboratory, the following rules must be obeyed:
- use smallest possible quantities,
- keep maximum distance from the experiment (vessels containing compounds not to be transported by hand, but with well fitting tongs or clamp instead),
- mechanical manipulation is to be used if possible, particularly for larger amounts,
- vessels never to be encased by hand (confinement),
- protective clothing is to be used (gloves, leather or Kevlar vest, ear protectors, face shield).

Additionally, one must bear in mind that although primary explosives are much more dangerous to handle – based on the probability of unexpected initiation due to their higher sensitivity – secondary explosives exhibit a much higher performance and if initiated, result in much larger damage than a primary explosive.

With regards to research into new energetic materials, a rule of thumb is that all synthetic attempts should first of all be carried out in 250 mg quantities while various sensitivity data is obtained (impact, friction and electrostatic sensitivity, thermal stability). Only when all of these values are known, can increasing the synthesis to 1 g be considered (only if the properties are suitable). Later, increasing it to 5 and perhaps eventually to 10 g quantities could be possible depending on the sensitivity data obtained. Further tests such as the long-term stability (DSC, long-term heat-flux measurements etc.) must then be carried out.

11.2 Protective Equipment

In addition to the clothing that every chemist should wear in the lab (lab coat, safety glasses, enclosed shoes, non-flammable clothing), one can be protected further during work with small quantities (see above) of explosives. Possibilities include:
- Suitable protective gloves (see below)
- Kevlar® wrist protectors
- Full-visor face shield with polycarbonate shield (to be worn over normal safety glasses)
- Ear protectors
- Shoes with conducting soles against electrostatics (it is important that the floor is also conducting and earthened, especially for handling primary explosives)
- Protective leather jacket (made from 2 mm thick leather) or Kevlar protective vest with groin protector.

Finally, only spatulas made out of conducting plastic instead of metal spatulas should be used. Furthermore, it is important to make sure that the lab equipment, and certainly the vessels for storing the energetic materials are also made out of conducting plastic and stored on a conducting surface.

Injuries to hands are one of the most common accidental injuries in a lab where energetic materials are handled. Safety gloves are available in different materials. In addition to the protective effect, the fine-feel also has to be taken into account. There are several opinions with respect to the most suitable protective gloves. Some chemists prefer thick leather gloves (welding gloves), while others prefer Kevlar gloves. According to the DIN EN 388 (mechanical risks), protective gloves have been classified into three groups based on their cut resistance (I = low, II = medium, III = high). Gloves from class III should always be chosen. Relatively

Fig. 11.1: Size and fragment distribution after an explosion with a quantity of 1 g Pb(N$_3$)$_2$ in a 10-ml glass vessel.

Fig. 11.2: Experimental set-up for investigating different protective gloves, scale unit 1 cm.

good protection is offered by the combination of two pairs of gloves: a tight-fitting pair made from Kevlar® ARMOR (a fiber developed by DuPont) or Kevlar® ES gloves and a second pair worn over the top made from Kevlar® with a woven steel core and a rubber or PVC coated palm side.

Different tests have been suggested to estimate the protection in an explosion. [66] For example, the explosion of 1 g of lead azide in a 10 ml glass vessel has been investigated. The results of the fragment distribution (size and direction dependent) are shown in Figure 11.1.

Tab. 11.1: Number of splinters and penetration depth using different gloves (exptl. set-up see Fig. 11.2 a and b).

	unprotected hand	Kevlar® glove		Kevlar®-glove with steel core		Kevlar®-glove with steel reinforcement		leather welding gloves	
	Exp. I	Exp. I	Exp. II	Exp. I	Exp. II	Exp. I	Exp. II	Exp. I	Exp. II
number of splinters									
in glove									
≥ 1 mm	0	10	13	17	18	10	14	5	4
< 1 mm	4	45	12	38	16	8	21	18	13
total	4	55	25	55	32	18	35	23	17
in hand									
≥ 1 mm	30[a]	14[b]	13	3	2	0	1	1[c]	0
< 1 mm	20	0	5	3	0	2	1	1	0
total	50	14	18	6	2	2	2	2	0
fragment distribution									
in glove									
≥ 1 mm	0 %	5 %	9 %	7 %	9 %	5 %	6 %	3 %	2 %
< 1 mm	0 %	4 %	1 %	3 %	1 %	1 %	2 %	1 %	1 %
total	0 %	2 %	2 %	4 %	2 %	1 %	2 %	1 %	2 %
in hand									
≥ 1 mm	13 %	9 %	9 %	1 %	1 %	0 %	0 %	1 %	0 %
< 1 mm	1 %	0 %	0 %	0 %	0 %	0 %	0 %	0 %	0 %
total	3 %	1 %	1 %	0 %	0 %	0 %	0 %	0 %	0 %
penetration depth (mm)									
deepest	15.0	7.0	8.6	5.0	6.0	1.0	6.0	5.0	–
average value	4.9	2.5	2.8	1.8	3.8	0.8	4.5	4.0	–

[a] 5 splinters > 5 mm
[b] 1 splinter penetrated the whole palm
[c] 1 splinter > 5 mm

Figure 11.2 shows a typical experimental set-up to investigate the protective effect of different gloves. Pictures a and b show the experimental set-up with a hand (gelatine in one-way gloves) at a distance of 10 cm from a 10 ml glass vessel containing 1 g $Pb(N_3)_2$. Figure 11.2c shows the perforations in an unprotected glove and Figure 11.2d shows the fragment impact on a hand protected with a Kevlar® glove. In Figure 11.2e, the penetration depth of a glass splinter is enlarged, despite the protection offered by the Kevlar glove (from Figure 11.2d, below right).

Table 11.1 shows the penetration depths of glass splinters with different types of gloves and an unprotected hand for comparison.

11.3 Laboratory Equipment

In addition to the personal protective equipment discussed above, one can/must equip the laboratory according to the current safety standards. Apart from the standards that every chemical laboratory must correspond to anyway, particular attention must be paid to the following:
- conducting floor covering which is earthened,
- humidity over 60 %, better 70 % (using a humidifier, in particular during winter in rooms with central heating),
- short-term "intermediate storage" of small quantities of energetic materials in appropriate protective containers,
- storing of larger quantities of energetic materials in a specially designated store-room and not in the laboratory,
- separate fridges/cold rooms for oxidizers (HNO_3, NTO, H_2O_2, MON) and fuels (MMH, UDMH),
- mechanical manipulators for the handling of larger quantities of energetic materials,
- autoclave rooms or bunkers for performing different tests (Koenen test, detonation velocity, ...).

Check your knowledge: See Chapter 14 for Study Questions.

12 Energetic Materials of the Future

Since the introduction of nitroglycerine as an explosive by Alfred Nobel in 1867 in the form of Dynamite, large advances have been made in the performance and reduction of sensitivities of highly energetic materials (Fig. 12.1). However, most of the formulations currently used are over 50 years old and do not fulfill all of today's requirements, in particular with regard to their performance, collateral damage, insensitivity, toxicity, compatibility with the environment and use in special operations.

Energetic materials are used mainly in explosives or propellant formulations. As we have already discussed, there are different parameters that characterize the performance of explosives and propellants and which are important in the development of new energetic materials. These include high densities (ρ, more important for explosives than for propellant charges), good oxygen balance (Ω) and high detonation and combustion temperatures (not for propellant charges). However, substances which are suitable for use as secondary explosives are not always suitable for use as propellants and vice versa. While the performance of explosives is particularly influenced by the heat of explosion (Q_{C-J}), the detonation pressure (p_{C-J}) and the detonation velocity (D), good rocket propellants possess a large specific impulse (I_{sp}) and thrust (F). For propellant charges, in addition to a high specific impulse, a high force (f), a low combustion temperature and the highest possible N_2/CO ratio in the products of the reaction are desirable.

When we consider the heat of explosion (Q_{C-J}), the detonation pressure (p_{C-J}) and the detonation velocity (D) as a measure that determines the performance of explosives, then Figure 12.1 shows that large advances have been made since the development of nitroglycerine.

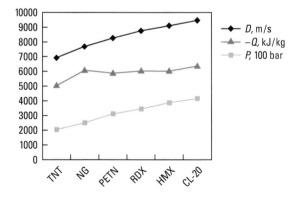

Fig. 12.1: Performance of explosives (Q, heat of explosion; D, detonation velocity; p, C-J detonation pressure).

Tab. 12.1: Nitrogen content and oxygen balance of some conventional explosives.

	nitrogen content / %	oxygen balance / %
TNT	18.5	−73.9
PETN	17.7	−10.1
RDX	37.8	−21.6
HMX	37.8	−21.6

Fig. 12.2: Structures of hydrazinium azotetrazolate (HZT) and triaminoguanidinium azotetrazolate (TAGzT).

Despite these advances, it appears to be difficult to make significant improvements to the performance with conventional C/H/N/O explosives based on nitro and nitrato compounds. This is partly due to, the fact that classic explosives suffer from the following three limitations (Tab. 12.1):
1. they are not nitrogen-rich enough (N rich > 60 % N),
2. the oxygen balance is often too negative,
3. to reach a good oxygen balance, different explosives have to be mixed.

More recent research on nitrogen-rich molecules (N > 60 %) has shown, that these compounds often have a higher energy content because of a much more positive enthalpy of formation in comparison with their C analogues. The first generation of nitrogen-rich compounds such as hydrazinium azotetrazolate (HZT) and triaminoguanidinium azotetrazolate (TAGzT) (Fig. 12.2) have already generated interest as components of propellant charges because of their low erosion properties. However, due to their poor oxygen balance, they are unsuitable for use as high-performance explosives (Tab. 12.2).

This could change with the introduction of the second generation of nitrogen-rich explosives which possess strongly oxidizing groups, and a (nearly) neutral oxygen balance. Examples of these are the compounds triaminoguanidinium

Tab. 12.2: Nitrogen content and oxygen balance of first and second generation N-rich compounds.

	nitrogen content / %	oxygen balance / %
1st generation		
HZT	85	−63
TAGZT	82	−73
2nd generation		
TAG-DN	57	−18
HAT-DN	58	0

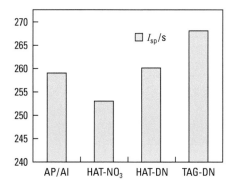

Fig. 12.3: Structures of triaminoguanidinium dinitramide (TAG-DN) and aminotetrazolium dinitramide (HAT-DN).

Fig. 12.4: Calculated specific impulses (I_{sp}^* / s) for AP/Al (70:30), HAT-NO$_3$, HAT-DN and TAG-DN (isobaric combustion at p = 70 bar).

dinitramide (TAG-DN) and aminotetrazolium dinitramide (HAT-DN) (Fig. 12.3, Tab. 10.3) [64, 65].

TAG-DN and HAT-DN are therefore potentially good candidates for use as explosives. But compounds with good oxygen balances are also suitable as components in rocket propellants. An increase in the specific impulse I_{sp}^* of only 20 s should be enough to allow almost a doubling of the payload. Figure 12.4 shows the calculated specific impulse (isobaric, p = 70 bar) for AP/Al (AP/Al) (70:30), HAT-NO$_3$, HAT-DN and TAG-DN.

Fig. 12.5: Detonation velocities (D) of classic CHNO and nitrogen-rich explosives.

Fig. 12.6: Energetic compounds (synthons) for "advanced oxidizers".

The calculated detonation velocities for the second generation of nitrogen-rich compounds (which are one of the criteria for the performance of explosives) are not only better than those of the first generation, but in the case of HAT-DN it is even better than the values for RDX and HMX (Fig. 12.5).

Possible directions for future research are (see also Tab. 12.3):
1. metallized nitrogen-rich compounds with positive oxygen balances,
2. metastable poly-nitrogen element modifications,
3. organometallic compounds with nitrogen-rich or poly-N-ligands,
4. boron-nitrogen compounds with good oxygen balances,
5. substitution of oxygen (in $-NO_2$ groups) by fluorine ($-NF_2$ groups),
6. new, better and perchlorate free oxidizers,
7. new synthetic, energy-rich polymers as nitrocellulose replacements,

Tab. 12.3: Future areas of research.

need	possible solution	research
downsizing of warheads: 4 → 1	increase of ΔU_{ex}: today: 5–6 kJ g^{-1} future: 10–20 kJ g^{-1}; up to 75 % of the mass of a warhead gets "wasted" (non-reactive structure material)	super high explosives, novel HEDMs: poly-N, poly-N$_2$O; reactive fragments: thermites, Al-PTFE composites, Al-Zn-Zr, Al-W and Al-U alloys
UxV's	increase of ΔU_{ex}: today: 5– 6 kJ g^{-1} future: 10–20 kJ g^{-1}; amalgamation of warhead and weapon into one no-return system	super high explosives, novel HEDMs: poly-N, poly-N$_2$O; reactive structures: novel materials
countermeasures against chemical WMDs	ADWs	peroxides, ozone, Cl-containing compounds, hypochlorites
countermeasures against biological WMDs	ADWs	peroxides, ozone, Cl-containing compounds, hypochlorites, F-containing compounds (HF formation)
more efficient fuels	nano	research concerning ageing problems, coating with strongly oxidizing groups (e.g. pernitro alcohols)
perchlorate-free oxidizers	new HEDOs	high-nitro compounds, compounds with peroxo groups
specialized GWT missions	thermobaric systems; Fuel Air Explosives (FAE)	coated nano-Al particles; strongly endothermic liquids

8. more effective, F, Cl and peroxide containing agent-defeat substances,
9. nano-materials (carbon nanotubes),
10. N$_4$, N$_8$ fixation in SWCNTs,
11. new high energy oxidizers (HEDO) (Fig. 12.6)

In this context it is worthwhile to stress that for certain applications various detonation parameters are of particular interest. For example, if the blast effect is the most important property, the detonation pressure (p_{C-J}) is one of the most relevant parameters. However, for shaped charge applications, high detonation velocities are required.

Since lethal weapons predominantly rely on overpressure and blast effect, new high-performance secondary explosives can help to down-size the weapon system. On the other hand, if weapon systems rely on the formation of fragments (fragmenting warheads), new more energetic high explosives can only marginally help to improve the performance. In this case, chemically reactive fragments could be an answer, since in conventional warheads up to 75 % of the mass correspondsto the non-reactive casing (steel casing). Research on chemically stable, but highly reactive materials could help to solve this problem. Such reactive structural materials (RSM) might be thermites, Al-PTFE composite materials or Al-Zn-Zr, Al-W or Al-U based alloys. Such RSMs should possess the following properties:
- High density
- High hardness
- Fast, highly exothermic reaction on hitting the target

A new system which might be extremely interesting in this context is the intermetallic phase ZrW_2. ZrW_2 is a high-density, very hard intermetallic compound that reacts/burns in a highly exothermic reaction with air at high temperatures. This intermetallic phase should provide a very suitable reactive material for warhead applications.

ZrW_2 adopts the $MgCu_2$ structure in the solid state, is very hard and also densely packed. One can also expect that this intermetallic compound should burn at very high temperature in a strongly exothermic reaction due to the zirconium content of 33 %. The heat of formation of ZrO_2 (−1100 kJ/mol) is high as that of CeO_2 (−1089 kJ/mol). For Cer misch metals it is very well-known that burning occurs with a strong white flame at high temperature and is very exothermic.

Metallized formulations are also suitable for special missions (e.g. caves, tunnels e.g. in GWT), since in enclosed spaces the oxygen in the air needed for breathing for survival is removed.

DFOX, which is a new secondary explosive by its own (Fig. 12.7) and possesses a positive oxygen balance (Ω_{CO} = +52 %, Ω_{CO_2} = +41 %) is unique in so far that, when aluminized, not only do the heat of detonation (Q_{ex}) and the detonation

Fig. 12.7: Synthesis of DFOX (1) from glyoxal and N_2O_5. DFOX has a higher oxygen content than NO_2, Li_2O_2, $C(NO_2)_4$, ADN, $KClO_4$, NH_4ClO_4 and tetranitroacetimidic acid.

Tab. 12.4: Detonation parameters for DFOX and RDX in aluminized formulations.

	DFOX	DFOX / Al 80/20	DFOX/Al 70/30	RDX	RDX / Al 80/20	RDX / Al 70/30
VoD / m s^{-1}	7678	7798	7156	8919	8201	7723
p_{C-J} / GPa	23.6	28.3	24.2	34.3	31.7	26.6
T_{ex} / K	2375	5358	6345	3749	4978	5391
Q_{ex} / kJ kg^{-1}	−2408	−7784	−10142	−5742	−8259	−9559

temperature (T_{ex}) increase significantly, but also do the detonation velocity (VoD) and detonation pressure (p_{C-J}) increase in the DFOX/Al 80/20 formulation whereas RDX shows "normal" behavior with decreasing VoD and p_{C-J} (Tab. 12.4). The DFOX/Al 70/30 mixture performs essentially equal to the RDX/Al 70/30 mixture, however, possesses a detonation temperature 1000 K higher and a released heat of detonation also by 500 kJ kg^{-1} higher than the corresponding aluminized RDX formulation.

Check your knowledge: See Chapter 14 for Study Questions.

13 Related Topics

13.1 Thermobaric Weapons

There are two related weapon systems in this category: the thermobaric weapons and the Fuel Air Systems, also known in German as aerosol bombs (**FAE, Fuel Air Explosive**). Both function according to the same principle.

An **aerosol bomb**, Fuel Air Explosive (FAE) or Fuel Air Bomb, is a weapon whose effect on the detonation of an aerosol or substance distributed as a dust cloud does not depend on an oxidizer being present in the molecule. A FAE bomb consists of a container with a flammable substance (e.g. ethylene oxide). Two explosive charges are used as detonators: the first explosion causes the fuel to be distributed as fine particles into the air as an aerosol. Following this, microseconds – milliseconds later the aerosol is detonated, which results in the release of high pressure.

The pressure wave which results from the explosive deflagration is considerably weaker than that of a comparable explosive such as RDX. However, the explosive deflagration occurs almost simultaneously in a large volume of 10–50 m diameter. The fuel can penetrate caves, tunnels or bunkers, which makes this weapon effective when used on targets in the global war against terror (GWT), unlike conventional explosive charges which only have a limited effect due to their poor pressure effect.

Furthermore, aerosol bombs have a considerably larger heat effect than conventional explosive charges. Therefore these bombs are more effective at killing people and at the destruction of other soft targets such as unarmed vehicles and other targets.

The explosion removes the oxygen from the air, because the explosive composition does not contain its own oxidizer and uses the oxygen present in the air instead. This is however not the deadly mechanism of the aerosol bombs. Much more significant is death by suffocation which is observed as the result of an aerosol bomb. The reason for this is not the lack of oxygen, but the damage which is done to the lungs through a so-called barotrauma. The negative pressure phase which occurs after the positive pressure phase (see Ch. 2.2) causes an expansion of the air in the lungs; resulting in damage. The properties of an aerosol bomb – long, relatively flat pressure wave with a corresponding distinct "partial vacuum" as well as the use of atmospheric oxygen are beneficial effects.

For **thermobaric weapons** (also known as **EBX** = enhanced blast explosives), in addition to a conventional explosion, a flammable substance (usually Al), with little or no oxidizer (e.g. oxygen) distributed in the air detonates immediately as a result of the explosion. This post-detonative reaction ("fireball" of Al with air) usually occurs within μs after the detonation of the high explosive. This causes the effect of the original explosion to be magnified which results in an even larger heat and pressure effect. It is interesting to note that when an EBX is detonated under anaerobic (experimental) conditions, the Al does not react and therefore does not contribute to the reaction enthalpy ($\Delta H_{comb.}$ or $\Delta H_{expl.}$).

The initial pressure wave and the subsequent underpressure, follow a phase in which the underpressure resulting from the explosion causes a flux of surrounding air to the centre of the explosion. The released, non-exploded, burning substance is as a result of the underpressure sucked into where the explosion occurred, and therefore enters all non air-tight objects (caves, bunker, tunnels) and burns these. Suffocation and internal damage can result to both humans and animals that are located out of the immediate vicinity: e.g. in deeper tunnels as a result of the pressure wave, the burning of oxygen, or the subsequent underpressure.

As discussed above, thermobaric weapons contain monopropellants or secondary explosives and energetic particles. Boron, aluminum, silicon, titanium, magnesium, zirconium and carbon can be considered to be energetic particles. The main advantage of thermobaric systems is that they release large quantities of heat and pressure, often in amounts larger than for only secondary explosives.

In thermobaric weapons, highly aluminized secondary explosives can be used instead of monopropellants. For example, RDX in combination with a binder and a large quantity of aluminum (fuel rich) can be used. Research is currently being undertaken to investigate energetic polymers which could possibly be metallized for possible applications.

In thermobaric weapons, first anaerobic detonation occurs within the microsecond time-frame, followed by a post-detonative combustion which is also anaerobic and occurs in the hundredth of a microsecond time-frame. Only then, the post detonative burning occurs, which lasts several milliseconds and generates strong heat radiation even at small shock-wave pressures (approx. 10 bar)

The following relationship gives a conservative approximation for the distance from the detonation of a thermobaric weapon, which can still experience large damage as a result of the heat and pressure-wave generated. D is the distance (in meters), C is a constant (0.15 for an assumed overpressure of 70 mbar), n is the so-called yield factor from the burning (normally 0.1) and E is the energy (in J) released on the explosion and combustion.

$$D = C(nE)^{0.33}$$

If the energy E is replaced by "mQ" in which m is the mass (in kg) of energetic material used and Q is the heat of combustion (in J kg^{-1}), the critical distance can be represented by:

$$D = 0.15 \text{ kg}^{-0.33} \text{ s}^{0.66} \text{ m}^{0.33} (0.1\, m\, Q)^{0.33}$$

Thermobaric weapons can be used as bombs or as shoulder-launch mobile systems. There are two fields of operation for which the thermobaric weapons are superior to other systems:
(i) Areas of conflict involving caves, tunnels or other difficult to access areas and deep, confined targets (HDBTs, hard and/or deeply buried targets).
(ii) Combat and destruction of chemical and biological (CB) weapons through the strong heat effect (see also Ch. 13.2).

Shock waves generated by the detonation of thermobaric weapons have a much longer duration than those which are generated by conventional high explosives. In addition, they also show a larger lethal radius. Under confinement, thermobaric detonations cause a series of reflected shock waves which maintain the fireball, and can extend its duration to between 10 and 50 ms, as exothermic recombination reactions occur. Further damage can result on cooling of the gases since this causes the pressure to drops sharply, which results in a partial under-pressure which is powerful enough to cause physical damage to people and structures. Consequently, thermobaric weapons are highly effective for use in the destruction of tunnels, bunkers, underground structures and buildings. A thermobaric explosion occurs in stages: in the initial stage, a cloud of flammable vapor or flammable metal is released (e.g. a finely powdered Al/AP mixture), which is then subsequently ignited. This effectively turns the air in the region of this cloud into a huge bomb. The resulting explosion has fragmentation, concussion, and incendiary elements. Actually, detonation of thermobaric explosives can be viewed as three discrete events merged together:
1. An initial anaerobic detonation reaction (microseconds duration) – defines the high pressure performance characteristics of the system: armor penetrating ability
2. Post detonation anaerobic combustion reactions (hundreds of microseconds in duration) – defines the system's intermediate pressure performance characteristics: wall/bunker breaching capability;
3. Post detonation aerobic combustion reaction (milliseconds in duration) – defines the personnel/material defeat capability of the system – impulse and thermal delivery.

Thermobaric effects are obtained by a long duration overpressure and heat due to the afterburning of the detonation products in air. Immediately after detonation occurs the ambient pressure increases, forming an overpressure pulse. The peak overpressure and positive phase duration determine the specific impulse of the blast wave. A negative phase succeeds the positive phase with its negative pressure (suction).

Today, cast composite thermobaric explosives most often consist of a high explosive (RDX, HMX), polymeric binder (e.g. HTPB), fuel component (metal powder for enhancement of the blast effect, e.g. Al, Mg) and an oxidizer (AP or AN).

An excellent paper on thermobaric detonations by Danica Simić et al. can be found in the literature (*Central European Journal of Energetic Materials*, **2016**, *13*(1), 161–182).

13.2 Agent Defeat Weapons

Agent defeat weapons (also known as agent defeat warheads, ADW) are airborne warheads, which can be used in the fight against chemical and biological (CB) weapons of mass destruction (WMD) to minimize the amount of collateral damage that occurs. Collateral damage can largely be minimized when the active agents in the chemical or biological weapons are destroyed or neutralized upon being attacked, before they can be spread into the environment.

An ADW is constructed similarly to a multiprojectile warhead (see Ch. 7.1), but is additionally combined with a thermobaric weapon (see Ch. 13.1). After penetration of the chemical or biological weapon system, the active substances are destroyed as a result of the large heat effect of the thermobaric explosion. In addition, some ADWs also contain Cl-releasing systems, which also provide further destruction through oxidation in particular of the biological warfare agents. Other chemical compounds which are suitable for fighting biological agents contain peroxide and hypochlorite groups as well as chlorine.

It has been shown that although Cl_2 is a better biocide than HCl, the biocidal activity of HF exceeds both Cl_2 and HCl. Only 200 ppm HF destroy most bacteria, including Anthrax spores. The most recent research in particular by Bob Chapman (NAWC, China Lake) has shown that ADW containing fluorinated payloads ($-NF_2$ groups) are effective in destroying biological threats. For example, tetrakis-(difluoramino)octahydro-dinitro-diazocine, HNFX (see Fig. 10.1) [67] can very quickly remove the dangers of Anthrax spores. The biocidal activity of HNFX is not due to heat or pressure, but due to the harsh conditions of exposure to biocidal detonation products (i.e. HF). A control experiment using HMX as the explosive showed an order of magnitude lower killing of spores (86 % spore survival). Also octa-F-PETA, i.e. $C(-CH_2-NF_2)_4$ has successfully been tested against Anthrax spores. A possible synthesis for octa-F-PETA is shown in Figure 13.1.

Alternatively, octa-F-PETA can also be prepared according to the procedure shown in Figure 13.2. Unfortunately, due to its high volatility (m.p. 40 42 °C; sublimes at ca. 30 °C) octa-F-PETA will not be employed as a main charge in ADWs.

A chlorine containing energetic material which also shows use as a potential agent defeat explosive is hexachloromelamine, $[-N=C(NCl_2)-]_3$ (Fig. 13.3).

This compound is known to be biocidal and can be prepared by direct chlorination of melamine (Fig. 13.1).

13.2 Agent Defeat Weapons

$$C(CH_2Br)_4 \xrightarrow{NaN_3} C(CH_2-N_3)_4 \xrightarrow{\text{reduction}} C(CH_2-NH_2)_4$$

$$\downarrow \text{select fluor}^®$$

$$\boxed{C(CH_2-NF_2)_4}$$

$$\uparrow F_2/N_2,\ 25\ °C,\ CH_3CN$$

$$\left[C\!-\!CH_2\!-\!HN\!-\!\underset{O}{\overset{O}{C}}\!-\!O\!-\!Et \right]_4$$

Fig. 13.1: Possible synthesis for octa-F-PETA.

$$C(CH_2NH_3^+Cl^-)_4 \xrightarrow[\text{aq. NaOH}]{EtOCOCl} C(CH_2NHCOOEt)_4$$

$$\downarrow F_2/N_2 \quad \substack{CH_3CN,\\ -25\ °C}$$

$$\boxed{C(CH_2NF_2)_4}$$

Fig. 13.2: Alternative synthesis for octa-F-PETA.

hexachloromelamine $\xleftarrow{Cl_2}$ melamine

Fig. 13.3: Synthesis of hexachloromelamine.

$$CatIO_3 \xrightarrow{40\%\ HF} CatIO_2F_2 \qquad Cat = NH_4^+\ (\mathbf{1a}) \\ CH_6N_3^+\ (\mathbf{1b})$$

$$\underset{HON}{\overset{Cl}{\diagdown}}\!\!\!=\!\!\!\underset{Cl}{\overset{NOH}{\diagup}} \xrightarrow[\text{DMF}]{NaN_3} \underset{HON}{\overset{N_3}{\diagdown}}\!\!\!=\!\!\!\underset{N_3}{\overset{NOH}{\diagup}} \\ \mathbf{2}$$

Fig. 13.4: Synthesis of difluoroiodate salts (**1a, b**) and diazidoglyoxime, **2**.

The killing efficiency R is usually defined as

$$R = \log(N_O + 1) - \log(N_E + 1)$$

Where N_O is the bacterial count fort he inoculated control and N_E the bacterial count after exposure to the detonation products. In control experiments it could be shown that the killing efficiency R for HMX is ca 0.06 whereas using HNFX it is ca. 3 for wet and up to 9 for dry spores.

In a more recent study it could be shown that the combination of an explosive (diazidoglyoxime, **2**, Fig. 13.4) with a strongly biocidal agent defeat ingredient (guanidinium difluoroiodate, **1b** generates I_2, I and HF), could also result in very high the killing efficiencies of $R = 6-8$ [68].

13.3 Nanothermites

Research into nanothermites is currently a hot topic in the field of energetic materials. For thermite-type reactions, a metal oxide is the oxidizer, and aluminum is the fuel. Thermite-type reactions on the nanoscale have been called many names including nanothermites, metastable intermolecular composites (MICs) or superthermites. The phrase "nanothermites" comes from the particle sizes used in these energetic mixtures, in contrast to the more familiar thermite type reaction in which the particle sizes are in the micron region. The term "metastable intermolecular composites" comes from the fact that the mixtures of metal oxide and aluminum are stable up to their ignition temperature, at which point self-propagating high-temperature synthesis (SHS) occurs and the thermodynamic products of a metal and aluminum oxide are produced. Finally, the term "superthermite" comes from the fact that thermites composed of nano-sized materials exhibit very different combustion characteristics when compared to those mixed with micron-sized precursors. Superior combustion velocities and explosive behavior compared to the usually observed deflagration often characterize thermites made with nanoscale precursors.

$$M_xO_y + Al \rightarrow M + Al_2O_3. \qquad (1)$$

While most metal oxides can be reduced with aluminum in an exothermic thermite reaction, and many have had their reactivity studied on the nanoscale (including Fe_2O_3, MnO_2, CuO, WO_3, MoO_3, and Bi_2O_3), discussion of all nanoscale thermite reactions performed in the literature would be too broad. Discussion will focus on iron[III] oxide, molybdenum[VI] oxide, and copper[II] oxide as they provide good examples of either differing synthetic methodologies tailoring properties, exemplary final product properties, or of a material that has a high potential for practical use. We will limit the discussion of thermite examples to those where aluminum is the

Tab. 13.1: Combustion rate as a function of contact points in an Si / Pb$_3$O$_4$ composite.

Si particle diameter (µm)	contact points (× 10^9)	combustion rate (mm / s)
2	30.2	257.4
4	8.7	100.6
5	6.1	71.5

fuel because of the practical uses of such mixtures. However other fuels such as Zr, Hf, Mg etc. have been used.

An energetic material's particle size traditionally ranges from 1–100 µm. As is known, generally smaller particles give higher reaction rates and nanothermites are not an exception. The reaction rate of a nanoscale thermite is generally several orders of magnitude larger than those on micron scales, and the much larger surface area can significantly change combustion behavior, as well as ignition behavior by increasing sensitivity. The changes in properties are all associated with reduced diffusion distances since the associated surface area increases.

The initial reactions in a binary fuel-oxidizer system such as a thermite reaction are assumed to be diffusion-limited solid-solid reactions. As a result, the rate of reaction and therefore the combustion velocity can be increased dramatically as the particle size decreases and particle contacts increase. A mathematical study done by Brown et al. compared observed combustion rates for a Si / Pb$_3$O$_4$ pyrotechnic mixture with the calculated fuel-oxidizer contact points for silicon particles of given sizes mixed with 5 µm Pb$_3$O$_4$. Contact points were calculated assuming the approximation that the fuel and oxidizer are hard spheres. Results from this study are presented in Table 13.1, and show that a small change in particle size has a very strong effect on fuel-oxidizer contact points, and a correspondingly large effect on the combustion velocity.

Despite this study not being on the nanoscale, it does show the dramatic effect particle size has on combustion rate, since doubling the particle size causes the combustion rate to be more than halved. The Si / Pb$_3$O$_4$ mixture has found use as a pyrotechnic delay mixture (slow burning mixture used for timing in pyrotechnics) as its combustion rate is easily tailored.

As particle sizes decrease, the sensitivity of the thermite mixture to impact and friction increases. The micron scale thermites are usually quite insensitive to impact and shock, but thermites on the nanoscale can be quite sensitive to both or one of the two depending on the metal oxide. This is exemplified in the work of Spitzer, where a tungsten(VI) oxide and aluminum thermite was prepared by mixing nano and micron aluminum with nano and micron WO$_3$. The results of this are presented in Table 13.2.

While increased sensitivity can make nano-thermites more dangerous to handle, the increased friction or impact sensitivity is also beneficial in some practical

Tab. 13.2: Sensitivities for nano vs. micron Al / WO₃ thermite.

Al diameter (nm)	WO₃ diameter (nm)	impact (J)	friction (N)	combustion rate (m/s)
1912	724	> 49 (insensitive)	> 353 (insensitive)	< 0.08
51	50	42 (insensitive)	< 4.9 (very sensitive)	7.3

applications such as percussion primers. However some nano-thermites have increased ESD sensitivity, which has no current practical application and is only a safety hazard. For example, a Bi_2O_3/Al nano-thermite has sufficient impact and friction sensitivity to be considered for use in ammunition primers, however it has an ESD sensitivity of 0.125 µJ (40 nm Bi_2O_3, 41 nm Al). This is a static potential, which is easily achieved by the human body. This makes handling hazardous, especially since the Bi_2O_3/Al thermite has a combustion velocity of over 750 m/s which means that this thermite explodes rather than burns. The increased ESD sensitivity of nano vs. micron composites is believed to be the result of the increased ability of high surface areas to develop charges.

A final property affected by the reduction in particle size from micron to nano scale is the ignition temperature of the thermites. For example, a thermite composed of 100 nm MoO_3 and 40 nm aluminum exhibits an ignition temperature of 458 °C whereas the same MoO_3 with 10–14 µm Al exhibits an ignition temperature of 955 °C. This indicates a change in mechanism between the two as the micron composite exhibits melting and volatilization of Al and MoO_3 (from DSC) before the thermite reaction occurs, while the nano-thermite reaction occurs before the Al melting could take place. This indicates that the reaction for the nano-composite is based on solid state diffusion, while for the micron composite the reaction is a gas (MoO_3)-liquid(Al) reaction. This change in mechanism may apply to other thermite systems as well, however detailed DSC data for many other systems has not been published so far.

In thermites using aluminum metal as the fuel, the passivation of the metal surface with oxide must be taken into account. For micrometer sized particles of aluminum, the oxide passivation layer is negligible, but on the nano-scale this passivation layer of alumina begins to account for a significant mass portion of the nanoparticles. In addition, the precise nature of the oxide layer is not the same for all manufacturers of aluminum nanoparticles, so the researcher must use TEM to measure oxide thickness to allow calculation of active aluminum content before stoichiometric calculations are carried out for the mixing of thermites. Table 13.3 shows details of some of the percentages of aluminum in aluminum nanoparticles and shows just how significant and inconsistent the oxide layer can be.

The effect of the aluminum oxide layer is also known to reduce the propagation of thermite reactions since alumina is an effective absorber of thermal energy. A

Tab. 13.3: Active aluminum content of aluminum nanoparticles.

average Al nanoparticle size (nm)	active aluminum content (%)
30	30
45	64
50	43
50	68
79	81
80	80
80	88

Tab. 13.4: CuO/Al thermites with highly oxidized (49 % Al) aluminum nanoparticles.

CuO	Al	combustion velocity (m/s)	mass burning rate (g/s)
micron	micron	220	2700
micron	nano (49 % Al)	200	1100
nano	micron	630	4850
nano	nano (49 % Al)	900	4000

study by Weismiller et al. of a 49 % active aluminum nanopowder in a thermite with copper oxide supports the idea that too much oxide can actually reduce thermite performance. Table 13.4 shows Weismiller's data which indicates the negative effect of a thick oxide layer on aluminum nanoparticles.

The combustion velocities given in Table 13.4 show that the use of nano CuO is more of a factor towards high velocities than the Al when the Al is highly oxidized, and even slightly decreases performance when the velocity of μm-CuO/μm-Al is compared to μm-CuO/nm-Al. The mass burning rate shows this even more starkly, as with a given particle size of CuO, the switch from micron to heavily oxidized nano aluminum decreases the mass burning rate.

In general, there are three common ways of preparing nano-thermites; these include arrested reactive milling, physical mixing, and sol-gel methodology. Changing from the microscale to the nano, these methods provide accessibility and use of different particle sizes, as well as materials with different degrees of contact between oxidizer and fuel.

Arrested reactive milling (ARM) is a technique for the preparation of nano-energetic materials by milling the metal oxide and aluminum in a ball or shaker mill. While it may or may not involve particle sizes on the nanoscale, the energetic composites prepared through this method have properties more similar to those of the nano-thermites as mixing is on the nanoscale; fuel and oxidizer may be contained in the same particle after milling. The particles produced by arrested reactive milling are in the range of 1–50 µm, and are composed of layers of aluminum and oxidizer on the scale of 10–100 nm. The size of the particles obtained is a function of the milling time, however due to the reactivity of the mixture, after a certain milling time (a function of initial particle sizes, milling media used, and type of metal oxide) when the particle size is reduced below a certain value, the milling causes ignition of the thermite mix. A liquid such as hexane is usually added to the mill to reduce static build up. The term arrested reactive milling comes from the fact that milling is stopped before the mix ignites, resulting in a useable thermite mixture. Advantages of this method are that the particles produced approach their maximum density. Aluminum is hidden within the particle matrix which reduces the presence of nonreactive alumina, the ability to start with non-nanoscale precursors, and milling time offers convenient control of the degree of intermixing and therefore the reactivity. Disadvantages include the fact that only few thermite mixtures can be prepared by this method as many other mixtures are too sensitive and ignite before sufficient intermixing can occur.

Physical mixing is the simplest and most common method for the preparation of nano-thermites. Nano-powders of metal oxide and aluminum are mixed in a volatile, inert liquid (to reduce static charge) and they are then sonicated to ensure good fuel-oxidizer mixing and to break up macroscale agglomerates. The liquid then evaporates and the thermite is ready to be used. Advantages of the physical mixing method is the simplicity and its wide range of applicability to many thermite systems. The only major disadvantage is the necessity of starting with nanoscale powders which may or may not be commercially available.

Sol-gel nanothermite preparations take advantage of the unique structural and mixing properties available from sol-gel chemistry. In these nanothermite mixtures, the aluminum nanoparticles reside in the pores of the metal oxide matrix, which is widely assumed to increase thermite power in comparison to physical mixing by huge reduction in diffusion distances between the fuel and oxidizer, and by increasing the contact area. The synthesis of energetic sol-gel thermites involves the addition of a suspension of aluminum nanopowder in a solvent to a metal oxide sol just prior to gellation. After gellation, the gel can be processed to an energetic xero or aero gel. Through the sol-gel methodology, interfacial area, pore size, and matrix geometry can be controlled which results in different, tunable energetic properties. In addition, the sol-gel methodology allows the molecular incorporation of organic agents to the matrix which further tune the thermite's properties by acting as a gas generating agent. Other than the ability to incorporate organics, advantages of the sol-gel method include its ability to form low density aero or xero gels, which brings with it the ability to form energetic surface coatings. Disadvantages include potential oxidation of the aluminum nanoparticles by water in the metal oxide gel before the solvent can be removed. This disadvantage can be overcome by using sol-gel chemistry to only prepare aero or xero-gel oxide precursors, which can then be followed by physical mixing with aluminum. But this option comes at the cost of reducing the high-level of interfacial contact achievable by sol-gel chemistry.

Due to the range of properties that nanothermites can be tailored to have, nanothermites find applications in a correspondingly large field. As research on nanothermites is a relatively new topic, many applications have only been proposed so far and testing in such areas has not yet been done. Many of the potential applications of nano-thermites are a direct result of their high energy densities, which are comparable to those of lithium batteries leading to applications of the power-generating variety such as microscale propulsion, energetic surface coatings and nano-scale welding. Other applications are a direct result of the pyrotechnic behavior of nano-thermites, and such materials may find use as gas generating agents for automobile air bags, exploding-on-contact missiles, environmentally friendly ammunition primers and electric igniters. The fastest thermites even have potential application as primary explosives. Their application in microscale propulsion, ammunition primers, and electric igniters has had more experimental work conducted with them than other potential applications for nano-thermites, and show very promising results.

Microscale propulsion involves the production of thrust on the microscale (< 1 mm). This is referred to as micropyrotechnics or microenergetics, and applications include rapid switching and propulsion for small spacecrafts. Performance energetic materials used for propulsion on the macroscale including RDX (hexahydro-1,3,5-trinitro-1,3,5-triazine, hexogen) or HMX (1,3,5,7-tetranitro-1,3,5,7-tetrazocane, octogen) are unable to function on the microscale, as they cannot sustain combustion in such small diameters since too much energy is lost to the combustion chamber which inhibits propagation through the energetic material. Nanothermites have no such problem as they have a much higher energy density, so energy lost to the chamber becomes insignificant.

An ammunition primer is the part of a round of ammunition that is impacted by the firing pin of the firearm. Traditional primers contain lead containing compounds including lead azide and lead styphnate, as these sensitive explosives detonate from the impact of the firing pin, and the resultant flame ignites the propellant in the cartridge, firing the bullet. Figure 13.5 shows a shotgun shell primer. The toxicity of lead means the use of such primers is both an environmental and personal hazard to the user. Nanothermites have been shown to be an effective replacement for the lead salts, as they have been tailored to have properties similar to the currently used lead azide/styphnate mixture.

Electric igniters find use throughout the energetic material industry in all areas of propellants, explosives and pyrotechnics; they are used wherever an electric current is required to initiate an energetic material. As a result of the precise timing they afford, they can be used for igniting everything from rockets to blasting caps, to fireworks displays. Electric igniters are also known as electric matches, and they consist of a flammable head of material around a resistive bridgewire, igniting

Fig. 13.5: Shotgun shell primer.

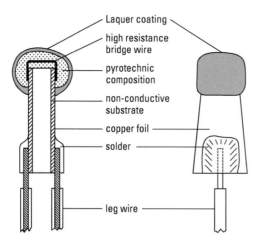

Fig. 13.6: Electric igniter/match diagram.

when a certain electrical current is passed through the match. Figure 13.6 shows a general diagram of an electric match. Like primers, commonly used electric matches contain toxic lead compounds including the dioxide, thiocyanate and nitroresorcinate. Therefore finding nontoxic replacements is desirable. Again, the use of nano-thermites has shown significant promise as non-toxic, green replacements for the lead compounds.

13.3.1 Example: Iron Oxide / Aluminum Thermite

On the macroscale, the iron oxide / aluminum thermite is used for the high temperature welding of iron railway rails. This is made possible through its slow rate of energy release and high combustion temperature. On the nanoscale, however, the iron oxide / aluminum thermite behaves very differently. In the literature, this thermite system is often synthesized by sol-gel methods as iron oxide sol-gel chemistry is well known, however, it has also been made by arrested reactive milling and physical mixing as well.

The iron oxide thermite is a very good example of the versatility of the sol-gel thermite methodology. Table 13.5 shows the huge effect the synthetic strategy used can have on the properties of the thermite.

From these numbers, the wide range of thermite properties that become available is apparent. The ease of tuning energetic properties is also apparent from the high velocity (320 and 895 m/s) xero and aerogels show as the xero and aerogel have different sensitivities to ESD and to impact. These differences are shown in Table 13.6.

The impact sensitivity of the aerogel thermite may make it suitable for use in primers, because all of the other preparatory methods for the iron oxide thermite produce thermites which are unsuitable for this application. The especially sensitive nature of aerogel thermites stems from the inability of the aerogel matrix to conduct heat, a characteristic which is only applicable to aerogels, due to their insulating property.

A further development on the synthesis-dependent application of the iron oxide thermite is that of Clapsaddle et al. Tailoring of energetic properties was obtained by the addition of organo-silicon precursors to the sol leading to aerogel oxidizers containing any desired percentage of organic-functionalized silica. The

Tab. 13.5: Effect of the synthetic methodology used to prepare the Fe_2O_3/Al thermite on the combustion velocity.

synthetic methodology	velocity (m/s)
arrested reactive milling (max mill time before ignition)	0.5
physical mixing (80 nm Al, oxidizer size unmentioned)	9
sol gel (aerogel oxide physically mixed with 80 nm Al)	80
sol gel (xerogel, 70 nm Al added before gelation)	320
sol gel (aerogel, 70 nm Al added before gelation)	895

Tab. 13.6: Impact and ESD sensitivities of aero- and xerogel Fe_2O_3/Al thermite.

type	impact (J)	ESD (mJ)
xerogel	36.6	> 1000
aerogel	21.9	30

silica functioned to reduce the combustion velocity, and the organic addition to increase the amount of gas released upon combustion. Research in this field is still in the initial stages, but the thermites may have application in the propulsive and gas-generating fields.

13.3.2 Example: Copper Oxide / Aluminum Thermite

Copper oxide and aluminum thermite is well known for its high combustion velocity even on the micron scale. This high combustion velocity is even more pronounced on the nano-scale where the thermites reach the highest combustion velocities known for a mixture of a metal oxide and aluminum.

Although in the literature CuO/Al thermites have not been prepared by reactive milling or sol-gel methods, unique to CuO/Al thermite is a new synthesis methodology by Gangopadhyay et al. called self-assembly. In this method, the authors coated copper nanorods (20 × 100 nm) with a coordinating polymer (P4VP) followed by coating the coated rods with 80 nm aluminum powder. The result of this self-assembly process is the formation of the thermite with the highest combustion velocity known of 2400 m/s. Table 13.7 gives the combustion velocity of the self-assembled thermite compared with physically mixed thermites.

Gangopadhyay et al. state that the reason for the increased combustion velocity is the highly improved interfacial contact between fuel and oxidizer. However the presence of an organic material binding the fuel and oxidizer will increase the combustion pressure which has been shown by Weismiller and Bulian to result in an increase in the combustion rates. The dependence of the combustion velocity on pressure has been thought of as resulting from the of decomposition of CuO to gaseous oxygen at 1000 °C which may be the active oxidizer.

Despite the reason for the high combustion velocity being under debate, the high combustion velocity is in the range of explosive velocities, so the CuO/Al thermite may have application as a primary explosive. The more easily prepared, physically mixed CuO nanothermite has been patented for use in ammunition primers. Sensitivity data for the CuO/Al thermite is unavailable, but it is expected to be relatively high. High impact sensitivity may be the reason why there are no literature reports of its preparation by arrested reactive milling.

Tab. 13.7: Combustion velocities of the CuO / Al thermite (80 nm Al, 2.2 nm oxide layer).

synthetic method	CuO morphology	combustion velocity (m/s)
physical mixing	micron	675
physical mixing	nanorods	1650
self assembly	nanorods	2400

13.3.3 Example: Molybdenum Trioxide / Aluminum Thermite

MoO_3/Al has been prepared exclusively by physical mixing and arrested reactive milling. Due to the availability of both MoO_3 and aluminum nanopowders, the preparation used exclusively for practical applications of these thermites is physical mixing. With combustion velocities ranging from 150–450 m/s MoO_3/Al thermites are perhaps some of the most widely studied nano-thermites, and extensive work has been reported on their practical applications in electric igniters, primers (patented), and microscale propulsion.

A study reported by Son et al. on the feasibility of the use of a MoO_3/Al nano-thermite (79 nm spherical Al, 30 × 200 nm sheet MoO_3) for microscale propulsion and microscale pyrotechnics shows great promise for the application of these materials. It was found that when confined in tubes with a diameter of 0.5 mm, the MoO_3/Al thermite would combust at rates as high as 790 m/s. This is an advantage over conventional energetic materials such as HMX which cannot propagate even in tubes under 1 mm.

Naud et al. have preformed extensive work on the use of the MoO_3/Al nano-thermite for application in electric matches. Electric matches prepared from this thermite mixture have been shown to have lower friction, impact, thermal and ESD sensitivities than the currently used matches, which contain toxic lead compounds making them both safer and more environmentally friendly to use.

13.4 Homemade Explosives

Usually explosives can be categorized as military, commercial and homemade (HME) explosives. HMEs are energetic formulations that can be created "at home". The term HME has been used to cover a wide range of materials from pure explosive compounds, such as triacetone triperoxide (TATP), that can be synthesized from readily available articles of commerce or pentaerythritol tetranitrate (PETN), as pentaerythritol can be purchased in bulk for paint use to home-made variants of explosive formulations, such as ammonium nitrate fuel oil (ANFO), that are used in very large commercial blasting operations.

A major difference between military and commercial explosives on the one hand and HMEs on the other hand is that we do look at "made in a factory with rigorous reproducibility" (military/commercial) vs. "made anywhere else" (HME).

HMEs such as black powder, triacetonetriperoxide (TATP), hexamethylene-triperoxidediamine (HMTD), chlorates, and perchlorates mixed with sugar or other fuels have come to be used by terrorists instead of commercial explosives.

13.5 Explosive Welding

Explosion welding or explosive welding (EXW) is a solid state joining process in which welding is accomplished by acceleration of one of the components at extremely high velocity through the use of chemical explosives [see also: Lancaster, J. F. (1999). *Metallurgy of welding* (6th ed.). Abington, Cambridge: Abington Pub., **1999**, ISBN 1-85573-428-1]. Explosive welding uses a controlled explosive detonation to force two metals together at high pressure. When an explosive is detonated on the surface of a metal, a high pressure pulse is generated which propels the metal at a very high rate of speed. If this piece of metal collides at an angle with another piece of metal, welding may occur. The composite system which results is joined by a durable, metallurgical bond. For welding to occur, a jetting action is required at the collision interface. This jet is the product of the surfaces of the two pieces of metals colliding. This cleans the metals and allows two pure metallic surfaces to join under extremely high pressure. The metals do not commingle; they are atomically bonded. Due to this fact, any metal can be welded to any metal (i.e. copper to steel; titanium to stainless). Typical impact pressures are millions of psi. Fig. 13.7 shows the explosive welding process.

Fig. 13.7: Schematic diagram of the explosive welding process (planar geometry).

This process is most commonly utilized to clad carbon steel plates with a thin layer of a corrosion resistant material (e.g. stainless steel, nickel alloy, titanium, or zirconium). Due to the nature of this process, the geometries which can be produced are very limited and must be simple. Typical geometries produced include plates, tubing, and tubesheets.

Explosion welding can produce a bond between two metals that cannot necessarily be welded by conventional means. It is important to point out that the process does not melt either metal, instead the surfaces of both metals are plasticized, causing them to come into contact, which is sufficiently intimate to create a weld.

13.6 Gas Generators (Air Bags)

An airbag system consists of/functions as follows:
- an electrical impulse heats the bridgewire in the igniter
- the primary charge of the igniter (pyrotechnic mixture *e.g.* zirconium-potassium-perchlorate, ZPP) ignites and produces hot gases
- the booster charge (*e.g.* titanium-hydride-potassium-perchlorate, THPP, or boron + potassium nitrate) is subsequently ignited and further hot gases are produced
- the resulting hot gas ignites the propellant (tablets)
- the propellant (main gas-generator, mostly N_2) burns, and the resulting gaseous combustion products which are produced are directed into the airbag and fill it (see Fig. 13.8)
- this all occurs within approximately 60 ms

Sodium azide was used as the propellant in the past, however it is not used nowadays due to its toxicity, and also because of security concerns (terrorism: NaN_3 from scrap + $Pb(NO_3)_2 \rightarrow ...$).

$$42\ NaN_3 + 2\ KNO_3 + 8\ Fe_2O_3 + 5\ SiO_2$$
$$\rightarrow 21\ Na_2O + K_2O + 8\ Fe + 8\ FeO + 5\ SiO_2 + 64\ N_2$$

The iron oxide lowers the burning temperature by about 600 K to 1330 °C, the silicon dioxide acts as a slag up-taker. Within 30 milliseconds the propellant is completely combusted and produces approximately 40 % nitrogen gas and 60 % slag (weight percent). Nowadays, modern propellants have a gas yield of 60–99 %.

Ammonium nitrate has also been used, however, this resulted in several accidents (explosion, not combustion) since in warm and moist areas (Florida) many phase transitions occurred with the corresponding density change → cracks, cavities ... (Fig. 13.9, Tab. 13.8). Currently, usually guanidinium nitrate (GN) or nitroguanidine (NiGu) are used as the main propellant together with a suitable oxidizer (e.g. basic copper nitrate). A typical gas-generating propellant composition could be:

    guanidinium nitrate (GN)   52 %
    SiO_2 3 %
    basic copper nitrate   45 %

Instead of basic copper nitrate, one can also use copper diammine dinitrate, e.g.:

    guanyl urea nitrate   38 %
    SiO_2 3 %
    Copper diammine dinitrate   59 %

Fig. 13.8: Air bag inflation device (gas generator) (left) and gas generator with autoignition tablet (right).

Fig. 13.9: Phase transitions in ammonium nitrate (AN) determined by DSC.

Gas-producing compounds develop a large volume of gas when they are ignited. One category of gas-producing compounds uses a compound based on Guanidine (HN = C(NH$_2$)$_2$) mixed with a sensitizing agent and/or oxidizer. For example, US Patent Nr. 2,165,236 from Holm reports on a gas-producing compound which contains Nitroguanidine in a binder. Part of the nitroguanidine can be replaced by guandinium nitrate (H$_2$NC(NH)NH$_2$ · HNO$_3$). Examples of typical binders are nitrocellulose and cellulose acetate.

Tab. 13.8: Summary of crystal structures of AN.

Phase	V	IV	III	II	I	
Crystal structure	orthorhombic	orthorhombic	orthorhombic	tetragonal	cubic	
Phase transition temperature / K		255	305	357	398	441(melt)
Density / g cm^{-3}	1.75	1.70	1.57	1.60	1.55	
Volume change / %		−2.8	+3.7	−1.7	+2.3	

Some advantages and disadvantages of different propellants in gas generators are summarized in Tab. 13.8a

Tab. 13.8a: Advantages and disadvantages of some common fuels and oxidizers for gas generatos.

fuel	formula	+	−
sodium azide	NaN$_3$	cheap	toxic
		reliable	may form HN$_3$
			potential of terrorist abuse
			does NOT produce CO, CO$_2$
			risk of formation of copper azide, Cu(N$_3$)$_2$
guanidinium nitrate	C(NH$_2$)$_3$]$^+$[NO$_3$]$^-$	good performance	rel. expensive
			Less energetic than NaN$_3$ and ATz
5-aminotetrazole	5-ATz	high energy per mass	expensive
			hygroscopic
di(ammonium)-5,5'-bis-1H-tetrazolate	BHT · 2 NH$_3$	high energy per mass	expensive
			hygroscopic (?)
oxidizer	**formula**	**+**	**−**
phase stabilized ammonium nitrate	[NH$_4$]$^+$[NO$_3$]$^-$, PSAN	cheap	hygroscopic
		high gas yield	multiple polymorphism near room temperature
		produces non-toxic gases: H$_2$O, N$_2$, O$_2$	endothermic decomposition: poor ignitability and burning performance
		safe to handle	

oxidizer	formula	+	−
strontium nitrate	$Sr(NO_3)_2$	no phase issues	more expensive than PSAN
		less hygroscopic than PSAN	lower gas yield than PSAN
basic copper nitrate	$Cu(OH)(NO_3)$, BCN	no phase issues	more expensive than PSAN
		less hygroscopic than PSAN	lower gas yield than PSAN
		more energetic	potentially hazardous

Usually the phase stabilization of ammonium nitrate (AN) is achieved by adding potassium nitrate (KN) in an amount of 9–10 w-% obtaining phase-stabilized ammonium nitrate (PSAN). KN has a slightly smaller cationic radius ($r(K^+)$ = 1.38 Å) in comparison with ammonium ($r(NH_4^+)$ = 1.48 Å). Therefore by replacing NH_4^+ ions with K^+ ions, the crystal structure of AN (originally phase IV) becomes close to phase III. KN doped PSAN is stable as phase III at room temperature and phase transition and volume expansions are suppressed over a wide temperature range (193~373 K).

13.7 Toxicity Measurements

The luminescent bacteria inhibition test using **vibrio fischeri** is a well-established method to determine the aquatic toxicity. The determined parameter is the effective concentration (EC_{50}) after 15 and 30 min. The inhibition of natural bioluminescence of these bacteria gives a first idea of the toxicity of compounds towards different ecosystems. It is a cost and time efficient experimental method, which does not involve animals.

The EC_{50} (effective concentration) values are usually determined using a spectrometer. The measurement is based on the bioluminescence of *Vibrio fischeri* NRRL-B-11177 bacteria strains, whereby the EC_{50} is the concentration level where the bioluminescence is decreased by 50%. Prior to the measurement, a ten-point dilution series was prepared according to DIN/EN/ISO 11348 (without G1 level) with a known weight of the compounds and a 2% NaCl stock solution. The measurements, which strictly have to be carried out at 15 °C, then started by determination of the bioluminescence of untreated reactivated bacteria. After 15 and 30 minutes exposure time with a specific amount of component, the bioluminescence is determined again. The compounds toxicity can then be classified as according to their EC_{50} values (non-toxic > 1.00 g L^{-1}; toxic 0.10–1.00 g L^{-1}; very toxic < 0.10 g L^{-1}).

To show a trend in the aquatic toxicity towards *Vibrio fischeri*, different types of energetic materials were tested (Fig. 13.10). The herein presented compounds are a small selection of possible primary- and secondary explosives, oxidizers and pyrotechnical materials with different chemical constitution. There are coordina-

Fig. 13.10: Overview of measured compounds 1–17.

tion (e.g. [Cu(dtp)₃](ClO₄)₂), neutral (e.g. azidoethanol and propyl-linked ditetrazoles) and ionic (e.g. polynitropyrazoles, -triazoles and tetrazoles) compounds, also different substitution patterns and energetic functionalities such as azido-, nitro-, flourodinitro- and nitramino-groups were investigated. Finally they were compared to commonly used materials, like RDX, ammonium perchlorate (AP) and azide salts.

Comparing 2-azidoethanol (**1**), 2,2,2-trinitroethanol (**2**), 3,3,3-trinitropropylammonium chloride (**3**) and 2-fluoro-2,2-dinitroethanol (**4**), **1** is the least toxic compound towards *Vibrio fischeri*. It is also relatively low in toxicity compared to ionic azide compounds NaN$_3$ and NH$_4$N$_3$. Once adding a trinitroalkyl moiety the toxicity drops dramatically and is even worse with a flourodinitroethyl moiety. This is consistent with former toxicity measurements using gram positive bacteria such as

Tab. 13.9: EC_{50} values of measured compounds after 15 and 30 minutes in g L^{-1} and their considered toxicity level after 30 minutes (+/o/−).

Compound	EC_{50} (15 min) [g L^{-1}]	EC_{50} (30 min) [g L^{-1}]	Toxicity level	Compound	EC_{50} (15 min) [g L^{-1}]	EC_{50} (30 min) [g L^{-1}]	Toxicity level
NaN_3	0.25	0.18	o	7a	0.75	0.74	o
NH_4N_3	0.26	0.15	o	7b	0.60	0.58	o
NH_4NO_3	10.49	6.39	+	8	0.13	0.07	−
$NH_4N(NO_2)_2$	7.25	4.50	+	8a	0.75	0.35	o
NH_4ClO_4	14.58	11.13	+	8b	>1.58	>1.58	+
RDX	0.33	0.24	o	10a	1.17	0.58	o
1	8.70	8.55	+	11a	0.32	0.24	o
2	0.29	0.22	o	12	13.90	10.30	+
3	<0.10	<0.10	−	13	0.81	0.79	o
4	0.002	0.001	−	14	0.36	0.36	o
5	0.27	0.19	o	15	0.44	0.35	o
5a	1.21	0.95	+	16	0.64	0.44	o
6a	0.70	0.43	o	17	0.34	0.28	o

Staphylococcus aureus and *Intestinal bacillus*. 2,4-Dintropyrozoles show toxicity towards the aquatic bacteria, though it is considered as moderate toxic. The findings were comparable, no matter if they were measured as neutral compound as 1-oxid or 3-amin, or measured as salts. The neutral nitraminotriazole **8** is very toxic towards *Vibrio fischeri*, whereas the hydroxylammonium salt **8a** is moderately toxic and the lithium salt **8b** is considered not toxic in this test. The substitution pattern can have a varying effect on the toxicity of tetrazoles. Tetrazoles **10a** and **11a** are both moderately toxic, whereby TKX-50 (salt **10a**) still is less toxic. Furthermore, it is even less toxic than the common used secondary explosive RDX. For tetrazoles **12–14**, which also have varying substitution patterns, the toxicity ranges from EC_{50} values of 0.36 g L^{-1} (**14**) to 10.30 g L^{-1} (**12**). Once adding the copper(II) metal, which is known for being toxic to microorganisms, the toxicities of the complexes of tetrazoles **12–14** increases partially dramatically. Nonetheless, complexes **15–17**, which could be used as potential primary explosives are less toxic compared to the measured azide salts.

Another well-established method for evaluating the harmful potential of a certain substance is the **AMES test**. This OECD accepted method uses bacteria to test if mutations can be caused in the DNA of the organism and is the minimum test carried out in the course of REACH (Registration, Evaluation and Authorization of Chemicals).

The Ames test is used to detect point mutations, which involves substitution, addition, or deletion of one or a few DNA base pairs. OECD Guidelines for Testing of Chemicals (No. 471) form the basis for the test, which employs auxotrophic strains of *Salmonella typhimurium* (TA98, TA100, TA1535, TA1537) and *Escheria coli*

Fig. 13.11: Compounds evaluated using AMES test and *Vibrio fischeri*.

Tab. 13.10: Results of experimental and *in silicio* AMES test and the considered mutagenicity/toxicity level (+/o/−).

Compound	AMES test	Vibrio fischeri
7a	n.a.	o
BDNAPM	−	n.a.
Li$_2$ANAT	o	+
TKX-50	+	o

(wp2[pKM101]+wp2 uvrA mixed 1 : 2). The biosynthesis of the corresponding amino acids is blocked by point mutations, made in the histidine (*Salmonella typhimurium*) or the tryptophan (*Escheria coli*) operon. The mutagenic potential is evaluated by detecting the appearance of the reverse mutants of the auxotrophic strains, making them prototrophs, able to grow in corresponding deficient media. In addition, rat liver S9 was chemically introduced to simulate the effect of metabolism, since certain compounds, like benzopyren, become mutagenic only after their metabolic conversion.

The mutagenic potential of salts **8b** (Li$_2$ANAT) and **10** (TKX-50) were evaluated using the AMES test with S9 mixture. The toxicological potential of salt **7a** was evaluated using *Vibrio fischeri*, due to its high water-solubility and because it is the water-soluble monomer of BDNAPM, a highly potential secondary explosive. BDNAPM (Figure 13.11) was also experimentally tested with the AMES test, but cannot be evaluated using *Vibrio fischeri* itself, because of its low water-solubility. All experimental findings were calculated using a smart algorithm and are listed in Table 13.10. The toxicity respectively mutagenic potential of the compounds in Table 13.9 and Table 13.10 are rated as non-toxic/mutagenic (+), moderately toxic/mutagenic (o) and toxic/mutagenic (−).

Check your knowledge: See Chapter 14 for Study Questions.

Literatur

C. Unger et al., *ICT Annual Conference* **2019**, P 85.

14 Study Questions

Answers can be found online under: www.degruyter.com/klapoetke-answers.

1. Please write suitable Lewis-type structures for the following molecules,
   FOX-7:
   PETN:
   RDX:

2. If 10 kg of a common high explosive is detonated, the minimum distance from the explosion which gives a chance of survival is
   ○ 2.2 m   ○ 4.3 m   ○ 8.6 m

3. Give examples for the following:
   – single base gun propellant:
   – double base gun propellant:
   – triple base gun propellant:

4. Why does Mg nowadays get replaced with Zr in smoke munition?

5. Name one commercially available computer code to calculate
   – enthalpies of formation:
   – detonation parameters:

6. Calculate the oxygen balance with respect to CO_2 of ethylene glycol dinitrate, $O_2N-O-CH_2-CH_2-O-NO_2$.

7. If a high explosive has a VoD of 9.000 m s^{-1}. What is its approximate Gurney velocity?

8. Which of the following materials is more suitable for a kinetic energy penetrator, and why?
   ○ SS-316   ○ DU
   reason:

9. In order to reduce the erosive effects of a gun propellant
   – the N_2/CO ratio should be   ○ high   ○ low
   – T_c should be   ○ high   ○ low

10. In a drophammer experiment, a mass of 5 kg hits the sample from a height of 50 cm. Calculate the impact energy.

11. Give five different nitrating agents.

12. For which application would hypergolic behaviour be desirable (give two applications)?

13. Name a possible replacement for TNT.

14. Name a possible replacement for Composition B.

15. How does the specific impulse I_{sp}^* depend on the combustion temperature (T_c) and the average molecular weight (M) of the combustion gases formed?

16. What is the relationship between the average thrust ($\overline{F}$) of a rocket and its specific impulse (I_{sp})?

17. Give one example each of a solid rocket propellant which belongs to the following classes and state whether the material is homogeneous or heterogeneous:
    − double-base:
    − composite:

18. As a "rule of thumb" by how much would one need to increase the specific impulse I_{sp}^* of a rocket in order to double its payload?

19. Which metals are clandestine signalling and illuminating formulations in this range 700–1000 nm usually based on?

20. How does the detonation pressure (p_{C-J}) scales with the loading density?

21. How can the enthalpies of sublimation ($\Delta H_{sub.}$) and vaporization ($\Delta H_{vap.}$) be estimated from the melting and boiling points of a substance?

22. How does the penetration depth of kinetic energy munition depend on the material of the penetrator?

23. How does the Gurney velocity ($\sqrt{2E}$) of an explosive depend on the detonation velocity (D = VoD)?

24. Describe three different methods suitable for the synthesis of N_2O_5.

25. Describe the laboratory scale synthesis of ADN starting from ammonia.

26. Which detonation product seems to be most effective for biological agent defeat explosives (e.g. against anthrax spores)?

27. Which general reaction describes the combustion of Al based nanothermites?

28. Name one lead-free tetrazolate based primary explosive.

29. Name one high explosive which has the same chemical composition ($C_aH_bN_cO_d$) as CL-20.

30. What are the environmental/health concerns of AP?

31. For which type of rocket/missile propulsion could chemical thermal propulsion (CTP) be very valuable?

32. In which pressure range do the combustion chambers of a solid rocket motor and a large calibre gun operate?

33. What are the most important performance parameters for
    a) high explosives (secondary explosives),
    b) gun propellants,
    c) rocket propellants?

34. Which two parameters influence the erosion problems in gun barrels the most?
35. Compare the perforation of a shaped charge device (SC) with that of an explosively formed projectile (EFP) for ideal standoff distances.
36. Which experimental test is suitable to assess the shockwave sensitivity of an explosive?
37. Which experimental test (IM test) would be suitable to assess the behaviour of an explosive when subjected to a fuel fire?
38. Compare the high explosives RDX and HNS. What are the most important advantages and disadvantages of the two compounds?
39. What prevents TATP from having any application in the defense world (two reasons)?
40. Which condition is a good/reasonable approximation (e.g. isochoric, isobaric) for the following:
    a) gun propellants,
    b) rocket propellants,
    c) secondary explosives?
41. What are the main light emitting species in visible signal flare formulations which emit in the following regions:
    a) red,
    b) green?
42. How does the maximum wavelength of a blackbody radiator (λ_{max}) depend on/change with the temperature? Which law describes this phenomenon?
43. How does the linear burn rate generally depend on the pressure?
44. How does the electrostatic discharge sensitivity (ESD) usually depend on the particle (grain) size?
45. Are shaped charges (SC) or explosively formed projectiles (EFP) more sensitive to the standoff distance?
46. Why do detonations or combustions often produce CO and not exclusively CO_2 despite a positive oxygen balance?
47. What are the three main "strategies" of designing CHNO based high explosives?
48. For each of the following, give one example of an explosive that only contains the following energetic groups,
    – nitro:
    – nitrate (nitrate ester):
    – nitramino:
    – nitrimino:
    – azide:
    – peroxo:

49. What are the disadvantages of using nitroglycerine for military applications?
50. According to Kamlet and Jacobs the detonation velocity (D) and the detonation pressure (p_{C-J}) can be estimated. Give the relevant equations.
51. Calculate the spark energy (ESD test) for a spark generated under the following conditions: C = 0.005 µF, V = 10000 V
52. Give two reagents which can be used to convert an amine into an N-oxide.
53. Give the structures of:

    MMH: UDMH: DMAZ:

    To which class(es) (in terms of application) do these compounds belong ? What might be a practical use of DMAZ and why?
54. A Hawker Siddeley Harrier is a single engine fighter plain with the capability of performing vertical take-off. How much thrust would the engine need to deliver to perform a vertical take-off if we assume a fighter mass of 17 t.
55. Name two types of electric detonators.
56. Give five advantages of an EBW detonator compared to a conventional hot-wire detonator.
57. Name a suitable replacement for tetrazene.
58. Name a suitable replacement for MMH in gelled hypergolic propellants.
59. The shock-wave energy and bubble-energy correspond to which important property?
60. Give two empirical relationships that correlate the VoD and pC-J with the specific impulse.
61. Name a possible replacement formulation for that of Composition-B that does not contain TNT.
62. Give one possible chemical synthesis of nitroform (NF) starting from commercially available ingredients.
63. Give one possible synthesis of trinitroethanol (TNE) starting from NF.

15 Literature

Additional Reading

General
Agrawal, J. P., *High Energy Materials*, Wiley-VCH, Weinheim, **2010**.
Agrawal, J. P.; Hodgson, R. D., *Organic Chemistry of Explosives*, Wiley-VCH, Weinheim, **2007**.
Akhavan, J., *The Chemistry of Explosives*, 2nd edn., The Royal Society of Chemistry, Cambridge, **2004**.
Bailey, A.; Murray, S. G., *Explosives, Propellants & Pyrotechnics*, Brassey's, London, **1989**.
Brinck, T. (ed.), Green Energetic Materials, Wiley, Weinheim, **2014**.
Cook, J. R., *The Chemistry and Characteristics of Explosive Materials*, 1st edn., Vantage Press, New York, **2001**.
Dolan, J. E.; Langer, S. S. (ed.), *Explosives in the Service of Man*, The Royal Society of Chemistry, Cambridge, **1997**.
Fair, H. D.; Walker, R. F., *Energetic Materials*, Vol. 1 & 2, Plenum Press, New York, **1977**.
Fedoroff, B. T.; Aaronson, H. A.; Reese, E. F.; Sheffield, O. E.; Clift, G. D., *Encyclopedia of Explosives*, U.S. Army Research and Development Command (TACOM, ARDEC) Warheads, Energetics and Combat Support Center, Picatinny Arsenal, New Jersey, **1960**.
Fordham, S., *High Explosives and Propellants*, 2nd edn., Pergamon Press, Oxford, **1980**.
Jonas, J. A.; Walters, W. P., *Explosive Effects and Applications*, Springer, New York, **1998**.
Keshavarz, M. H.; Klapötke, T. M., *Energetic Compounds*, De Gruyter, Berlin/Boston, **2017**.
Keshavarz, M. H.; Klapötke, T. M., *The Properties of Energetic Materials*, De Gruyter, Berlin/Boston, **2018**.
Keshavarz, M. H., *Liquid Fuels as Jet Fuels and Propellants*, Nova, New York, **2018**.
Keshavarz, M. H., *Combustible Organic Materials*, De Gruyter, Berlin/Boston, **2018**.
Klapötke, T. M. (ed.), *High Energy Density Materials*, Structure and Bonding, Vol. 125, Springer, Berlin, **2007**.
Klapötke, T. M., in: E. Riedel (ed.), *Moderne Anorganische Chemie*, 3. edn., ch. 1, Nichtmetallchemie, Walter de Gruyter, Berlin, **2007**.
Klapötke, T. M., *Energetic Materials Encyclopedia*, 2. Aufl., De Gruyter, Berlin/Boston, **2019**.
Koch, E.-C., *Sprengstoffe, Treibmittel, Pyrotechnika*, 2. Aufl., De Gruyter, Berlin/Boston, **2019**.
Köhler, J.; Meyer, R.; Homburg, A., *Explosivstoffe*, 10. Auflage, Wiley-VCH, Weinheim, **2008**.
Kubota, N., *Propellants and Explosives*, Wiley-VCH, Weinheim, **2002**.
Olah, G. A.; Squire, D. R., *Chemistry of Energetic Materials*, Academic Press, San Diego, **1991**.
Urbanski, T., *Chemistry and Technology of Explosives*, Vol. 1–4, Pergamon Press, Oxford, **1964/1984**.
Venugopalan, S., *Demystifying Explosives: Concepts in High Energy Materials*, Elsevier, Amsterdam, **2015**.

Synthetic Chemistry
Agrawal, J. P.; Hodgson, R. D., *Organic Chemistry of Explosives*, Wiley, Hoboken, NJ, **2007**.
Ang, H. G.; Santhosh, G., *Advances in Energetic Dinitramides*, World Scientific, New Jersey, London, Singapore, **2007**.
Bailey, P. D.; Morgan, K. M., *Organonitrogen Chemistry*, Oxford Science Publications, Oxford, **1996**.
Boyer, J. H., *Nitroazoles*, VCH, Weinheim, **1986**.
Feuer, H.; Nielsen, A. T., *Nitro Compounds*, VCH, Weinheim, **1990**.
Feuer, H., *Nitrile Oxides, Nitrones, and Nitronates in Organic Synthesis*, Wiley, Hoboken, NJ, **2008**.

Hammerl, A.; Klapötke, T. M., *Pentazoles*. In: Katritzky, A.R.; Ramsden C. A.; Scriven, E. F. V.; Taylor, R. J. K. (ed.); *Comprehensive Heterocyclic Chemistry III*, Elsevier, Oxford, **2008**; Vol. 6, p. 739–758.

Ledgard, J., *The Preparatory Manual of Explosives*, 3rd edn., J. B. Ledgard, **2007**.

Nielsen, A. T., *Nitrocarbons*, VCH, Weinheim, **1995**.

Ono, N., *The Nitro Group in Organic Synthesis*, Wiley-VCH, New York, **2001**.

Torssell, K. B. G., *Nitrile Oxides, Nitrones, and Nitronates in Organic Synthesis*, VCH, Weinheim, **1988**.

Gao, H.; Shreeve, J. M., *Chem. Rev.*, 2011, 111, 7377–7436.

Detonation Physics

Dremin, A. N., *Toward Detonation Theory*, Springer, Berlin, New York, **1999**.

Fickett, W.; Davis, W. C., *Detonation Theory and Experiment*, Dover Publications, Mineola, New York, **1979**.

Hidding, B., *Untersuchung der Eignung von Silanen als Treibstoffe in der Luft- und Raumfahrt* (Diplomarbeit), UniBW München, **2004**.

Holian, B. L.; Baer, M. R.; Brenner, D.; Dlott, D. D.; Redondo, A.; Rice, B. M.; Sewell, T.; Wight, C. A., *Molecular Dynamics Simulations of Detonation Phenomena*, ITRI Press, Laurel, MD, **2004**.

Liberman, M. A., *Introduction to Physics and Chemistry of Combustion*, Springer, Berlin, **2008**.

Penney, W. G.; Taylor, G.; Jones, H.; Evans, W. M.; Davies, R. M.; Owen, J. D.; Edwards, D. H.; Thomas, D. E.; Adams, C. A.; Bowden, F. P.; Ubbelohde, A. R.; Courtnev-Pratt, J. S.; Paterson, S.; Taylor, J.; Gurton, O. A., *Proc. Royal Soc.* **1950**, *204 A*, 3.

Zel'dovich, Y. B., Raizer, Y. P., *Physics of Shock Waves and High-Temperature Hydrodynamic Phenomena*, Dover, Mineola, New York, **2002**.

Pyrotechnics

Conkling, J. A., *Chemistry of Fireworks*, Marcel Dekker, New York, **1985**.

Hardt, A. P., *Pyrotechnics*, Pyrotechnica Publications, Post Falls, **2001**.

Koch, E.-C., *Journal of Pyrotechnics* **2002**, *15*, 9–23.

Koch, E.-C., *Journal of Pyrotechnics* **2005**, *21*, 39–50.

Koch, E.-C., *Propellants, Explosives, Pyrotechnics* **2001**, *26(1)*, 3–11.

Koch, E.-C., *Propellants, Explosives, Pyrotechnics* **2002**, *27(5)*, 262–266.

Koch, E.-C., *Propellants, Explosives, Pyrotechnics* **2002**, *27*, 340.

Koch, E.-C., *Propellants, Explosives, Pyrotechnics* **2004**, *29(2)*, 67–80.

Koch, E.-C., *Propellants, Explosives, Pyrotechnics* **2006**, *31(1)*, 3–19.

Koch, E.-C., *Propellants, Explosives, Pyrotechnics* **2008**, *33(3)*, 165–176.

Koch, E.-C., *Propellants, Explosives, Pyrotechnics* **2008**, *33(5)*, 333–335.

Koch, E.-C., *Propellants, Explosives, Pyrotechnics* **2009**, *34*, 6–12.

Russell, M. S., *The Chemistry of Fireworks*, The Royal Society of Chemistry, Cambridge, **2000**.

Steinhauser, G.; Klapötke, T. M., *Angew. Chem.* **2008**, *120*, 2–21; *Intl. Ed.* **2008**, *47*, 2–20.

Steinhauser, G.; Klapötke, T. M., *J. Chem. Educ.* **2010**, *87(2)*, 150–156.

Koch, E.-C., *Metal-Fluorocarbon Based Energ. Materials*, Wiley-VCH, Weinheim, **2012**.

Crow, J. M., *Chem. World* **2012**, 46–49.

Material Sciences

Teipel, U. (ed.), *Energetic Materials*, Wiley-VCH, Weinheim, **2005**.

Engineering and Technology

Cooper, P. W.; Kurowski, S. R., *Technology of Explosives*, Wiley-VCH, New York, **1996**.

Cooper, P. W., *Explosives Engineering*, Wiley-VCH, New York, **1996**.

Rocket Propulsion
Kubota, N., *Propellants and Explosives*, Wiley, VCH, Weinheim, **2002**.
Jacobs, P. W. M.; Whitehead, H. M., *Chem. Rev.* **1969**, *69(4)*, 551–590.
Schmidt, E. W., *Hydrazine and its Derivatives*, 2. Aufl., Band 1 und 2, Wiley, New York, **2001**.
Yang, V.; Brill, T. B.; Ren, W.-Z. (ed.), *Solid Propellant Chemistry, Combustion, and Motor Interior Ballistics*, Vol. 185, AAIA, Reston, VA, **2000**.
Ulas, A.; Kuo, K. K., *Comb. Flame*, **2001**, *127*, 1935–1957.
Beckstead, M. W.; Puduppakkama, K.; Thakre, P.; Yang, V., *Progress in Energy and Combustion Science*, **2007**, *33* 497.
DeLuca, L. (ed.), Chemical Rocket Propulsion – A Comprehensive Survey of Energetic Materials, Springer, ISBN: 978-3-319-27746-2, **2017**.
UY, H. E. A., "Double-base Propellants, Solid Rocket Propulsion Technology", **2012**, 369.
Naminosuke, K., "Propellant and Explosives", in Thermochemical Aspects of Combustion in Germany, Wiley, **2007**.
Epishina, M. A.; Ovchinnikov, I. V.; Kulikov, A. S.; Makhova, N. N.; Tartakovsky, V. A., *Mendeleev Communications*, **2011**, *21*, 21.
Ermakov, A.; Bulatov, P.; Vinogradov, D.; Tartakovskii, V., *Russ. J. Org. Chem.*, **2004**, *40*, 1062.

Gun Propulsion
Jahnuk, H., *Prop. Expl. Pyrotech.* **1976**, *1*, 47–50.
Dürschner, M. G., *Prop. Expl. Pyrotech.* **1976**, *1*, 81–85.
Zentner, B. A.; Reed, R., *Calculated Performance of Gun Propellant Compositions Containing High Nitrogen Ingredients*, 5th Intl. Gun Propellant Symp., Picatinny Arsenal, NJ, November 19–21, **1991**, IV–278.

Weapons
Carlucci, D. E.; Jacobson, S. S., *Ballistics – Theory and Design of Guns and Ammunition*, CRC Press, Boca Raton, Fl., **2008**.
Dathan, H., *Waffenlehre für die Bundeswehr*, 4th edition, Mittler, Herford, Bonn, **1980**.
Heard, B. J., *Handbook of Firearms and Ballistics*, 2nd edition, Wiley-Blackwell, Chichester, **2008**.

Testing Methods
Kneisl, P., *13th NTREM*, **2010**, 184–193.
Sućeska, M., *Test Methods for Explosives*, Springer, New York, **1995**.

Gurney Velocity
Backofen, J. E., *New Trends in Research of Energetic Materials*, **2006**, 9th Seminar, 76–89.
Backofen, J. E.; Weickert, C. A., in: Furnish, M. D.; Thadhani, N. N.; Horie, Y. (eds.), *Shock Compression of Condensed Matter*, **2001** Amer. Inst. Phys. **2002**, 954–957.
Cooper, P. W., *Explosives Engineering*, Wiley-VCH, New York, Chichester, Weinheim, Kap. 27, **1996**, 385–403.
Gurney, R. W., *Internal Report BRL-405*, **1943**, Ballistic Research Laboratory, Aberdeen, MD.
Hardesty, D. R.; Kennedy, J. E., *Combust. Flame* **1977**, *28*, 45.
Jaansalu, K. M.; Dunning, M. R.; Andrews, W. S., *Prop. Expl. Pyrotech.* **2007**, *32*, 80.
Kamlet, M. J.; Finger, M., *Combust. Flame* **1979**, *34*, 213.
Keshavarz, M. H., *Prop. Expl. Pyrotech.* **2008**, *33*, 316.
Koch, A.; Arnold, N.; Estermann, M., *Prop. Expl. Pyrotech.* **2002**, *27*, 365.
Reaugh, J. E.; Souers, P. C., *Prop. Expl. Pyrotech.* **2004**, *29*, 124.
Stronge, W. J.; Xiaoqing, M.; Lanting, Z., *Int J. Mech. Sci.* **1989**, *31*, 811.

Computational Chemistry (quantum mechanics)

Politzer, P.; Murray, J. S. N., *Energetic Materials*, Part 1: Decomposition, Crystal and Molecular Properties, in: *Theor. Comput. Chem.*, Vol. 12, Elsevier, Amsterdam, **2003**.

Politzer, P.; Murray, J. S. N., *Energetic Materials*, Part 2, Energetic Materials, Part 2: Detonation, Combustion, in: *Theor. Comput. Chem.*, Vol. 13, Elsevier, Amsterdam, **2003**.

Rice, B. M.; Hare, J. J.; Byrd, E. F. C., *J. Phys. Chem.* **2007**, *III*, 10874–10879.

Seminario, J. M.; Politzer, P., Modern Density Functional Theory: A Tool for Chemistry, in: *Theor. Comput. Chem.*, Vol. 2, Elsevier, Amsterdam, **1995**.

Politzer, P. A.; Murray, J. S., *Energetic Materials*, Part 1 & 2, Elsevier, Amsterdam, **2003**.

Nanothermites

Brown, M. E.; Taylor, S. J.; Tribelhorn, M. J., *Propellants, Explosives, Pyrotechnics*. **1998**, *23*, 320–327.

Bulian, C. J.; Kerr, T. T.; Puszynski, J. A., *Proceedings of the International Pyrotechnics Seminar* **2004**, 31st 327–338.

Clapsaddle, B. J.; Zhao, L.; Gash, A. E.; Satcher, J. H.; Shea, K. J.; Pantoya, M. L.; Simpson, R. L., *Res. Soc. Symp. Proc. Vol. 800*. **2004**, AA2.7.1–AA2.7.6.

Clapsaddle, B. J.; Zhao, L.; Prentice, D.; Pantoya, M. L.; Gash, A. E.; Satcher, J. H.; Shea, K. J.; Simpson, R. L., *International Annual Conference of ICT* **2005**, 36th (Energetic Materials), 39/1–39/10.

Comet, M.; Pichot, V.; Spitzer, D.; Siegert, B.; Ciszek, F.; Piazzon, N.; Gibot, P., *International Annual Conference of ICT* **2008**, 39th, V38/1–V38/8.

Gash, A. E.; Satcher, J. H.; Simpson, R. L.; Clapsaddle, B. J., *Mat. Res. Soc. Symp. Proc. Vol. 800*. **2004**, AA2.2.1–AA2.2.12.

Naud, D. L.; Hiskey, M.; Son, S. F.; Busse, J. R.; Kosanke, K., *Journal of Pyrotechnics* **2003**, *17*, 65–75.

Pantoya, M. L.; Granier, J. J., *Propellants, Explosives, Pyrotechnics*, **2005**, *30(1)*, 53–62.

Perry, W. L.; Smith, B. L.; Bulian, C. J.; Busse, J. R.; Macomber, C. S.; Dye, R. C.; Son, S. F., *Propellants, Explosives, Pyrotechnics*, **2004**, *29*, 99–105.

Piercey, D. G.; Klapötke, T. M., *Central Europ. J. Energ. Mat.* **2010**, in press.

Prakash, A.; McCormick, A. V.; Zachariah, M. R., *Chem. Mater.* **2004**, *16(8)*, 1466–1471.

Puszynski, J. A.; Bichay, M. M.; Swiatkiewicz, J. J., United States Patent Application Number 20060113014. June 1, 2006.

Puszynski, J. A.; Bulian, C. J.; Swiatkiewicz, J. J., *Journal of Propulsion and Power* **2007**, *23(4)*, 698–706.

Puszynski, J. A., *Mat. Res. Soc. Symp. Proc. Vol. 800*. **2004** AA6.4.1–AA6.4.10.

Sanders, V. E.; Asay, B. W.; Foley, T. J.; Tappan, B. C.; Pacheco, A. N.; Son, S. F., *Propulsion and Power*. **2007**, *23(4)*, 707–714

Schoenitz, M.; Ward, T.; Dreizin, E. L., *Res. Soc. Symp. Proc. Vol. 800*. **2004**, AA2.6.1–AA2.6.6.

Shende, R.; Subramanian, S.; Hasan, S.; Apperson, S.; Thiruvengadathan, R.; Gangopadhyay, K.; Gangopadhyay, S.; Redner, S.; Kapoor, D.; Nicolich, S.; Balas, W., *Propellants, Explosives, Pyrotechnics*. **2008**, *33(2)*, 122–130.

Son, S. F., *Mat. Res. Soc. Symp. Proc. Vol. 800*. **2004**, AA5.2.1–AA5.2.12.

Son, S. F.; Yetter, R. A.; Yang, Y., *Journal of Propulsion and Power* **2007**, *23(4)*, 643–644.

Son, S. F.; Asay, B. W.; Foley, T. J.; Yetter, R. A.; Wu, M. H.; Risha, G. A., *Propulsion and Power*. **2007**, *23(4)*, 714–721.

Spitzer, D.; Comet, M.; Moeglin, J.-P.; Stechele, E.; Werner, U.; Suma, A.; Yves, *International Annual Conference of ICT* **2006**, 37th (Energetic Materials), 117/1–117/10.

Tillotson, T. M.; Gash, A. E.; Simpson, R. L.; Hrubesh, L. W.; Satcher, J. H.; Poco, J. F., *Journal of Non-Crystalline Solids*, **2001**, *285*, 338–345.

Umbrajkar, S. M.; Schoenitz, M.; Dreizin, E. L., *Propellants, Explosives, Pyrotechnics*. **2006**, *31(5)*, 382–389.

Walker, J.; Tannenbaum, R., *Res. Soc. Symp. Proc. Vol. 800*. **2004**, AA7.8.1–AA7.8.10.

Walter, K. C.; Pesiri, D. R.; Wilson, D. E., *Propulsion and Power* **2007**, *23(4)*, 645–650.

Weismiller, M. R.; Malchi, J. Y.; Yetter, R. A.; Foley, T. J., *Chemical and Physical Processes in Combustion*, **2007**, 595, 600.

Primary Explosives

Cartwright, M.; Wilkinson, J., *Propellants, Explosives, Pyrotechnics* **2011**, *36*, 530–540.

Cartwright, M.; Wilkinson, J., *Propellants, Explosives, Pyrotechnics* **2010**, *35*, 326–332.

Fronabarger, J. W.; Williams, M. D.; Sanborn, W. B.; Bragg, J. G.; Parrish, D. A.; Bichay, M., *Propellants, Explosives, Pyrotechnics* **2011**, *36*, 541–550.

Ugryumov, I A.; Ilyushin, M. A.; Tselinskii, I. V.; Kozlov, A. S., *Russian Journal of Applied Chemistry* **2003**, *76*, 439–441.

Fronabarger, J. W.; Williams, M. D.; Sanborn, W. B.; Parrish, D. A.; Bichay, M., *Propellants, Explosives, Pyrotechnics* **2011**, *36*, 459–470.

Fischer, D.; Klapötke, T. M.; Piercey, D. G.; Stierstorfer, J., *J. Energ. Mat.* **2012**, *30*, 40–54.

Mehta, N.; Damavarapu, R.; Cheng, G.; Dolch, T.; Rivera, J.; Duddu, R.; Yang, K., *Proceedings of the International Pyrotechnics Seminar* **2009**, *36*, 175–180.

Matyáš, R.; Pachman, J., *Primary Explosives*, Springer, Berlin, **2012**.

Ahmad, S. R.; Cartwright M., *Laser Ignition of Energetic Materials*, Wiley, Weinheim, **2015**.

Ilyshin, M. A.; Tselinsky, I. V.; Shugalei, I. V., *Centr. Europ. J. Energ. Mat.* **2012**, *9(4)*, 293–327.

Thermally Stable Explosives

Baird, T.; Drummond, R.; Langseth, B.; Silipigno, L., *Oilfield Review*, **1998**, 50.

Sikder, A. K.; Sikder, N., *Journal of Hazardous Materials*, **2004**. *112*, 1.

Wang, X., *Current situation of study on insensitive composite explosives in USA*. Huozhayao Xuebao, **2007**, *30*, 78–80.

Albright, D., *South Africa and the affordable bomb*, in: *Bulletin of the Atomic Scientists*, Jul **1994**, Educational Foundation for Nuclear Science, Inc., 37–47.

Nuclear notebook, in *Bulletin of the Atomic Scientists*, May **1988**, Educational Foundation for Nuclear Science, Inc., 55.

Rieckmann, T.; Völker, S.; Lichtblau, L.; Schirra, R., *Chem. Eng. Sci.*, **2001**, *56*, 1327.

Coburn, M. D.; Jackson, T. E., *J. Heterocycl. Chem.*, **1968**, *5*, 199.

Meyer, R.; Köhler, J.; Homburg, A., *Explosives*, 7th edn., **2008**, Wiley, Weinheim.

Sitzmann, M. E., *J. Energ. Mater.* **1988**, *6*, 129.

Agrawal, J. P., *Progress in Energy and Combustion Science*, **1998**, *24*, 1.

Agrawal, J. P., *Propellants, Explos. Pyrotech.*, **2005**, *30*, 316.

Agrawal, J. P., *Central European J. of Energ Materials*, **2012**, *9*, 273.

Agrawal, J. P., Prasad, U. S.; Surve, R. N., *New. J. Chem*, **2000**, *24*, 583.

Sikder, N.; Sikder, A. K.; Agrawal, J. P., *Ind. J. Eng. & Mater Sci.*, **2004**, *11*, 516.

Mousavi, S.; Esmaelipour, K.; Keshavaraz, M. H., Central *Europ. J. of Energ. Mat.* **2013**, *10*, 455.

Klapötke, T. M.; Witkowski, T., *Chem. Plus Chem.* **2016**, *81*, 357.

Gottfried, J. L.; Klapötke, T. M.; Witkowski, T. G., *Prop. Expl. Pyrotech.*, in press: DOI: 10.1002/prep.201600257.

Zhang, X.; Xiong, H.; Yang, H.; Cheng, G., *Propellants Explos. Pyrotech.*, **2017**, *42*, in press.

Snyder, C. J.; Myers, T. W.; Imler, G. H.; Chavez, D. E.; Parrish, D. A.; Veauthier, J. M.; Scharff, R. J., *Propellants Explos. Pyrotech.* **2017**, *42*, in press.

Cited Literature

[1] (a) Beck, W.; Evers, J.; Göbel, M.; Oehlinger, G.; Klapötke, T. M., *Z. Anorg. Allg. Chem.* **2007**, *633*, 1417–1422.
(b) Beck, W.; Klapötke, T. M., *J. Mol. Struct.* (THEOCHEM) **2008**, *848*, 94–97.
(c) Karaghiosoff, K.; Klapötke, T. M.; Michailovski, A.; Holl, G., *Acta Cryst. C* **2002**, *C 58*, o580–o581.
(d) Gökçınar, E.; Klapötke, T. M.; Bellamy, A. J., *J. Mol. Struct.* (THEOCHEM), **2010**, *953*, 18–23.
(e) Majano, G.; Mintova, S.; Bein, T.; Klapötke, T. M., *Advanced Materials* **2006**, *18*, 2240–2243.
(f) Majano, G.; Mintova, S.; Bein, T.; Klapötke, T. M., *J. Phys. Chem. C*, **2007**, *111*, 6694–6699.
(g) Matyushin, Y. N.; KonkovaT. S.; Korchatova, L. I., *To the Question of the Enthalpy of Formation for the Nitrocellulose*, ICT, **1998**, Karlsruhe, Germany, 197-1–179-8.

[2] (a) Evers, J.; Klapötke, T. M., Mayer, P.; Oehlinger, G., Welch, J., *Inorg. Chem.* **2006**, *45*, 4996–5007.
(b) Crawford, M.-J.; Evers, J.; Göbel, M.; Klapötke, T. M.; Mayer, P.; Oehlinger, G.; Welch, J. M., *Propellants Explosives and Pyrotechnics* **2007**, *32*, 478–495.

[3] Huynh, M. H. V.; Hiskey, M. A.; Meyer, T. J.; Wetzler, M., *PNAS*, **2006**, *103*, 5409–5412. See also: Klapötke, T. M.; Miro Sabate, C.; Welch, J. M., *J. Chem. Soc., Dalton Trans.*, **2008**, 6372–6380.

[4] Geisberger, G.; Klapötke, T. M.; Stierstorfer, J., *Eur. J. Inorg. Chem.* **2007**, 4743–4750.

[5] SERDP information: Cleanup CU-1164: http://www.p2pays.org/ref/19/18164.pdf (2/21/03).

[6] The official SERDP source for perchlorate information: http://www.serdp.org/research/er-perchlorate.cfm

[7] SERDP & ESTCP Annual Symposium **2007**: http://www.serdp.org/Symposium/.

[8] Göbel, M.; Klapötke, T. M., *Z. Anorg. Allg. Chem.* **2007**, *633*, 1006–1017.

[9] Klapötke, T. M.; Penger, A.; Scheutzow, S.; Vejs, L., *Z. Anorg. Allg. Chem.*, **2008**.

[10] Karaghiosoff, K.; Klapötke, T. M.; Sabate, C. Miro, *Chemistry Europ. J.*, **2008**.

[11] Doherty, R. M., Novel Energetic Materials for Emerging Needs, 9th-IWCP on Novel Energetic Materials and Applications, Lerici (Pisa), Italy, September 14–18, **2003**.

[12] Fischer, N.; Klapötke, T. M.; Scheutzow, S.; Stierstorfer, J., *Centr. Europ. J. Energ. Mat.* **2008**, *5*, 3–18.

[13] Steinhauser, G.; Klapötke, T. M., *Angew. Chem.* **2008**, *120*, 2–21; *Intl. Ed.* **2008**, *47*, 2–20.

[14] Klapötke, T. M.,Tetrazoles for the safe detonation, *Nachrichten aus der Chemie*, **2008**, *56(6)*, 645–648.

[15] Klapötke, T. M.; Stierstorfer, J.; Tarantik, K. R., Neue, umweltverträglichere Formulierungen für Pyrotechnika, *Die aktuelle Wochenschau, GDCh*, 15.09.**2008**, Woche 37: http://www.aktuelle-wochenschau.de/woche37/woche37.html.

[16] (a) Halford, B., *Chem. Eng. News*, **2008**, June 30, 14–18.
(b) Sabatini, J. J.; Poret, J. C.; Broad, R. N., *Angew. Chem. Int. Ed.*, **2011**, *50*, 4624–4626.

[17] Kamlet, M. J.; Jacobs, S. J., *J. Chem. Phys.* **1968**, *48*, 23.

[18] Kamlet, M. J.; Ablard, J. E., *J. Chem. Phys.* **1968**, *48*, 36.

[19] Kamlet, M. J.; Dickinson, C., *J. Chem. Phys.* **1968**, *48*, 43.

[20] Sućeska, M., Proceedings of 32nd Int. Annual Conference of ICT, July 3–6, Karlsruhe, **2001**, S. 110/1–110/11.

[21] Curtiss, L. A.; Raghavachari, K.; Redfern, P. C.; Pople, J. A., *J. Chem. Phys.* **1997**, *106*, 1063.

[22] Byrd, E. F. C.; Rice, B. M., *J. Phys. Chem. A.* **2006**, *110*, 1005.

[23] Rice, B. M.; Pai, S. V.; Hare, J., *Combustion and Flame* **1999**, *118*, 445.

[24] Linstrom, P. J.; Mallard, W. G. (eds.), NIST Chemistry WebBook, NIST Standard Reference Database Number 69, June **2005**, National Institute of Standards and Technology, Gaithersburg MD, 20899 (http://webbook.nist.gov).

[25] Westwell, M. S.; Searle, M. S.; Wales, D. J.; Williams, D. H., *J. Am. Chem. Soc.* **1995**, *117*, 5013.
[26] Jenkins, H. D. B.; Roobottom, H. K.; Passmore, J.; Glasser, L., *Inorg. Chem.* **1999**, *38(16)*, 3609–3620.
[27] Jenkins, H. D. B.; Tudela, D.; Glasser, L., *Inorg. Chem.* **2002**, *41(9)*, 2364–2367.
[28] Jenkins, H. D. B.; Glasser, L., *Inorg. Chem.* **2002**, *41(17)*, 4378–88.
[29] Jenkins, H. D. B., Chemical Thermodynamics at a Glance, Blackwell, Oxford, **2008**.
[30] Ritchie, J. P.; Zhurova, E. A.; Martin, A.; Pinkerton, A. A. , *J. Phys. Chem. B*, **2003**, *107*, 14576–14589.
[31] Sućeska, M., Calculation of Detonation Parameters by EXPLO5 Computer Program, Materials Science Forum, **2004**, 465–466, 325–330.
[32] Sućeska, M., *Propellants, Explos., Pyrotech.* **1991**, *16*, 197.
[33] Sućeska, M., *Propellants Explos., Pyrotech.* **1999**, *24*, 280.
[34] Hobbs, M. L.; Baer, M. R., Calibration of the BKW-EOS with a Large Product Species Data Base and Measured C-J Properties, Proc. of the 10th Symp. (International) on Detonation, ONR 33395-12, Boston, MA, July 12–16, **1993**, p. 409.
[35] (a) Fischer, N.; Klapötke, T. M.; Scheutzow, S.; Stierstorfer, J., *Central Europ. J. of Energ. Mat.*, **2008**, *5(3–4)*, 3–18;
(b) Gökçinar, E.; Klapötke, T. M.; Kramer, M. P., *J. Phys. Chem.*, **2010**, *114*, 8680–8686.
[36] Trimborn, F., *Propellants Explos., Pyrotech.* **1980**, *5*, 40.
[37] Held, M., *NOBEL-Hefte*, **1991**, *57*, 14.
[38] Sućeska, M. , Intl. J. Blasting and Fragmentation, 1997, 1, 261.
[39] (a) Politzer, P.; Murray, J. S., *Computational Characterization of Energetic Materials*, in: Maksic, Z. B.; Orville-Thomas, W. J. (eds.), Pauling's Legacy: Modern Modelling of the Chemical Bond, Theoretical and Computational Chemistry, Vol. 6, **1999**, 347–363;
(b) Politzer, P.; Murray, J. S.; Seminario, J. M.; Lane, P.; Grice, M. E.; Concha, M. C., *J. Mol. Struct.* **2001**, *573*, 1.
(c) Murray, J. S.; Lane, P.; Politzer, P., *Molec. Phys.* **1998**, *93*, 187;
(d) Murray, J. S.; Lane, P.; Politzer, P., *Mol. Phys.* **1995**, *85*, 1.
(e) Murray, J. S.; Concha, M. C.; Politzer, P., *Molec. Phys.* **2009**, *107(1)*, 89;
(f) Politzer, P.; Martinez, J.; Murray, J. S.; Concha, M. C.; Toro-Labbe, A., *Mol. Phys.* **2009**, *107(19)*, 2095–2101;
(g) Pospisil, M.; Vavra, P. ; Concha, M. C.; Murray, J. S.; Politzer, P., *J. Mol. Model.* **2010**, *16*, 985–901.
[40] (a) Rice, B. M.; Hare, J. J.; Byrd, E. F., *J. Phys. Chem. A*, **2007**, *111(42)*, 10874.
(b) Rice, B. M.; Byrd, E. F. C., *J. Mater. Res.*, **2006**, *21(10)*, 2444.
[41] Byrd, E. F.; Rice, B. M., *J. Phys. Chem. A*, **2006**, *110(3)*, 1005.
[42] Rice, B. M.; Hare, J., *Thermochimica Acta*, **2002**, *384(1–2)*, 377.
[43] Rice, B. M.; Pai, S. V.; Hare, J., *Combustion and Flame*, **1999**, *118(3)*, 445.
[44] Rice, B. M.; Hare, J. J., *J. Phys. Chem. A*, **2002**, *106(9)*, 1770.
[45] Engelke, R., *J. Phys. Chem.*, **1992**, *96*, 10789.
[46] Glukhovtsev, M. N.; Schleyer, P. v. R., *Chem. Phys. Lett.*, **1992**, *198*, 547.
[47] Eremets, M. I.; Gavriliuk, A. G.; Serebryanaya, N. R.; Trojan, I. A.; Dzivenko, D. A.; Boehler, R.; Mao, H. K.; Hemley, R. J., *J. Chem. Phys.* **2004**, *121*, 11296.
[48] Eremets, M. I.; Gavriliuk, A. G.; Trojan, I. A.; Dzivenko, D. A.; Boehler, R. , *Nature Materials* **2004**, *3*, 558.
[49] Eremets, M. I.; Popov, M. Y.; Trojan, I. A.; Denisov, V. N.; Boehler, R.; Hemley, R. J., *J. Chem. Phys.* **2004**, *120*, 10618.
[50] Klapötke, T. M.; Holl, G., *Green Chemistry*, **2001**, G75.
[51] Klapötke, T. M.; Holl, G., *Chemistry in Australia*, **2002**, 11.

[52] Klapötke, T. M.; Stierstorfer, J., *Central Europ. J. of Energ. Mat.*, **2008**, *5(1)*, 13.
[53] Klapötke, T. M.; Stierstorfer, J., *J. Chem. Soc., Dalton Trans.*, **2008**, 643–653.
[54] Göbel,M.; Klapötke,T. M., *Adv. Funct. Mat.*, **2009**, *19*, 347–365.
[55] (a) Singh, R. P.; Verma, R. D.; Meshri, D. T.; Shreeve, J. M., *Angew. Chem. Int. Ed.*, **2006**, *45*, 3584.
(b) Sikder, A. K.; Sikder, N., *J. Haz. Mater.*, **2004**, *A112*, 1.
[56] (a) Joo, Y.-H.; Twamley, B.; Grag, S.; Shreeve, J. M., *Angew. Chem.* **2008**, *120*, 6332; *Angew. Chem. Int. Ed.* **2008**, *47*, 6236;
(b) Yoo, Y.-H.; Shreeve, J. M., *Angew. Chem.* **2009**, *48(3)*, 564– 567.
(c) Klapötke, T. M.; Mayer, P.; Stierstorfer, J.; Weigand, J. J., *J. Mat. Chem.* **2008**, *18*, 5248–5258.
(d) Saly, M. J.; Heeg, M. J.; Winter, C. H., *Inorganic Chemistry* **2009**, *48(12)*, 5303–5312.
[57] (a) Chambreau, S. D.; Schneider, S.; Rosander, M.; Hawkins, T.; Gallegos, C. J.; Pastewait, M. F.; Vaghjiani, G. L., *J. Phys. Chem. A*, **2008**, *112*, 7816.
(b) Schneider, S.; Hawkins, T.; Ahmed, Y.; Rosander, M.; Hudgens, L.; Mills, J., *Angew. Chem. Int. Ed.*, **2011**, *50*, 5886–5888.
[58] (a) Amariei, D.; Courthéoux, L.; Rossignol, S.; Kappenstein, C., *Chem. Engeneering and Processing*, **2007**, *46*, 165.
(b) Singh, R. R.; Verma, R. D.; Meshri, D. T.; Shreeve, J. M., *Angew Chem.* **2006**, *118(22)*, 3664–3682.
(c) Gao, H.; Joo, Y.-H.; Twamley, B.; Zhou, Z.; Shreeve, J. M., *Angew. Chem.* **2009**, *121(15)*, 2830–2833.
(d) Zhang, Y.; Gao, H.; Joo, Y.-H.; Shreeve, J. M., *Angew. Chem. Int. Ed.*, **2011**, *50*, 9554–9562.
[59] (a) Altenburg, T.; Penger, A.; Klapötke, T. M.; Stierstorfer, J., *Z. Anorg. Allg. Chem.* **2010**, *636*, 463–471.
(b) Klapötke, T. M.; Stierstorfer, J., *Helv. Chim. Acta*, **2007**, *90*, 2132–2150.
(c) Bolton, O.; Matzger, A. J., *Angew. Chem. Int. Ed.*, **2011**, in press.
(d) Fischer, N.; Fischer, D.; Klapötke, T. M.; Piercey, D. G.; Stierstorfer, J., *How High Can You Go? Pushing the Limits of Explosive Performance*, *J. Mater. Chem*, **2012**, *22(38)*, 20418–20422.
(e) Fischer, N.; Klapötke, T. M.; Matecic Musanic, S.; Stierstorfer, J.; Sućeska, M., *TKX-50, New Trends in Research of Energetic Materials*, Part II, Czech Republic, **2013**, 574–585.
(f) Golubev, V. ; Klapötke, T. M.; Stierstorfer, J., TKX-50 and MAD-X1 – A Progress Report, *40th Intl. Pyrotechnics Seminar*, Colorado Springs, CO, July 13–18, **2014**.
[60] Millar, R. W.; Colclough, M. E.; Golding, P.; Honey, P. J.; Paul, N. C.; Sanderson, A. J.; Stewart, M. J., *Phil. Trans. R. Soc. Lond. A*, **1992**, *339*, 305.
[61] Harris, A. D.; Trebellas, J. C.; Jonassen, H. B., *Inorg. Synth.*, **1967**, *9*, 83.
[62] Borman, S., *Chem. Eng. News*, **1994**, Jan. 17, 18.
[63] Christe, K. O.; Wilson, W. W.; Petrie, M. A.; Michels, H. H.; Bottaro,J. C.; Gilardi, R., *Inorg. Chem.*, **1996**, *35*, 5068.
[64] Klapötke, T. M.; Stierstorfer, J., *Phys. Chem. Chem. Phys.*, **2008**, *10*, 4340.
[65] Klapötke, T. M.; Stierstorfer, J.; Wallek, A. U., *Chem. Mater.*, **2008**, *20*, 4519.
[66] Klapötke, T. M.; Krumm, B.; Mayr, N.; Steemann, F. X.; Steinhauser, G., *Safety Science*, **2010**, *48*, 28–34.
[67] Chapman, R. D.; Gilardi, R. D.; Welker, M. F.; Kreutzberger, C. B., *J. Org. Chem.* **1999**, *64*, 960–965.
[68] Fischer, D.; Klapötke, T. M.; Stierstorfer, J., *Z. Anorg. Allg. Chem.* **2011**, *637*, 660–665.
[69] Langlet, A.; Latypov, N.; Wellmar, U.; Goede, P., *Pyrotechics*, **2004**, *29*, No. 6.

[70] Schöyer, H. F. R.; Welland-Veltmans, W. H. M.; Louwers, J.; Korting, P. A. O. G.; van der Heijden, A. E. D. M.; Keizers, H. L. J.; van der Berg, R. P., *J. Propuls. Power*, **2002**, *18*, 131.
[71] Schöyer, H. F. R.; Welland-Veltmans, W. H. M.; Louwers, J.; Korting, P. A. O. G.; van der Heijden, A. E. D. M.; Keizers, H. L. J.; van der Berg, R. P., *J. Propuls. Power*, **2002**, *18*, 138.
[72] Shechter, H.; Cates Jr., H., *J. Org. Chem.*, **1961**, *26*, 51.
[73] Feuer, H.; Swarts, W. A., *J. Org. Chem.*, **1962**, *27*, 1455.
[74] Ovchinnikov,I. V.; Kulikov, A. S.; Epishina, M. A.; Makhova, N. N.; Tartakovsky, V. A., *Russ. Chem. Bull. Int. Ed.*, **2005**, *54*, 1346.
[75] Metelkina, E. L., Novikova, T. A., *Russ. J. Org. Chem.*, **2002**, *38*, 1378.
[76] Klapötke, T. M.; Krumm, B.; Scherr, M.; Spieß, G.; Steemann, F. X., *Z. Anorg. Allg. Chem.*, **2008**, *8*, l244.
[77] Göbel, M.; Klapötke, T. M.; Mayer, P. M., *Z. Anorg. Allg. Chem.*, **2006**, *632*, 1043.
[78] Welch, D. E., *Process for Producing Nitroform*, US 3491160, **1970**.
[79] Frankel, M. B.; Gunderloy, F. C.; Woolery, D. O., US 4122124, **1978**.
[80] Langlet, A.; Latypov, N.; Wellmar, U., *Method of preparing Nitroform*, WO 03/018514 A1, 2003.
[81] Behrend, R.; Meyer, E.; Rusche, F., *Liebigs Ann. Chem.*, **1905**, *339*, 1.
[82] M. H. Keshavarz, H. Motamedoshariati, R. Moghayadnia, H. R. Nazari, J. Azarniamehraban, A new computer code to evaluate detonation performance of high explosives and their thermochemical properties, part I., *J. Haz. Mat.* **2009**, *172*, 1218–1228.
[83] M. H. Keshavarz, M. Jafari, H. Motamedoshariati, R. Moghayadnia, Energetic Materials Designing Bench (EMDB), Version 1.0, Malek-Aschtar University, Schahinschahr, **2017**.
[84] J. L. Gottfried, T. M. Klapötke, T. G. Witkowski, Estimated Detonation Velocities for TKX-50, MAD-X1, BDNAPM, BTNPM, TKX-55 and DAAF using the Laser-induced Air Shock from Energetic Materials Technique, *Prop. Expl. Pyrotech.* **2017**, *42*, 353–359.
[85] M. H. Keshavarz, T. M. Klapötke, M. Suceska, Prop. Expl. Pyrotech., in press.

Computer Codes

Sućeska, M., *EXPLO5* program, version 5.04, Zagreb, Croatia, **2010**.
Fraunhofer Institut Chemische Technologie, ICT, ICT Thermodynamic Code, Version 1.00, **1998–2000**.
Frisch, M. J. et al., Gaussian 03, Revision B04, Gaussian Inc., Wallingford, CT, **2004**.
Fried, L. E.; Glaesemann, K. R.; Howard, W. M.; Souers, P. C., CHEETAH 4.0 User's Manual, Lawrence Livermore National Laboratory, **2004**.
Lu, J. P., Evaluation of the Thermochemical Code–CHEETAH 2.0 for Modelling Explosives Performance, DSTO-DR-1199, Defence Science and Technology Organization (Australian Government), Edinburgh, **2001**.
HyperChem, Release 7.0, Molecular Modeling System for Windows, Hypercube Inc., **2002**.
Kee, R. J.; Rupley, F. M.; Miller, J. A.; Coltrin, M. E.; Grcar, J. F.; Meeks, E.; Moffat, H. K.; Lutz, A. E.; Dixon-Lewis, G.; Smooke, M. D.; Warnatz, J.; Evans, G. H.; Larson, R. S.; Mitchell, R. E.; Petzold, L. R.; Reynolds, W. C.; Caracotsios, M.; Stewart, W. E.; Glarborg, P., CHEMKIN Collection, Release 3.5, Reaction Design, Inc., San Diego, CA, **1999**.
McBride, B.; Gordon, S.,NASA-Glenn Chemical Equilibrium Program CEA, , **1999**.
N. F. Gavrilov, G. G. Ivanova, V. N. Selin, V. N. Sofronov, UP-OK program for solving continuum mechanics problems in one-dimensional complex, VANT. Procedures and programs for numerical solution of mathematical physics problems, (3), **1982**, 11–21.
ANSYS Autodyn® Academic Research, Release 15.0, Help System, Equations of State, ANSYS, Inc., **2015**.

WEB Pages

ARL	http://www.arl.army.mil/www/default.htm
SERDP	http://www.serdp.org/
ESTCP	http://www.estcp.org/
WIWEB	http://wiweb-erding.de/
ARDEC	http://www.pica.army.mil/PicatinnyPublic/organizations/ardec/index.asp
NAWCWD	http://www.navair.navy.mil/nawcwd/
NSWC	http://www.ih.navy.mil/
ICT	http://www.ict.fraunhofer.de/
DTRA	http://www.dtra.mil/index.cfm
DTIC	http://www.dtic.mil/dtic/
EPA	http://www.epa.gov/
HEMSI	http://www.hemsichd.org
RTO	http://www.rta.nato.int
NIOSH	http://www.cdc.gov/NIOSH/
NIST	http://webbook.nist.gov/chemistry/
IMEMG	http://www.imemg.org/imemg-home.html
MSIAC	http://www.msiac.nato.int/
FHG	http://www.ict.fraunhofer.de/
SYSTAG	http://www.systag.ch/A4d_TAV_1m24.pdf
OZM	http://www.ozm.cz/en/

Journals

Propellants, Explosives, Pyrotechnics: http://www3.interscience.wiley.com/journal/26213/home
Central European Journal of Energetic Materials: http://www.ipo.waw.pl/wydawnictwa/cejem-glowna.html
Journal of Energetic Materials: http://www.marinkas.com/
Journal of Pyrotechnics: http://www.jpyro.com/
The American Institute of Aeronautics and Astronautics: http://www.aiaa.org/content.cfm?pageid=318&pubid=2
Journal of Hazardous Materials: http://www.sciencedirect.com/science/journal/03043894
Journal of Energetic Materials: http://www.tandf.co.uk/journals/titles/07370652.html
International Journal of Energetic Materials and Chemical Propulsion (IJEMCP): http://www.begellhouse.com/journals/17bbb47e377ce023.html
Engineering Science and Military Technologies http://ejmtc.js.iknito.com/

Conferences

Gordon Research Conference: Energetic Materials, every other year (2008, 2010, ...), usually June, in New Hampshire (USA): http://www.grc.org/conferences.aspx?id=0000071
ICT International Annual Conference, annualy, June/July, in Karlsruhe (Germany): http://www.ict.fraunhofer.de/ http://www.ict.fraunhofer.de/EN/VuM/ICT_Jahrestagung/index.jsp
New Trends in Research of Energetic Materials (NTREM), annualy, April, in Pardubice (Czech Republic): http://www.ntrem.com/
PARARI – Australian Ordnance Symposium, every other year (2007, 2009, ...), usually November, (Australia): http://www.adi-limited.com/
International Pyrotechnics Seminar, every other year (2008, 2010, ...), July, CO (USA): http://www.ipsusa.org/
High Energy Materials Society of India (HEMSI), biannually (2011, 2013, ...), usually November (India): http://www.hemsichd.org/index.html

Insensitive Munitions (IM) and Energetic Materials Technology Symposium (usually every 2 years): http://www.imemts2015.com/

Biennial APS Conference on Shock Compression of Condensed Matter – SCCM (every other year): http://msl.cas.usf.edu/sccm-2015/

International Conference IPOEX: Explosives Research – Application – Safety (yearly), Institute of Industrial Organic Chemistry, Ustroń Zawodzie, Poland

16 Appendix

*I am one of those who think like Nobel,
that humanity will draw more good
than evil from new discoveries.*
Marie Curie

Important reaction types in organic nitrogen chemistry

$$R-NH_2 \xrightarrow{HNO_2,\ aq.} [R-N_2]^+ \xrightarrow{H_2O} R-OH$$

$$R-NH_2 \xrightarrow{HNO_2,\ aq.} [R-N_2]^+ \xrightarrow{NaN_3} R-N_3$$

$$R_2NH \xrightarrow{HNO_2,\ aq.} R_2N-N=O$$

$$Ph-NH_2 \xrightarrow{HNO_2,\ aq.} [Ph-N_2]^+ \xrightarrow{PhH,\ -H^+} Ph-N=N-Ph$$

$$Ph-NH_2 \xrightarrow{HNO_2,\ aq.} [Ph-N_2]^+ \xrightarrow{NaN_3} Ph-N_3$$

$$Ph-NH_2 \xrightarrow{HNO_2,\ aq.} [Ph-N_2]^+ \xrightarrow{NaBF_4} [Ph-N_2]^+[BF_4]^- \xrightarrow{NaN_3} Ph-N_5 \xrightarrow{-N_2} PhN_3$$

$$Ph-NH-NH-Ph \xrightarrow{HgO,\ -H_2O,\ -Hg} Ph-N=N-Ph$$

$$Ph-NH_2 \xrightarrow[\text{diazotation}]{HNO_2,\ aq.} [Ph-N_2]^+ \xrightarrow[\text{Sandmeyer}]{Cu^IX} Ph-X$$

$$NaN_3 + R-Br \longrightarrow R-N_3 \xrightarrow{H_2/Pd-C} R-NH_2$$

$$Ph-CH_2-Br + NaN_3 \longrightarrow Ph-CH_2-N_3$$

$$Ph-CH_2-N_3 \xrightarrow{H_2/Pd-C} Ph-CH_2-NH_2 + N_2$$

$$Ph-C(O)-Cl + NaN_3 \longrightarrow Ph-C(O)-N_3$$

$$Ph-CH_2-CN \xrightarrow{H_2/cat.\ or\ LiAlH_4} Ph-CH_2-CH_2-NH_2$$

$$R-CN \xrightarrow{H_2/cat.\ or\ LiAlH_4} R-CH_2-NH_2$$

$$R_1R_2HC-NO_2 \xrightarrow{H_2/cat.\ or\ LiAlH_4} R_1R_2HC-NH_2$$

$$R_1R_2HC-CN \xrightarrow{H_2/cat.\ or\ LiAlH_4} R_1R_2HC-CH_2-NH_2$$

$$R-NO_2 \xrightarrow{H_2/cat.\ or\ LiAlH_4} R-NH_2$$

$$R-C(O)-NH_2 \xrightarrow{PCl_5} R-CN$$

$$R-C(O)-NH_2 \xrightarrow{LiAlH_4} R-CH_2-NH_2$$

$$HMTA \xrightarrow{HNO_3} RDX \quad (+\ HMX)$$

$$C(-CH_2-OH)_4 \xrightarrow{HNO_3} PETN$$

$$R-C(O)-R \xrightarrow{N_2H_4} R_2C=N-NH_2$$

$$R-X \xrightarrow{N_2H_4} R-NH-NH_2 \quad (X = Hal, OMe)$$

$$RC(O)R + H_2NR' \longrightarrow RC(=NR')R$$

$$R_2NH + CH_2O \xrightarrow{H^+} [R_2N=CH_2]^+ \longleftrightarrow [R_2N-CH_2]^+$$
$$\text{Mannich} \quad \text{iminium ion (reacts with C nucleophiles)}$$

$$R_2N-C(O)H + POCl_3 \longrightarrow [R_2N=CHCl]^+ \longleftrightarrow [R_2N-CHCl]^+$$
$$\text{Vilsmeyer iminium ion}$$

$$R-COOH \xrightarrow{SOCl_2} R-C(O)-Cl$$

$$R-C(O)-NH-R' \xrightarrow{N_2H_4} R-C(O)-NH-NH_2 + R'-NH_2$$
$$\text{hydrazinolysis}$$

$$R-C(O)-OR \xrightarrow{N_2H_4} R-C(O)-NH-NH_2 \xrightarrow{HNO_2} R-C(O)-N_3$$

$$R-X + KCN \longrightarrow R-CN \quad (X = Cl, Br, I, OTs, ...)$$

$$Ph-C(O)-NH_2 \xrightarrow{PCl_5} Ph-CN$$

$$Ph-Cl \xrightarrow{N_2H_4} Ph-NH-NH_2$$

$$R-C(O)-Cl \xrightarrow{N_2H_4} R-C(O)-NH-NH_2$$

$$R-C(O)-NH-NH_2 + NO^+ \xrightarrow{-H^+} R-C(O)-NH-NH-N=O$$
$$\longrightarrow R-C(O)-NH-N=N-OH$$
$$\longrightarrow R-C(O)-N_3 + H_2O$$

$$R-C(O)-N_3 \xrightarrow{T,\ -N_2} R-N=C=O$$

$$R-C(O)-R \xrightarrow{N_2H_4} R_2C=N-NH_2$$

$$R_3N \xrightarrow{PhCO_3H \text{ or } H_2O_2/AcOH} R_3N^+-O^- \quad \text{(also with pyridines instead of } R_3N\text{)}$$

$$R_2N-H \xrightarrow{MSH \text{ or } DPA,\ Base} R_2N-NH_2$$

MSH = mesitylsulfonylhydroxylamine

DPA = diphenylphosphenylhydroxylamine

TOHA =

HOSA = $H_3N^+-O-SO_3^-$

N amination: TOHA = O-tosylhydroxylamine

C amination: TMJI = trimethylhydrazinium iodide

Curing of a diol (HTPB) with isocyanate binder

$$O=C=N-R^1-N=C=O \; + \; H-O-R^2-OH \longrightarrow O=C=N-R^1-\underset{H}{N}-\overset{O}{\underset{\|}{C}}-O-R^2-OH$$

$$\longrightarrow \longrightarrow -\overset{O}{\underset{\|}{C}}-\underset{H}{N}-R^1-\underset{H}{N}-\overset{O}{\underset{\|}{C}}-O-R^2-O-\overset{O}{\underset{\|}{C}}-\underset{H}{N}-R^1-\underset{H}{N}-\overset{O}{\underset{\|}{C}}-O-R^2-O-$$

Important reaction types in inorganic nitrogen chemistry

azide transfer reagents: NaN_3
                                         AgN_3
                                         Me_3SiN_3

HN_3 synthesis:

$$RCO_2H + NaN_3 \xrightarrow[\text{melt}]{\sim 140\,°C} RCO_2Na + HN_3\uparrow$$

stearic acid

$$HBF_4 + NaN_3 \xrightarrow{Et_2O} NaBF_4\downarrow + HN_3\uparrow$$

$$H_2SO_4 + NaN_3 \longrightarrow HN_3 + NaHSO_4$$

NaN_3 synthesis:

$$NaNO_3 + 3\,NaNH_2 \xrightarrow{100\,°C,\;NH_3(l)} NaN_3 + 3\,NaOH + NH_3$$

$$N_2O + 2\,NaNH_2 \xrightarrow{190\,°C} NaN_3 + NaOH + NH_3$$

$$ {\scriptstyle HC(OR)_3,\; NaN_3\; H^+}$$

$$R'-NH_2 \longrightarrow R'-N_4CH$$

hydrazine synthesis:

Raschig process (SNPE):

$$NaOCl_{aq.} + NH_{3\,aq.} \xrightarrow{0\,°C} NH_2Cl_{aq.} + NaOH$$

$$NH_2Cl_{aq.} + NaOH + NH_3 \xrightarrow{130\,°C} N_2H_4 + H_2O + NaCl$$

$$NH_2Cl + H_2N(CH_3) \xrightarrow[-NaCl\,/\,-H_2O]{NaOH} H_2N-NH(CH_3)$$
$$\text{MMH}$$

laboratory-scale preparation of anhydrous hydrazine:

$$N_2H_4 \cdot H_2O \xrightarrow{BaO} N_2H_4 \xrightarrow{Na} N_2H_4 \text{ anh.}$$

$$(HOCN)_3 + N_2H_4/H_2O \longrightarrow (HOCN)_3 \cdot N_2H_4 \xrightarrow{200\,°C} (HOCN)_3 + N_2H_4$$

ammonium perchlorate (AP):

$$NH_3 + HClO_4 \longrightarrow AP$$

$$R-NH-NH_2 + HNO_2\,(NaNO_2/HCl) \longrightarrow R-N_3$$

$$R-C(O)-Cl + N_2H_4 \longrightarrow R-C(O)-NH-NH_2 + HCl$$

or:

$$R-C(O)-OMe + N_2H_4 \longrightarrow R-C(O)-NH-NH_2 + MeOH$$

$$R-C(O)-NH-NH_2 + HNO_2 \longrightarrow R-C(O)-N_3 \longrightarrow R-NCO + N_2$$

$$R-NH_2 + H_2O_2/H_2SO_4 \longrightarrow R-NO_2$$

ammonium nitrate (AN):

$$NH_3 + HNO_3 \longrightarrow AN$$

N-oxide formation:

$$\underset{R^2}{\overset{R^1}{N}} \xrightarrow{\text{Oxidation Reagent (OR)}} \underset{R^2}{\overset{R^1}{\overset{\oplus}{N}}}-O^{\ominus}$$

OR e.g. Oxone
MCP BA
H_2O_2/H_2SO_4
$H_2O_2/F_3C-COOH$
Ozone
HO-F*CH_3CN ($F_2+H_2O+CH_3CN$)

For organic azides the following rule of thumb applies: organic, covalently bound azide compounds can be handled and are not expected to be explosive if the number (#) of N atoms is smaller than the number of C atoms and if the ratio of (C atoms + O atoms) divided by the number of N atoms is greater than three:

$$\#(N) < \#(C)$$

$$[\#(C) + \#(O)] / \#(N) > 3$$

(a) P. A. S. Smith, Open-Chain Nitrogen Compounds, Bd. 2, Benjamin, New York, **1966**, 211–256;
(b) J. H. Boyer, R. Moriarty, B. de Darwent, P. A. S. Smith, *Chem. Eng. News* **1964**, *42*, 6.

Laboratory-scale synthesis of ADN starting from potassium sulfamate and cyanoguanidine:

$$NH_2\text{-}SO_3\text{-}K \xrightarrow{1.\,HNO_3/H_2SO_4} H^+\left[N(NO_2)_2\right]^-$$
$$\text{HDN}\quad\bigg|\,H_2N\text{-}C(=NH)\text{-}NH\text{-}CN$$

$$\left[H_2N\text{-}C(=NH_2)\text{-}NH\text{-}CO\text{-}NH_2\right]^+\left[N(NO_2)_2\right]^- \xrightarrow[\Delta\,50°C]{KOH/H_2O/EtOH}$$
$$\text{GUDN}$$

$$NH_4^+\left[N(NO_2)_2\right]^- \xleftarrow{\text{Ion exchange}} K^+\left[N(NO_2)_2\right]^- + H_2N\text{-}C(=NH)\text{-}NH\text{-}CO\text{-}NH_2$$
$$\text{ADN}\qquad\qquad\qquad\qquad\qquad\text{KDN}$$

(according to: M. Mahdavi, K. Ghani, K. Nosratzadegan, K. Barati, *J. Energ. Mat.*, in press)

Tab. I: Technical terms in German and English language.

German	English
Abbrandgeschwindigkeit	burn rate
Abgaswolke (Flugzeug, Rakete)	plume
Aufschlagzünder oder Perkussionszünder	percussion primer
Binder	binder
Brandwaffe	incendiary device
Brennkammer	combustion chamber
Brennstoff, Treibstoff	fuel
Detonationsgeschwindigkeit	velocity of detonation, detonation velocity
Dreibasig	triple-base
Einbasig	single-base
Elektrostatische Empfindlichkeit	electrostatic sensitivity
Elektrostatisches Potential	electrostatic potential
Energetischer Stoff	energetic material
Expansionsdüse	(expansion) nozzle
Fallhammer	drophammer
Flachladung	flat cone charge
Gasgenerator	gas generator
Hohlladung	shaped charge
Ionische Flüssigkeit	ionic liquid
Köder-Munition oder Täuschkörper	decoy flares / countermeasure munition
Leuchtsatz	illuminating device
Mündungsfeuer	muzzle flash
Oxidator	oxidizer
Plastifizierungsmittel	plasticizer

Tab. I (continued)

German	English
Primärer Explosivstoff	primary explosive
Projektilbildende Ladung	explosively formed projectile
Pyrotechnische Mischung	pyrotechnic composition
Raketentreibstoff	rocket propellant
Raucherzeuger	smoke generator
Reibeapparat	friction tester
Reibempfindlichkeit	friction sensitivity
Schlagempfindlichkeit	impact sensitivity
Schmelzgießen	melt cast
Schwarzpulver	blackpowder
Sekundärer Explosivstoff	secondary explosive, high explosive
Signalfackel	signal flare
Spezifische Energie	force
Sprenggelatine	blasting gelatine
Sprengkapsel (Schockwelle)	blasting cap / detonator / primer
Stahlhülsentest	steel shell test
Täuschkörper	decoy flare
Treibladungspulver	gun propellant
Universalbombe	general purpose bomb
Verdämmung	confinement
Verzögerungssatz	delay composition
Zünder (durch Hitze)	initiator
Zweibasig	double-base

Tab. II: Abbreviations.

AA	ammonium azide
ADN	ammonium dinitramide
ADW	agent defeat warhead
AF	azidoformamidinium
AF-N	azidoformamidinium nitrate
AG	aminoguanidinium
AG-N	aminoguanidinium nitrate
ALEX	electrically exploded aluminum, nano material (20–60 nm)
AN	ammonium nitrate
ANFO	ammonium nitrate fuel oil
AT, 5-AT	aminotetrazole, 5-aminotetrazole
BI	bullet impact
BP	black powder (gun powder)
BTA	bis(tetrazolyl)amine
BTAT	bis(trinitroethyl)-1,2,4,5-tetrazine-3,6-diamine
BTH	bistetrazolylhydrazine
BTTD	bis(trinitroethyl)-tetrazole-1,5-diamine

Tab. II (continued)

C-J point	Chapman-Jouguet point
CL-20	HNIW, hexanitrohexaazaisowurtzitane
CTP	chemical thermal propulsion
DADNE	1,1-diamino-2,2-dinitro-ethene
DADP	diacetone diperoxide
DAG	diaminoguanidinium
DAG-Cl	diaminoguanidinium chloride
DAT	diaminotetrazole
DDNP	diazodinitrophenole
DDT	deflagration to detonation transition
DINGU	dinitroglycoluril
DNAN	2,4-dinitroanisole
EDD	explosive detection dog
EFP	explosively formed projectile
EPA	environmental protection agency
ESD	electrostatic discharge sensitivity
ESP	electrostatic potential
FAE	fuel-air explosives
FCO	fast cook-off
FI	fragment impact
FLOX	fluorine – liquid oxygen
FOX-12	guanylurea dinitramide
FOX-7	DADNE, diaminodinitroethene
FOX-7	1,1-diamino-2,2-dinitro-ethene
GUDN	guanylurea dinitramide
GWT	global war on terror
H_2BTA	bis(tetrazolyl)amine
HA	hydrazinium azide
HAA	hydroxylammonium azide
HAN	hydroxylammonium nitrate
HAT-DN	aminotetrazolium dinitramide
HME	homemade explosive
HMTA	hexamethylenetetramine
HMTD	hexamethylene triperoxide diamine
HMX	octogen, high melting explosive, her/his Majesty's explosive
HN	hydrazinium nitrate
HNFX-1	tetrakis(difluoramino)octahydro dinitro diazocine
HNS	hexanitrostilbene
HTPB	hydroxy-terminated polybutadiene
Hy-At	hydrazinium aminotetrazolate
HZT	hydrazinium azotetrazolate
IED	improvised explosive device
IM	insensitive munition
IRFNA	inhibited red fuming nitric acid
LOVA	low-vulnerability ammunition
LOX	liquid oxygen

Tab. II (continued)

MEKP	methylethylketone-peroxide
MF	mercury fulminate
MMH	monomethylhydrazine
MOAB	massive ordnance air blast bomb
MON-XX	mixture of dinitrogen tetroxide (NTO) with XX % NO
MTV	magnesium-Teflon-Viton
Na$_2$ZT	sodium azotetrazolate
NC	nitrocellulose
NG	nitroglycerine
NIOSH	National Institute for Occupational Safety and Health
NQ	nitroguanidine
NT	nitrotetrazole
NTO	5-nitro-1,2,4-triazol-3-one
NTP	nuclear thermal propulsion
ONC	octanitrocubane
PA	picric acid
PBX	polymer bonded explosive
PETN	nitropenta, pentaerythritoltetranitrate
RDX	hexogen, research department explosive, Royal demolition explosive
RFNA	red fuming nitric acid
SC	shaped charge
SCI	shaped charge impact
SCJ	shaped charge jet
SCO	slow cook-off
SCRAM-Jet	supersonic combustion RAM jet
SR	sympathetic reaction
STP	solar thermal propulsion
TAG	triaminoguanidine
TAG-Cl	triaminoguanidinium chloride
TAG-DN	triamonoguanidinium dinitramide
TAGzT	triaminoguanidinium azotetrazolate
TART	triacetyltriazine
TATB	triaminotrinitrobenzene
TATP	triacetone triperoxide
TMD	theoretical maximum density (in the crystal)
TNAZ	1,3,3-trinitroazetidine
TNT	trinitrotoluene
TO	triazol-5-one
TTD	trinitroethyl-tetrazol-1,5-diamine
TTD	trinitroethyl-1H-tetrazole-1,5-diamine
UDMD	unsymmetrically substituted dimethylhydrazine
UXO	unexploded ordnance
VOD	velocity of detonation
WFNA	white fuming nitric acid
WMD	weapon of mass distruction
ZT	azotetrazole

Tab. III: CBS-4M calculated gas-phase enthalpies of formation.

M	M	$\Delta_f H°(g, M)$ / kcal mol^{-1}	$\Delta_f H°(g, M)$ / kcal mol^{-1}, lit. value
1-H-nitrotetrazole, 1-H-NT	CHN_5O_2	+ 87.1	
1-Me-aminotetrazolium, 1-MeHAT$^+$	$C_2H_6N_5^+$	+224.0	
1-Me-nitriminotetrazole, 1-MeHAtNO$_2$	$C_2H_4N_6O_2$	+ 86.5	
1-Me-nitrotetrazole, 1-Me-NT	$C_2H_3N_5O_2$	+ 80.7	
2-H-nitrotetrazole, 2-H-NT	CHN_5O_2	+ 84.0	
2-Me-nitrotetrazole, 2-Me-NT	$C_2H_3N_5O_2$	+ 74.5	
aminoguanidinium, AG$^+$	$CH_7N_4^+$	+161.0	
aminotetrazolium, HAT$^+$	$CH_4N_5^+$	+235.0	
ammonium, NH_4^+	NH_4^+	+151.9	+147.9[a]
azidoformidinium, AF$^+$	$(H_2N)_2CN_3^+$, $CH_4N_5^+$	+235.4	
azidotetrazolate, CN_7^-	CN_7^-	+114.1	
bistetrazolylamine, BTA	$C_2H_3N_9$	+178.2	
diaminotetrazolium, HDAT$^+$	$CH_5N_6^+$	+252.2	
dinitramide, DN$^-$	$N(NO_2)_2^-$, $N_3O_4^-$	− 29.6	
guanizinium, Gz$^+$	$C_2H_7N_6$	+207.7	
hydrazinium, $N_2H_5^+$	$N_2H_5^+$	+185.1	+184.6[b]
NG-A	$C_3H_7N_5O_7$	− 13.1	
nitrate, NO_3^-	NO_3^-	− 74.9	− 71.7[c,d]
nitriminotetrazole, H$_2$AtNO$_2$	$CH_2N_6O_2$	+ 95.0	
nitroglycerine, NG	$C_3H_5N_3O_9$	− 67.2	
perchlorate, ClO_4^-	ClO_4^-	− 66.1	
triaminoguanidinium, TAG$^+$	$C(NH-NH_2)_3^+$, $CH_9N_6^+$	+208.8	

[a] D. A. Johnson, *Some thermodynamic aspects of inorganic chemistry*, Cambridge Univ. Press, 2nd edn., Cambridge, **1982**, appendix 5.

[b] P. J. Linstrom and W. G. Mallard, Eds., *NIST Chemistry WebBook*, NIST Standard Reference Database Number 69, June **2005**, National Institute of Standards and Technology, Gaithersburg MD, 20899 (http://webbook.nist.gov).

[c] J. A. Davidson, F. C. Fehsenfeld, C. J. Howard, *Int. J. Chem. Kinet.*, **1977**, *9*, 17.

[d] D. A. Dixon, D. Feller, C.-G. Zhan, J. S. Francisco, *International Journal of Mass Spectrometry*, **2003**, *227(3)*, 421.

Tab. IV: Molecular volumes[e,f,g].

	V_M / Å^3	V_M / nm^3
aminotetrazolium cation, [HAT]$^+$	69	0.069
ammonium, NH_4^+	21	0.021
azide, N_3^-	58	0.058
Ba^{2+}	12.3	0.0123
chloride, Cl$^-$	47	0.047
diaminotetrazolium cation, [HDAT]$^+$	93	0.093

Tab. IV (continued)

	V_M / Å³	V_M / nm³
dinitramide, DN⁻	89	0.089
guanizinium, [Gz]⁺	117.6	0.118
H₂O, hydrate water	25	0.025
H₂O, structural water	14	0.014
hydrazinium, N₂H₅⁺	28	0.028
Mg²⁺	2.0	0.00199
nitrate, NO₃⁻	64	0.064
nitriminotetrazolate anion, [HAtNO₂]⁻	136	0.136
nitriminotetrazolate dianion, [AtNO₂]²⁻	136	0.136
perchlorate, [ClO₄]⁻	89.0	0.089
Sr²⁺	8.6	0.00858
triaminoguanidinium, [TAG]⁺	108	0.108
[1-MeHAT][NO₃]	208	0.208
[2-MeHat][DN]	206.4	0.206
[2-MeHat]⁺	117.4	0.117
[AF][DN]	174	0.174
[AF][DN]·H₂O	199	0.199
[AG][CN₇]	201.8	0.202
[CH₃N₄][DN]	161.3	0.161
[Gz][DN]	206.6	0.206
[HAT][DN]	172	0.172
[HAT][NO₃]	133	0.133
[HDAT][DN]	182	0.182
[N₂H₅][CN₇]	151.6	0.152
[N₂H₅][CH₂N₅]	125.7	0.126
[N₂H₅]₂[OD]	224.4	0.224
[NH₄][1,5-BT]	164	0.164
[NH₄][CN₇]	132.3	0.132
[NH₄][DN], ADN	110	0.110
[NH₄][NO₃], AN	77	0.077
[NH₄]₂[OD]	195.0	0.195
[Sr][AtNO₂]·2 H₂O	172.5	0.173
[TAG][1-Me-AtNO₂]	262	0.262
[TAG][Cl]	153.5	0.154
[TAG][DN]	215	0.215
[TAG][HAtNO₂]	244	0.244
[TAG][NO₃]	174.1	0.174
[TAG]₂[AtNO₂]	352	0.352

[e] H. D. B. Jenkins, H. K. Roobottom, J. Passmore, L. Glasser, *Inorg. Chem.*, **1999**, *38(16)*, 3609.
[f] H. D. B. Jenkins, D. Tudela, L. Glasser, *Inorg. Chem.*, **2002**, *41(9)*, 2364.
[g] H. D. B. Jenkins, L. Glasser, *Inorg. Chem.*, **2002**, *41(17)*, 4378.

Tab. V: Classification of some weapon systems and war-heads.

weapon	description	example
bomb	war-head respectively HE in a shell, solely ballistic	MK 80 series
rocket	bomb + propulsion system, solid rocket motor	Hydra 70 (70 mm)
missile	guided rocket, guidance: sensor, IR seeker, radar, GPS	AIM (air intercept missile), e.g. AIM-9-Sidewinder
strategic missile	missile for larger targets, target: military base or city; large often intercontinental; war-head often nuclear; propulsion: AP / Al = high I_{sp}	Peacekeeper
tactical missile	missile for tactical use, target: localized in battlefield; often conventional war-head with HE; only occasionally nuclear, relatively small system; 3 types: 1. high I_{sp}: AP / Al 2. smoke-reduced: AP only 3. minimum signature (smoke): NG / NC or NG / NC / NQ	Cruise Missile
ballistic missile	no propulsion system for most of the flight (except launch), free fall (ballistic) to target, leaves atmosphere during flight, liquid or solid boosters	SCUD (Russian)
cruise missile	remains in atmosphere, own propulsion system during entire flight, smaller than ballistic missile, for tactical purpose	Tomahawk

Tab. VI: Classification of low- and high-explosives.

	low-explosive	high-explosive
example	black powder, gun powder	RDX, HMX
initiation	flame, spark	detonation, shock-wave
confinement	necessary	not necessary
propagation	particle-to-particle burning (thermal)	shock-wave
velocity	subsonic	supersonic
effect	bursting, deformation, cracking	shattering
chemical analysis after reaction	uncombusted particles	traces (25–100 ppb), MS, LC/MSMS (electrospray)

Collection of formulas

Berthelot-Rot value:
$$B_R \text{ [kJ m}^{-3}\text{]} = \rho_0^2 \, V_0 \, Q_v$$

Burn rate:
$$r = \beta(T) \, p^\alpha \qquad \alpha < 1 \text{ deflagrating explosive}$$
$$\alpha > 1 \text{ detonating explosive}$$

Brisance:
$$B = \rho \times F \times D \qquad D = \text{VoD}$$

Specific energy:
$$F = p_{ex} \qquad V = n \, RT$$

Distance rule:
$$D = c w^{0.33} \approx 2 \, w^{0.33} \qquad w: \text{ mass of high expl.}$$
$$D: \text{ distance}$$

Projectile velocity:
$$v = \sqrt{\frac{2 \, m \, Q \, \eta}{M}} \qquad \begin{array}{l} m: \text{ mass of propellant[g]} \\ Q: \text{ heat of combustion [J g}^{-1}\text{]} \\ M: \text{ mass of projectile [g]} \\ \eta: \text{ constant} \end{array}$$

Specific impulse:
$$I_{sp} = \frac{F \cdot t_b}{m} = \frac{1}{m} \cdot \int_0^{t_b} F(t) dt$$

$$I_{sp}^\star = \frac{I_{sp}}{g}$$

$$I_{sp} = \sqrt{\frac{2 \, y \, R \, T_c}{(y-1)M}}$$

$$y = \frac{C_p}{C_V}$$

$$I_{sp}^* = \frac{1}{g}\sqrt{\frac{2yRT_c}{(y-1)M}}$$

$$F = I_{sp}\frac{\Delta m}{\Delta t}$$

$$I_{sp} \propto \sqrt{\frac{T_c}{M}}$$

$$F_{impulse} = \frac{dm}{dt}v_e$$

$$F = F_{impulse} + F_{pressure} = \frac{dm}{dt}v_e + (p_e - p_a)A_e$$

Planck's rule:

$$W_\lambda = \frac{2\pi h c^2}{\lambda^5}\frac{1}{e^{\frac{hc}{\lambda kT}} - 1}$$

Rule of Wien:

$$\lambda_{max} = 2897.756 \text{ µm K } T^{-1}$$

VoD for infinitely large diameter:

$$D = D_\infty \left(1 - A_L\frac{1}{d}\right)$$

Density at room temp.:

$$d_{298\,K} = \frac{d_T}{1 + \alpha_V(298 - T_0)} \quad \text{with} \quad \alpha_V = 1.5 \cdot 10^{-4} \text{ K}^{-1}$$

Kamlet-Jacobs equations:

$$p_{C-J}[\text{kbar}] = K\rho_0^2\Phi$$

$$D[\text{mm µs}^{-1}] = A\Phi^{0.5}(1 + B\rho_0)$$

$$\Phi = N(M)^{0.5}(Q)^{0.5}$$
$$K = 15.88$$
$$A = 1.01$$
$$B = 1.30$$

N: number of moles gas per gram explosive
M: mass of gas in gram per mol of gas
Q: heat of explosion in cal per gram

Oxygen balance for $C_aH_bN_cO_d$:

$$\Omega_{CO_2} = \frac{\left[d-(2a)-\left(\frac{b}{2}\right)\right] \times 1600}{M}$$

$$\Omega_{CO} = \frac{\left[d-a-\left(\frac{b}{2}\right)\right] \times 1600}{M}$$

Power index:

$$PI = \frac{Q \times V_0}{Q_{PA} \times V_{PA}} \times 100 \qquad \text{PA: picric acid}$$

Explosion temperature:

$$T_{ex} = \frac{Q_{ex}}{\Sigma C_v} + T_i$$

Atomization method:

$$\Delta_f H^\circ (g, m) = H^\circ_{(molecule)} - \Sigma H^\circ_{(atoms)} + \Sigma \Delta_f H^\circ_{(atoms)}$$

$H^\circ_{(molecule)}$: from CBS-4M

$\Sigma H^\circ_{(atoms)}$: from CBS-4M

$\Sigma \Delta_f H^\circ_{(atoms)}$: from NIST

enthalpy of sublimation:

$$\Delta H_{sub.} \, [\text{J mol}^{-1}] = 188 \, T_m \, [\text{K}]$$

enthalpy of vaporization:

$$\Delta H_{sub.} \, [\text{J mol}^{-1}] = 90 \, T_b \, [\text{K}]$$

lattice energies & enthalpies:

$$(1 \, \text{Å}^3 = 10^{-3} \, \text{nm}^3)$$

$$U_L = |z_+| \, |z_-| \, \nu \, (\alpha/(V_M)^{0.33} + \beta)$$

(V_M in nm³)

	α [kJ/mol]	β [kJ/mol]
MX	117.3	51.9
MX$_2$	133.5	60.9
M$_2$X	165.3	−29.8

$$\Delta H_L(M_pX_q) = U_L + [p\,(n_M/2 - 2) + q\,(n_X/2 - 2)]\,RT$$

n_M, n_X: 3 monoatomic ions

5 linear polyatomic ions

6 non-linear polyatomic

Nobel-Abel equation:

$$p\,(v - b_E) = n\,RT$$

b_E: covolume

Spark energy:

$$E = \frac{1}{2}CU^2$$

Gurney velocity:

$$\sqrt{2E} = \frac{3\sqrt{3}}{16}D \approx \frac{D}{3.08} \quad [\text{in km s}^{-1}]$$

Cylindrical charge: $\quad \dfrac{V}{\sqrt{2E}} = \left(\dfrac{M}{C} + \dfrac{1}{2}\right)^{-0.5}$

Spherical charge: $\quad \dfrac{V}{\sqrt{2E}} = \left(2\dfrac{M}{C} + \dfrac{3}{5}\right)^{-0.5}$

Sandwich: $\quad \dfrac{V}{\sqrt{2E}} = \left(2\dfrac{M}{C} + \dfrac{1}{3}\right)^{-0.5}$

M: mass of metal confinement

C: mass of explosive

Enthalpy of formation vs. energy of formation:

$$H_m = U_m + \Delta n\, RT$$

e.g. $\quad 7\,C + 5/2\,H_2 + 3/2\,N_2 + 3\,O_2 \rightarrow C_7H_5N_3O_6\,(s)$

$$\Delta n = -7$$

$$U_m = H_m - \Delta n\, RT$$

conversion of electronic energies (Gaussian) into enthalpies:

$$p\Delta V = \sum v_i\, RT \qquad \Delta U^{tr} = \sum v_i\, 3/2\, RT$$

$$\Delta U^{rot} = \sum v_i\, (F_i^{rot}/2)\, RT \qquad \Delta U^{vib} = \sum v_i\, zpe$$

Author

Thomas M. Klapötke, born in 1961, studied chemistry at TU Berlin and received his PhD in 1986 under the supervision of Hartmut Köpf. After his postdoctoral studies in Fredericton, New Brunswick with Jack Passmore he finished his habilitation at TU Berlin in 1990. From 1995 until 1997 Klapötke was Ramsay Professor of Chemistry at the University of Glasgow in Scotland. Since 1997 he is Professor and holds the Chair of Inorganic Chemistry at LMU Munich. In 2009 Klapötke was appointed a permanent visiting Professor for Mechanical Engineering and Chemistry at the Center of Energetic Concepts Development (CECD) in the University of Maryland (UM), College Park. In 2014 Klapötke was appointed an Adjunct Professor at the University of Rhode Island. Klapötke is a Fellow of the Royal Society of Chemistry (C.Sci., C.Chem. F.R.S.C., U.K.), a member of the American Chemical Society (ACS) and the Fluorine Division of the ACS, a member of the Gesellschaft Deutscher Chemiker (GDCh), a Life Member of the International Pyrotechnics Society (IPS), a Life Member of the National Defense Industrial Association (NDIA, USA) and was appointed a Honorary Fellow of the Indian High Energy Materials Society (HEMSI) in 2011. Most of Klapötke's scientific collaborations are between LMU and the US Army Research Laboratory (ARL) in Aberdeen, MD and the US Army Armament Research, Development and Engineering Center (ARDEC) in Picatinny, NJ. Klapötke also collaborates with the US Army Engineer Research and Development Center (ERDC) in Champaign, IL and several industrial partners in Germany and the USA. From 2011–2012 Klapötke was a Technical Team Member of NATO's Munitions Related Contamination team (AVT-197). He is executive editor of *Zeitschrift für Anorganische and Allgemeine Chemie* (ZAAC, Wiley), the subject Editor in the area of explosives synthesis of the *Central European Journal of Energetic Materials* (CEJEM) and an editorial board member of *Propellants, Explosives and Pyrotechnics* (PEP), *Journal of Energetic Materials, Chinese Journal of Explosives and Propellants* and the *International Journal of Energetic Materials and Chemical Propulsion* (IJEMCP). Klapötke has published over 800 scientific papers in international peer reviewed journals, 32 book chapters, 14 books and holds 13 patents.

Index

acetylene 277
additive 113
adiabatic compression 173–174
adiabatic condition 141
adiabatic flame temperature 51
ADN 35, 144, 153, 335–336
ADNQ 301
ADW 360
Aerosil 200 88
aerosol 111–112, 357
aerosol bomb 357
Aestus engine 88
agent defeat warhead 360
agent defeat weapon 360
AIM-9 105
AIM-9-Sidewinder 410
air bag 15, 35, 367
Air bag 374
air intercept missile 410
air-breathing engine 79, 82
ALEX 82
aluminium hydride 82
aluminium octanoate 88
aluminized formulation 163
aluminum 35, 64, 160, 362
aluminum hydride 89, 335
aluminum metal 364
aluminum nanoparticles 365–366
aluminum thermite 363, 369
amidosulfonic acid 337
aminotetrazole 153, 284
aminotetrazolium dinitramide 351
ammonium azide 297
ammonium dinitramide 35, 144
ammonium nitrate 4, 138, 403
ammonium perchlorate 35, 160, 402
ammunition primer 367
ANFO 4
anhydrous hydrazine 402
Anthrax spores 360
antimony sulphide 28
ANTTO 12
apex angle 236, 239
APX 301
Ariane V 88
armor steel 236
armor steel plate 236
arrested reactive milling 365

Atlas rocket 88
atomisization energy 144
AWD 360
azide transfer reagents 401
azides, organic 403
azidopentazole 279
azotetrazolate 122

Bachmann process 6
back-bone 275
ball mill 365
ballistic cap 237
ballistic missile 410
ballistite 3
BAM drop hammer 187
BAM regulations 207
α-band 106
β-band 106, 110
band spectrum 97
barotrauma 357
basis set 143
BDNAPM 312, 319
Becker-Kistiakowsky-Wilson-Neumann parameters 147
Becker-Kistiakowsky-Wilson's equation of state 147
Bernoulli equation 243
Berthelot-Rot 43
beryllium 64, 88
binder 11–12, 81–82, 91, 109, 111, 337
biocid 360
biocidal activity 360
biocidal detonation product 360
biological agent 360
biological threat 360
biological weapon 359–360
bipropellant 67, 85–87, 296
bistetrazolylamine 290
bistetrazolylhydrazine 71
bis(trinitroethyl)-1,2,4,5-tetrazine-3,6-diamine 291–292
BKW-EOS 147
blackbody 106, 110, 383
blackpowder 1–2, 69, 92–93
blast wave 62, 318
blasting caps 92
blasting gelatine 3
boiling point 145

bomb 250–251, 255, 410
bond energy 277
booster charge 55
borate 291
boron carbide 101
bridge wire 91
brisance 60, 244
Brockman process 6
BTAT 291, 295
BTNPM 312, 319
BTTN 13
bubble energy 265
building 64
building block 331
bullet impact 217
bunker 347, 358
burn rate 44, 69, 93, 109
burning rate 109–110, 125
butane-1,2,4-trioltrinitrate 13

cadmium azide 57
cage strain 275
camouflage 111
capactive resistance 210
carbon back-bone 275
carbon fiber 113
carbon soot 106
casting 338
Castor oil 88
cave 358
CBS-4M 143–144, 161, 165, 408
CBS method 143
cellulose 88
cesium 122
cesium pentazolate 282
Chapman-Jouguet condition 46
Chapman-Jouguet Point 130
Characterization 42
CHEETAH 147
chemical thermal propulsion 89
chemical weapon 359–360
chlorate 115
chloroprene 114
CHNO-explosive 138
chromaticity diagram 100, 123
C-J point 48, 130, 147–149
CL-20 2, 10, 15–17, 275, 291, 300
classification 55
cluster bomb 251
CMDB 83

C nitration 332
co-crystallization 302
collateral damage xii, 360
colorant 91, 99, 104
colored NICO 154
colored smoke 115–116
combustion 44–45, 49, 151
combustion chamber 74–75, 86, 89, 151, 162
combustion efficiency 82, 111
combustion gas 72
combustion parameter 167
combustion pressure 113
combustion temperature 69, 154, 168
combustion velocity 363, 365
compatability 103
complete basis set 143
composite material 113
composite modified double-base propellant 83
composite propellant 79
composition B 65, 255
computational methods 143
computer calculation 269
conducting floor 347
confined charge 63
confinement 64, 250, 253, 343
contamination 25, 57
continuous spectrum 97
cook-offtest 81
copper complex 57
copper(I) 5-nitrotetrazolate 26
copper(II) oxide 362
cordite 3
countermeasure 108
covalent radii 277
co-volume 154, 158
covolume constant 147
Cowan-Fickett's equation of state 147
critical diameter 127, 248
cruise missile 410
cryogenic solid 89
crystal density 272
Crystal structure 375
CTP 89–90
cubic 375
CuO/Al thermites 370
cyanogen azide 289
cylinder energy 310
cylindrical charge 251, 255
cylindrical grain 160

DAAF 22–23, 318
DADNE 19
DADP 24
DAGL 34, 361
damage 343
dantacol 11
DBX-1 26
DDNP 56
DDT 45, 52, 125
decomposition temperature 82, 214
decoy flare 104, 106–107, 109
decoy munition 92
deeply buried target 359
deflagration 44, 52
deflagration-to-detonation transition 45, 52, 125
delay composition 92–93, 95
Delta rocket 88
density 61, 81, 291
depleted uranium 243
detonation 44–45, 52, 125–127, 174, 358
detonation behavior 130
detonation chamber 247–248
detonation gas 141
detonation parameter 143, 147, 150–151
detonation pressure 60–62, 133, 135, 137, 147, 284–285, 301
detonation product 142, 149
detonation temperature 141, 143
detonation velocity 47, 60–61, 127, 133, 135, 137, 143, 149–150, 244–246, 248–250, 284–285, 301, 317, 347, 352
detonation wave 46, 249
detonator 91–92, 125
detoxification 65
Dewar benzene 279
DHE 11
diaminotetrazole 284
diaminotetrazolium dinitramide 286
diaminotetrazolium nitrate 214
diamond cell 282
diamond-cell technique 281
diazide structure 278
diazodinitrophenol 55, 93
diazodinitroresorcinate 93
diazotation 397
diazotization 333
dicyanamide 290
difluoramino 360
difluoroamine 332

diisocyanate 337
dimethylaminoethylazide 86
DINGU 20
dinitramide 99–100, 297
dinitramide ion 335
dinitraminic acid 336
dinitroanisole 4, 67
dinitrogen tetroxide 86
dinitroglycoluril 20
dinitroguanidine 300–301
dinitrourea 20
diphenylphosphenylhydroxylamine 400
distance 64, 343
disulfuric acid 337
DMAZ 86
DNAM 20, 22
DNAN 4, 67–68
DNQ 300
doorway modes 173
double-base propellant 69, 79, 85, 167
DPO 324
drop height 187, 271
drop weight 187
drop-hammer 43, 187
DSC 103–104, 212, 214, 301, 343, 364
DSC plot 19, 214
dye 115–116
dynamic compression 126
dynamite 4, 349

ear protectors 64, 343
EBW 183–185
EBX 358
ecological impact 65
EDD 24
EFI 183–185
EFP 239, 241
EGDN 11
ejection velocity 78
electric bridge wire detonator 183
electric igniter 367
electric initiator 92
electric match 91, 368
2-electron bond 276
2-electron bond energy 277
electron correlation 144
electron density 270
electrophilic substitution 332
electrostatic balance parameter 273
electrostatic discharge sensitivity 210

electrostatic potential 165–166, 269, 271–272
electrostatic sensitivity 187, 210, 284, 343
emulsion explosives 4
emulsions 4
energetic binder 11
energetic filler 72
energetic material 42, 55
energy and momentum conversation 151
energy limit 143
enhanced blast explosives 357
enthalpy 139, 144, 146, 164
enthalpy of combustion 139
enthalpy of detonation 139
enthalpy of explosion 139
enthalpy of formation 275, 408
enthalpy of sublimation 144
enthalpy of vapourization 145–146
environment 65, 104
epichlorohydrine 16
equilibrium flow 162
ERA 65
erosion 69, 71, 167, 350
ESD 104, 211, 369
ESD sensitivity 212, 364
ESD test apparatus 210
ESP 165, 269
ethylenglycoldinitrate 11
ethylenoxide 357
exhaust gases 106
exhaust velocity 162
expansion nozzle 152, 162
EXPLO5 147–153, 159, 162–163, 167, 249, 301
exploding foil initiator 183
explosion pressure 62
explosion stimuli 187
explosion temperature 60, 137, 141
explosive detection dog 24
explosive lens 241
explosive power 141, 280
explosive welding 372
explosively formed penetrator 239
explosively formed projectile 239
Explosive-Power 141
extruder 338
extruding 337–338
extruding processes 133
EXW 372

face shield 64, 343
FAE bomb 357

far IR 112, 114–115
far IR region 112
fast cook-off-test 217
FCO 216
fiberoptic technique 245
figure of merit 119
fire 49
firework 92–93
first fire 91–92
flame front 52
flare 122
FlexyTSC safety calorimeter 214
floor 347
flow behavior 88
FLOX 88
fluorine 88, 360
fluorocarbon 113
force 61, 154, 244
Formulation 68
formulations 7
FOX-7 18–19, 21
FOX-12 18–19
F-PETN 361
fragment distribution 344
fragment velocity 250–251
fragmentation 250
fragmenting munitions attack 217
fragmenting warhead 241
free energy 144
free fall test 220
friction 174, 343
friction apparatus 207
friction force 207
friction sensitivity 43, 187, 212, 219, 284
friendly fire 65
frozen flow 162
fuel 79–80, 86, 109, 283, 297
Fuel Air Systems 357
fuel-oxidizer system 363
fume-cupboard 343
functional group 331
fybrid rocket motors 89

G2MP2 143
G3 143
GAP 11, 72, 81
gap test 217–218
gas bubble 173
gas-dynamic propulsion step 253
gasless delay composition 93

Gaussian 143
gel propellants 88
gelatine dynamite 3
gelling agent 88
general purpose bomb 255
global minimum 279
global war against terror 357
global war on terror xii
glove 64, 343–344, 346
glycidylazid polymer 11
granule 337
graphite 106, 114
graphite fluoride 109
green 60, 104
green propellants 297
grenade 114, 250–251
greybody 105–106
ground water 104
guanidinium nitrate 5
guanylurea 19
guanylurea dinitramide 19
GUDN 19
Guhr Dynamite 3
guidance system 105
guided rocket 410
gun barrel 71, 168
gun propellant 42, 72, 154, 167
gun propulsion 71
gunpowder 69
Gurney model 250
Gurney velocity 245, 250–253, 255
GWT xii, 65, 357

Hammond's postulate 165
HAN 35
hand grenade 95
HAT-DN 351
hazard potential 213
H_2BTA 290
HC 114
HDAT nitrate 214
HDBT 359
heat capacity 142
heat effect 359–360
heat exchanger 90
heat flow rate 214
heat generating pyrotechnic 92
heat of explosion 60, 137, 284
heat radiation 358
heat-seaking missile 105

heavy metal 100, 104
heterogeneous explosives 129
hexachlorethane 114
hexachloromelamine 360
hexamethylentetramine 333
hexamine 122
hexanitrobenzene 300
hexanitrohexaazaisowurtzitane 15
hexanitrostilbene 2, 6
hexogen 2, 6
high explosive 60, 275, 302, 410
high-nitrogen formulation 168
high-nitrogen propellant 167
high-performance explosive 279
high-speed camera 245
high-speed streak camera 245
HME 24
HMTD 24
HMX 2, 6, 72, 138, 252
HNF 35, 296
HNFX 331, 360
HNIW 2, 15
HNS 2, 6, 57, 138, 323
HN_3 synthesis 401
homemade explosive 24
homogeneous explosives 129
hotspot 173–174
howitzer 168
HTPB 79–81, 89, 337, 401
Hugoniot adiabate 126, 148–149
Hugoniot-Adiabaten 46
humidifier 347
humidity 111, 347
h_{50} value 272
hybrid engine 89
hybridization 277
hydrazine 85–86, 153, 297, 402
hydrazine synthesis 402
hydrazinium aminotetrazolate 153
hydrazinium azide 297
hydrazinium azotetrazolate 350
hydrazinium nitrate 297
hydrazinium nitroformate 35
hydrazinium salt 290
hydraziniumnitroformate 296
hydrazinolysis 399
hydro carbon 88
hydrodynamic detonation theory 150
hydrogen peroxide 86
hydroxyl terminated polybutadiene 80

hydroxylammonium azide 297
hydroxylammonium nitrate 35, 297
hypergolic mixture 86
hypergolic propellant 86
hypergolicity 88
hypersurface 269–270, 279

ICT code 163
ICT program 162
ideal gas constant 147
ideal gas law 147
ignitation temperature 43
ignition 113, 173–174
ignition temperature 173, 364
IHC 218
illuminant 96–97, 99, 122
illumination 95
IM 216
imidazolium salt 296
impact 174, 187, 343
impact linking 173
impact sensitivity 43, 166, 187, 212, 219, 269, 271–272, 284, 369
impetus 154
impulse 63, 72
IMX-101 67
incandescence 97
incendiary device 86, 92
inflator 15
inhalation 285
inhalation toxicity 296
initiating charge 27
initiation 173–174
initiator 91
inorganic nitrogen chemistry 401
insensitive munition 19, 60, 216
intensity ratio 109
intercontinental rocket 86
inter-crystalline friction 174
interim hazard classification 218, 220
interior ballistics 158
internal energy 140
ionic liquid 296
ionization potential 161
IR 104
IR detector 106
IR emitter 106
IRFNA 87
iron carbide 71
iron oxide 369

iron oxide thermite 369
iron(III) oxide 362
IR signiature 106
isobaric combustion 151
isobaric combustion temperature 152
isobaric condition 152
isobaric heat of combustion 152
isochoric combustion 151, 154
isochoric combustion processes 153
isochoric heat of combustion 153
isocyanate 79
isocyanate binder 401
isopropylnitrate 86
isotherm safety calorimeter 214

jet 235, 237
jet engine 78
jet formation 235, 237
jet velocity 238

Kamlet and Jacobs equations 133
KDN 286
KDNP 26–27
Kevlar 64
Kevlar vest 343
Kieselguhr 3
kinetic energy munition 236, 243–244
Kneisl-Test 214
Koenen test 43, 212, 291, 347

lactose 122
LASEM 317
laser excitation 320
laser-induced air shock 317
laser-induced plasma 319
laser-supported detonation wave 318
lattice energy 145, 296
lattice enthalpy 145–146, 161
launch platform 238
lead azide 25–26, 55, 91, 367
lead styphnate 25–26, 55, 91–92, 367
lead-free 26
leather welding gloves 346
light-generating pyrotechnics 92
line spectrum 97
liquid propellant 88
liquid rocket propellant 85
LLM-105 20, 22
loading density 61, 133, 158, 244
loading voltage 210

long-term stability 214
LOVA 72
low frequency vibrational modes 173
low-explosive 410
low-vulnerability ammunition 72
LOX 87
luminescence 97

Mach Stem 63
magnesium 64
magnesiumphosphide 113
main charge 91
Mannich reaction 291–292, 399
manometric pressure bomb 158
mass flow 113
mCPBA 24
mechanical manipulation 343
mechanical manipulator 64, 347
MEKP 24
melt-cast 337
melt-castable component 337
melting 337
melting point 145, 338
mercury fulminate 2–3, 25, 55
mesitylsulfonylhydroxylamine 400
metal oxide 362
metal spatulas 344
metal-lined shaped charge 235, 237
metastable intermolecular composites 362
metrioltrinitrate 13
micron composite 364
mid IR 112, 114–115
milling 365
minimal population localization 144
minimum smoke missile 80
missile 105, 251, 410
missile design 81
missile propellant 160
mixed acid 332–333, 337
MK 80 bomb 255
MK 80 series 95
MMH 86–88, 296
MOAB 7, 62
molecular volume 145–146, 408
molybdenum(VI) oxide 362
MON-25 87
monomethylhydrazine 86
monopropellant 67, 85, 297
MoO_3/Al nanothermite 371
MSIAC 217

MTN 13
MTV 92, 106, 108–110
MTX-1 57
multi-channel analyzer 245–246
multi-perforated grain 160
multiprojectile warhead 360
Munitions Safety Information Analysis Center 217
muzzle flash 69

N_2/CO ratio 71, 168
nano-Al 82
nano-composite 364
nano material 82
nanoparticle 364
nanopowder 365
nanoscale 362–363, 369
nanoscale powder 366
nanoscale thermite 362–363
nano-sized material 362
nanothermite 367
nanothermites 362
NaN_3 synthesis 401
NATO angle 243
near infrared 122
near IR region 112
NENA 11–12
NICO 154
NILE 154, 167
NIR 122–123
N_6 isomers 278
N_8 isomers 279
nitramine 273, 300
nitraminotetrazolate 57
nitrate 100, 113
nitrating agent 332, 334
nitration 332
nitration reaction 332
nitrato ester 161
nitric acid 332–334
nitrimino group 332
nitriminotetrazole 287, 290, 331, 333
nitrocellulose 3, 69, 96
nitro-compounds 271
Nitroform 37
nitroform 40, 296
nitrogen triazide 277
nitrogen-rich 71, 280, 284, 291, 350
nitrogen-rich compound 122
nitrogen-rich ligands 99

nitrogen-rich molecules 275
nitroglycerine 2, 69, 138, 144, 349
nitroguadine 69
nitroguanidine 2, 4, 138
nitromethane 86
nitronium 161
nitronium triflate 332
nitropenta 2
nitrosonium 161
nitrotetrazolate 57
nitrotetrazolate salt 289
nitrotetrazole 287, 289
N_6 molecule 277
N—N bond order 277
N nitration 332
^{15}N NMR spectroscopy 289
N≡N triple bond 276
Nobel-Abel equation 158
non-hypergolic mixture 86
normal-casting 337
N-oxide 24
nozzle 74–75, 151, 162
nozzle plate 212
nozzle throat 151
NTO 15, 17
NTP 89
nuclear thermal propulsion 89
Nytox 89

obscurant 111
octa-F-PETA 360
octanitrocubane 15
octogen 2, 6
Octol 65
ONC 15–17, 300
O nitration 332
Oppauer salt 43
optical fiber 245, 248–249
organic nitrogen chemistry 397
orthorhombic 375
output charge 27
overexpanding nozzle 75
oxalic acid 163
oxalic acid dinitrate ester 164, 167
oxide layer 364
oxide thickness 364
oxidizer 35, 82, 86, 88, 91, 109, 153, 163–164, 275, 283, 297, 332
oxidizer gel 89
oxidizing group 283

Oxone 24–25
oxygen balance 35, 64, 81, 137–138, 160, 163, 283, 291, 332, 350
oxygen-rich 291
ozonation 334–335
ozone 334

parachute 95, 122
Pardubice 11
particle size 363–364
particulate matter 104
PATO 8, 323
payload 280
PBAN 81
PBX 11, 338
penetration behavior 236
penetration depth 237, 243
penetration process 236
penetrator 243
pentasilane 82
pentazolate anion 283
pentazole 284
perchlorate 35, 96, 99, 104, 286
percussion primer 91
performance 60, 65, 286, 291, 300, 349
perturbation-theory 143
PETN 2, 6, 28, 91, 138
phase transition 10
Phase transition temperature / K 375
phenylpentazole 279
phosphides 114
phosphine 113
phosphonium salt 296
phosphorus 112–113
photodiode 245–246
picrates 4
picric acid 2, 4–5, 138, 141
Planck's rule 105
plane 105–106
plasma emission 319
plasticizer 11–12
plastiziser 337
plunger assembly 187
PMF 109
poly-AMMO 11
poly-BAMO 11
polycarbonmonofluoride 109
polychloroprene 113
poly-GLYN 11
polyisobutylene 11

polymer 106
polymer binder 338
polymer bound explosive 338
polymer-bonded explosive 11
polymeric nitrogen 282
polymorph 10, 17, 19, 300
poly-NIMMO 11
polynitro-compound 294
polynitrogen 277, 280
polynitrogen compound 276
poly(pyrazolyl)borate 291
polyvinylalcohol 113
porcelain peg 207
positive potential 271
post-detonation 64
potassium 122
potassium dinitramide 286
potassium nitroformate 39
potential 271
potential energy hypersurface 164
Power-Index 141
pressing 337–338
pressure 61, 281
pressure-coordination rule 281
primary explosive 25, 42, 55, 57, 67, 91, 302, 343
primer 92, 367
primer composition 91
prismane 279
processing 337
progressive burning 160
projectile 95, 237
propagation 125, 127
propellant 35, 79, 90–91
propellant charge 69–70, 72, 91, 140
propulsion system 410
protective clothing 64, 343
PTFE 106, 108–109
pulsed engine 86
pulse-mode 88
PVA 113
PVC 96, 99
pyrophoric effect 236
pyrosulfuric acid 337
pyrotechnic 43, 67, 92, 99, 367
pyrotechnic composition 122
pyrotechnic delay mixture 363
pyrotechnic mixture 91, 95, 103, 106
pyrotechnic reaction 91
PYX 8, 323

quantities 343
quantum chemical calculation 269
quantum mechanical calculation 143, 279

RADAR detection 335
RADEX cell 214
radiant intensity 123
radiation intensity 107
Raleigh line 148
RAM jet 78
Raschig process 402
Rayleigh line 47, 130
RDX 2, 6, 65–66, 72, 138, 252, 283, 300, 333
reaction vessel 343
reactive armor 251
red fuming nitric acid 88
red phosphorus 112, 114
reducing agent 91
regression rate 89
regressive burning 160
re-ignition 86
remediation 65
residual volume 147
resistance heating 91
RFNA 88–89
ring strain 275
rocket 74, 78, 86, 410
rocket equation 78
rocket fuel 43, 140
rocket launch 78
rocket motor 72, 161–162
rocket propellant 72, 79, 151–152, 335
rule of Wien 106
ryogenic 86

safety standards 347
salt mixture 297
Sandmeyer reaction 397
sandwich charge 251
satellite 86
schock adiabate 126, 148
SCO 81
SCRAM jet 78, 82
secondary explosive 15, 42, 60–61, 65, 67, 91, 137, 275, 343
semicarbazid 16
semi-solid 88
Semtex 11
Semtin 11
sensitivity 269, 302

sensitizer 55
SF_6 90
shaped charge 235, 238, 241, 244–245, 250
shaped charge weapon attack 217
shear stress 88
shell primer 367
shielding 112–114
shock adiabate 46, 127, 130, 148
shock energy 265
shock front 62, 150
shockwave 52, 62–63, 91, 125, 127, 173
Sidewinder 105
signal flare 92, 95, 122
signiature 106
silane 82
silicon 114
silver acetylide nitrate 56
silver nitraminotetrazolate 57
single-base propellant 69, 167
SINOXID 92
SINTOX 92
slow cook-off-test 217
slug 239
small arms attack 217
small-scale-burning test 219
smoke 91, 111–113
smoke grenade 93
smoke munition 92, 111, 114
smoke pot 114
smoke signal 111
smoke-free combustion 280, 284
smoke-generating pyrotechnics 92
sodium azide 290
solar thermal propulsion 89
solid grain design 159
solid propellant 79, 88
solid rocket booster 160
solid rocket motor 76, 81, 163, 410
solid rocket propellant 67, 160–161
solid state reaction 91
soot 106
SORGUYL 20
Space Shuttle 88
spark energy 210
spatulas 344
specific energy 61, 153–154, 244
specific impulse 72–74, 77, 86, 88–90, 152, 162–163, 280, 351
specific radiant emittance 105
specific volume 126, 149

spherical charge 251
Springall-Roberts rules 138, 141
stab detonator 27
stabilizer 83
π-stacking 25
standard enthalpy of formation 144
stand-off distance 237, 239
stationary state 125
steady-state model 148
steel-sleeve test 43, 212, 291
storage 347
STP 89
strategic missile 410
streak camera 246
sulfamic acid 337
sulfuric acid 332
sulphur hexafluoride 90
Summary of crystal structures of AN. 375
superthermite 362
surface potential 166
sustainable synthesis 284

tactical missile 161, 410
tactical missile rocket motor 160
TAG-DN 351
TAGNF 35
TAGzT 71–72, 167, 350
tail fin 243
TATB 2, 6, 9, 175, 319, 323
TATP 23–24, 371
Taylor wave 48
Teflon 106
TEM 364
terephthalic acid 118
tetracene 28
tetragonal 375
tetranitromethane 39, 293
tetrazene 55, 92–93
tetrazine 24, 291
tetrazole 99, 284, 286, 288–289, 291
tetrazole skeleton 286
tetryl 4
TEX 17–18
theoretical maximum density 133
thermal conductivity 113
thermal sensitivity 187
thermal stability 211, 219, 284, 291
thermite 106, 362, 364, 366, 369, 371
thermite performance 365
thermite reaction 362–364

thermobaric explosion 360
thermobaric weapon 357–358, 360
thermodynamic data 143
thermodynamic values 137
Thermodynamics 137
Thixatrol 88
thrust 73–74, 76, 86, 152, 162
TIGER 147
time critical target xii
TKX-50 305–306, 318
TKX-55 318, 324, 329
TMD 125, 133
TMTZ 88
TNAZ 15–17, 335
TNC-NO$_2$ 36
TNE 293
TNM 293
TNT 2, 4, 65, 137–139, 252, 255
torpedo 64, 90
TORPEX 64
total impulse 76
toxicity 60, 65, 88, 113
training ground 66
transfer charge 27
transition state 165, 278
triacetone triperoxide 23
triacetyltriazine 333
triaminoguanidinium azotetrazolate 350
triaminoguanidinium dinitramide 351
triaminotrinitro benzene 6
triaminotrinitrobenzole 2
triazine triazide 28
triazole 284
trinitroazetidine 335
trinitroethanol 291, 293
trinitroethyl 284, 291
trinitroethyl group 291
trinitromethane 293, 296
trinitrotoluene 2, 4
triple bottom line 104
triple-base propellant 69, 167
tristimulus function 98
Tritonal 255

Trouton's rule 145
TTA 34–35
TTTO 312
tubular grain 160
tunnel 358

UDMH 86–88, 296
ultrasonic activation 332
underexpanding nozzle 75
underwater 64
unexploded munition 65
unexploded ordnance 65
UN guidelines 211, 213
upper stage 86
uranium 243
UXO 65

vacuum stability test 311
vapor pressure 81, 88, 296
Vilsmeyer-Reaktion 399
viscosity 88
Viton 110
volume 60
Volume change / % 375
volume-based sensitivity 273

warhead 251, 360, 410
wavefront 241
weapons of mass destruction 360
welding 372
WFNA 89, 296
White, Johnson and Dantzig's free energy
    minimization technique 147
WMD 360

X-ray diffraction 286, 294–295

yield factor 119

zero-point energy 143
zinc peroxide 93
Zone Model 50
zwitterionic structure 300